How to Design and Implement Powder-to-Tablet Continuous Manufacturing Systems

Expertise in pharmaceutical process technology series

How to Design and Implement Powder-to-Tablet Continuous Manufacturing Systems

Edited by

Fernando J. Muzzio

Sarang Oka

ELSEVIER

ACADEMIC PRESS

An imprint of Elsevier

Academic Press is an imprint of Elsevier
125 London Wall, London EC2Y 5AS, United Kingdom
525 B Street, Suite 1650, San Diego, CA 92101, United States
50 Hampshire Street, 5th Floor, Cambridge, MA 02139, United States
The Boulevard, Langford Lane, Kidlington, Oxford OX5 1GB, United Kingdom

Library of Congress Cataloging-in-Publication Data
A catalog record for this book is available from the Library of Congress

British Library Cataloguing-in-Publication Data
A catalogue record for this book is available from the British Library

ISBN: 978-0-12-813479-5

For information on all Academic Press publications visit our website at
https://www.elsevier.com/books-and-journals

Publisher: John Fedor
Acquisitions Editor: Erin Hill-Parks
Editorial Project Manager: Timothy Bennett
Production Project Manager: Sreejith Viswanathan
Cover Designer: Mark Rogers

Typeset by TNQ Technologies

I dedicate this book to the hundreds of students, postdocs, and faculty and industry collaborators who did the work described here.
and to Daisy. For everything.
Long Branch, NJ, 2/5/2022

Fernando J. Muzzio

To the contributors. To Fernando, to mentors, past and present.
To Aai, Baba, and Shankali.
Princeton, NJ, 2/11/2022

Sarang Oka

Contents

1. Introduction

Sarang Oka and Fernando J. Muzzio

2. Characterization of material properties

*Sonia M. Razavi, Sarang Oka, M. Sebastian Escotet-Espinoza,
Yifan Wang, Tianyi Li, Mauricio Futran and Fernando J. Muzzio*

3.   Loss-in-weight feeding
Tianyi Li, Sarang Oka, James V. Scicolone and Fernando J. Muzzio

4.   Continuous powder mixing and lubrication
Sarang Oka and Fernando J. Muzzio

5.   Continuous dry granulation
*Nirupaplava Metta, Bereket Yohannes, Lalith Kotamarthy,
Rohit Ramachandran, Rodolfo J. Romañach and Alberto M. Cuitino*

8.   Continuous tableting

*Sonia M. Razavi, Bereket Yohannes, Ravendra Singh,
Marcial Gonzalez, Hwahsiung P. Lee, Fernando J. Muzzio and
Alberto M. Cuitiño*

**9.   Continuous film coating within continuous oral
solid dose manufacturing**

Oliver Nohynek

10. Role of process analytical technology in continuous manufacturing

Joseph Medendorp, Andrés D. Román-Ospino and Savitha Panikar

11. Developing process models of an open-loop integrated system

Nirupaplava Metta and Marianthi Ierapetritou

16. **Orkambi: a continuous manufacturing approach to process development at Vertex**

Stephanie Krogmeier, Justin Pritchard, Eleni Dokou, Sue Miles, Gregory Connelly, Joseph Medendorp, Michael Bourland and Kelly Swinney

17. **Outlook—what comes next in continuous manufacturing (and in advanced pharmaceutical manufacturing)**

Fernando J. Muzzio

Contributors

Michael Bourland, Vertex Pharmaceuticals, Boston, MA, United States

Gregory Connelly, Vertex Pharmaceuticals, Boston, MA, United States

Alberto M. Cuitiño, Department of Mechanical and Aerospace Engineering, Rutgers, The State University of New Jersey, Piscataway, NJ, United States

Eleni Dokou, Vertex Pharmaceuticals, Boston, MA, United States

M. Sebastian Escotet-Espinoza, Engineering Research Center for Structured Organic Particulate Systems (C-SOPS), Department of Chemical and Biochemical Engineering, Rutgers, The State University of New Jersey, Piscataway, NJ, United States; Oral Formulation Sciences and Technology, Merck & Co., Inc., Rahway, NJ, United States

Mauricio Futran, Janssen Supply Chain, The Janssen Pharmaceutical Companies of Johnson and Johnson, Raritan, NJ, United States

Marcial Gonzalez, School of Mechanical Engineering, Purdue University, West Lafayette, IN, United States

Douglas B. Hausner, Thermo Fisher Scientific, Waltham, MA, United States

Marianthi Ierapetritou, Department of Chemical and Biomolecular Engineering, University of Delaware, Newark, DE, United States

Lalith Kotamarthy, Department of Chemical and Biochemical Engineering, Rutgers University, Piscataway, NJ, United States

Stephanie Krogmeier, Vertex Pharmaceuticals, Boston, MA, United States

Hwahsiung P. Lee, Department of Mechanical and Aerospace Engineering, Rutgers, The State University of New Jersey, Piscataway, NJ, United States

Tianyi Li, Drug Product Development, J-Star Research, Cranbury, NJ, United States; Department of Chemical and Biochemical Engineering, Rutgers University, Piscataway, NJ, United States; Engineering Research Center for Structured Organic Particulate Systems (C-SOPS), Department of Chemical and Biochemical Engineering, Rutgers, The State University of New Jersey, Piscataway, NJ, United States

Joseph Medendorp, Vertex Pharmaceuticals, Boston, MA, United States

Nirupaplava Metta, Applied Global Services, Applied Materials Inc., Santa Clara, CA, United States; Automation Products Group, Applied Materials, Logan, UT, United States; Department of Chemical and Biochemical Engineering, Rutgers University, Piscataway, NJ, United States

Sue Miles, Vertex Pharmaceuticals, Boston, MA, United States

Christine M.V. Moore, Organon & Co., Jersey, NJ, United States

Shashank Venkat Muddu, Department of Chemical and Biochemical Engineering, Rutgers, The State University of New Jersey, Piscataway, NJ, United States

Fernando J. Muzzio, Engineering Research Center for Structured Organic Particulate Systems (C-SOPS), Department of Chemical and Biochemical Engineering, Rutgers, The State University of New Jersey, Piscataway, NJ, United States

Oliver Nohynek, Driam USA Inc., Coating Technology, Spartanburg, SC, United States

Sarang Oka, Hovione, Drug Product Continuous Manufacturing, East Windsor, NJ, United States; Engineering Research Center for Structured Organic Particulate Systems (C-SOPS), Department of Chemical and Biochemical Engineering, Rutgers, The State University of New Jersey, Piscataway, NJ, United States

Savitha Panikar, Drug Product Continuous Manufacturing, Hovione, LLC., East Windsor, NJ, United States

Justin Pritchard, Vertex Pharmaceuticals, Boston, MA, United States

Rohit Ramachandran, Department of Chemical and Biochemical Engineering, Rutgers, The State University of New Jersey, Piscataway, NJ, United States

William Randolph, Janssen Supply Chain, The Janssen Pharmaceutical Companies of Johnson and Johnson, Raritan, NJ, United States

Sonia M. Razavi, Engineering Research Center for Structured Organic Particulate Systems (C-SOPS), Department of Chemical and Biochemical Engineering, Rutgers, The State University of New Jersey, Piscataway, NJ, United States

Eric Sánchez Rolón, Janssen Supply Chain, The Janssen Pharmaceutical Companies of Johnson and Johnson, Raritan, NJ, United States

Rodolfo J. Romañach, Department of Chemistry, University of Puerto Rico-Mayagüez campus, Mayagüez, PR, United States

Andrés D. Román-Ospino, Rutgers University, Piscataway, NJ, United States

James V. Scicolone, Department of Chemical and Biochemical Engineering, Rutgers University, Piscataway, NJ, United States

Ravendra Singh, Engineering Research Center for Structured Organic Particulate Systems (C-SOPS), Department of Chemical and Biochemical Engineering, Rutgers, The State University of New Jersey, Piscataway, NJ, United States

Stephen Sirabian, Equipment & Engineering Division, Glatt Air Techniques, Inc., Ramsey, NJ, United States

Kelly Swinney, Vertex Pharmaceuticals, Boston, MA, United States

Zilong Wang, Manufacturing Intelligence, Global Technology and Engineering, Pfizer Global Supply, Pfizer Inc., Peapack, NJ, United States

Yifan Wang, US Food and Drug Administration, Center for Drug Evaluation and Research, Office of Pharmaceutical Quality, Silver Spring, MD, United States

Bereket Yohannes, Department of Mechanical and Aerospace Engineering, Rutgers, The State University of New Jersey, Piscataway, NJ, United States; Drug Product Development, Bristol-Myers Squibb Company, New Brunswick, NJ, United Staes

About the *Expertise in Pharmaceutical Process Technology Series*

Numerous books and articles have been published on the subject of pharmaceutical process technology. While most of them cover the subject matter in depth and include detailed descriptions of the processes and associated theories and practices of operations, there seems to be a significant lack of practical guides and "how to" publications.

The *Expertise in Pharmaceutical Process Technology* series is designed to fill this void. It comprises volumes on specific subjects with case studies and practical advice on how to overcome challenges that the practitioners in various fields of pharmaceutical technology are facing.

Format

- The series volumes will be published under the Elsevier Academic Press imprint in both paperback and electronic versions. Electronic versions will be full color, while print books will be published in black and white.

Subject matter

- The series will be a collection of hands-on practical guides for practitioners with numerous case studies and step-by-step instructions for proper procedures and problem solving. Each topic will start with a brief overview of the subject matter and include an exposé, as well as practical solutions of the most common problems along with a lot of common sense (proven scientific rather than empirical practices).
- The series will try to avoid theoretical aspects of the subject matter and limit scientific/mathematical exposé (e.g., modeling, finite elements computations, academic studies, review of publications, theoretical aspects of process physics or chemistry) unless absolutely vital for understanding or justification of practical approach as advocated by the volume author. At

best, it will combine both the practical ("how to") and scientific ("why") approach, based on *practically proven* solid theory − model − measurements. The main focus will be to ensure that a practitioner can use the recommended step-by-step approach to improve the results of his or her daily activities.

Target audience

- The primary audience includes pharmaceutical personnel, from R&D and production technicians to team leaders and department heads. Some topics will also be of interest to people working in nutraceutical and generic manufacturing companies. The series will also be useful for those in academia and regulatory agencies. Each book in the series will target a specific audience.

The *Expertise in Pharmaceutical Process Technology* series presents concise, affordable, practical volumes that are valuable to patrons of pharmaceutical libraries as well as practitioners.

Welcome to the brave new world of practical guides to pharmaceutical technology!

Foreword

Continuous manufacturing is certainly not a new manufacturing mode in the process and related industries. Chemical engineering has been firmly rooted in continuous—steady-state—processing for perhaps a century with integrated continuous processes dating back even further to developments such as the Solvay process. Moreover, the notion that the continuous mode might or ought to be employed in pharma manufacturing also had been a topic of speculation and discussion for quite some time—beginning some 30 years ago. Indeed, there are a number of former and current industry thought leaders who might and some do claim to have been among the first to observe that pharmaceutical manufacturing could benefit substantially from a transition from batch to continuous mode. However, the real and perceived barriers to such a transition were many, including the substantial capital investment in existing batch facilities, the reluctance to accept the risk of manufacturing innovation, and the perception that there were substantial regulatory barriers to overcome. As a result, although there were exploratory projects initiated within some companies, it took initiatives from the academic community, where the cost of failure of an R&D activity was not as high, to begin more intense investigation of the technical challenges of continuous pharmaceutical manufacturing and, in the course of doing so, raise the visibility of this technology. Fortunately, there were a few pharmaceutical companies willing to co-sponsor these initial steps. Moreover, some 20 years ago, there were encouraging signals from the regulatory side that manufacturing innovation would be welcomed.

The senior editor of this book, Professor Fernando Muzzio, undertook to organize one of these pioneering collaborative partnerships in the form of the NSF Engineering Research Center for Structured Organic Particulate Systems (CSOPS), which at its peak drew in some 40 supporting industry members. Several of my colleagues and I at Purdue University as well as selected faculty from the University of Puerto Rico and New Jersey Institute of Technology also joined that effort. While CSOPS focused on solid oral dosage products, a parallel project under the banner of MIT-Novartis pursued continuous manufacture of both drug substance and drug product. Since then, other such consortium/center activities in the United States and abroad have emerged and continuous process research has certainly flourished, aided by increasing funding from industry and government, including the FDA. Moreover, adoption of continuous manufacturing has progressed beyond the early adopter major pharmaceutical companies and is poised to grow substantially. It is thus fitting to take stock of what has been learned to date, to compile the knowledge on how to design and implement continuous tablet manufacturing systems, and thus hopefully to help to accelerate the growth of continuous pharmaceutical manufacturing across the industry. It is also appropriate that tablet manufacture be the focus as it has advanced further than the adoption of continuous in drug substance and biologics products manufacture. It is laudable that Professor Muzzio and the team of contributors he has assembled, largely from the Rutgers contingent of CSOPS, have undertaken the substantial challenge of actually attempting to compile that knowledge.

In order to meet the very ambitious goal of conveying "how do design and implement powder to tablet continuous manufacturing systems," it is necessary to cover a wide range of topics:

- the materials processed and their properties,
- the main unit operations, including their characterization and modeling,
- the process analytical technology, required to monitor the performance of the unit operations and critical quality attributes of the materials these operations produced,
- the methodology and software for assembling and solving integrated process flowsheet models composed of suitable individual unit operations models and the optimization of such flowsheet models to support process development decisions,
- the design and tuning of the process control systems required to maintain product quality during operation,
- The regulatory issues specific to continuous operations.

The authors of the 17 chapters of this book have collectively covered this array of topics and have done so to varying depth and degree of success. The book also describes case studies of successful industrial development of continuous SOD processes, namely the pioneering work of the Janssen and Vertex teams on the Prezista and Orkambi products. These case studies, particularly, the very thorough exposition offered by the Janssen team, are very valuable additions to conveying the "how to" to the reader.

The design of a continuous pharmaceutical tableting line is inherently an integrated product and process design activity. The properties of the drug substance and desired performance of the drug product drive the decisions on formulation, the manufacturing route, and the operating condition. It is thus appropriate that the book leads with a chapter on properties of pharmaceutical materials, their measurement, and compilation into databases that can power material selection, equipment, and product performance decisions. The power of physical properties databases and estimation methods to enable predictive process models and their use in process and product design has been amply demonstrated in the industries working with gases and liquids. There is much to be done to bring the particulate solids domain to that level. This chapter sets the stage for such future efforts in material databases and property prediction models as well as the next stage which is predictive design of product formulations and associated processes.

The next seven chapters of the book cover the portfolio of unit operations that are used to configure the three main pathways for producing SOD, namely, direct compression, dry granulation, and wet granulation—based continuous lines. These chapters are necessarily quite different in scope and content because the degree of prior use of these unit operations in the industry is so different. For instance, loss in weight feeders and blenders have been introduced specifically to enable continuous operation: reliable and accurate feeding and blending are essential to assuring product content uniformity and operation under state of control. Thus, the chapters dedicated to them require more in-depth treatment. Moreover, the physical realization of these unit operations is quite different from these processing functions in the batch mode. The authors of these chapters do a fine job of connecting performance to material properties and equipment design features such as screw design and blade configuration and outlining methodology for characterizing these devices. These two chapters certainly deliver the "how to" component.

By contrast, some others of the unit operations have been widely used in batch manufacture although they are inherently continuous in nature (e.g., roll compactor and integrated milling step, twin screw granulator and tablet press). Thus, those chapters require less review of basic processing characteristic and elaboration of existing models and thus the focus could shift to the new aspects relevant to continuous lines. In these chapters the authors have made a good effort to strike a balance between the known and the new elements. Thus, roll compactor chapter balances the review of compactor models with more detailed discussion of population model development for the integrated milling step and review of characterization methods of granules. The wet granulation chapter focuses on residence time distribution, which is essential to tracking material flow in continuous operation and offers useful example case studies in determining RTD. The unit operations chapters covering high shear wet granulation, fluid bed drying, and coating involve unit operations also in use in batch tablet manufacture and thus are more familiar, but their continuous adaptations have unique features that need elaboration. The descriptions of these unique features are generally well done but the "how to" element could have been given more emphasis.

Sensors are critical to establishing that the operation of any process, batch or continuous, is in a state of control but is especially important in the continuous case given the relatively short residence times and fast dynamics of the operations used in SOD manufacture. The dictum—if you cannot measure it, you cannot control it—really does capture the reality of continuous pharmaceutical operations. The authors have very competently covered this critical topic, going beyond just describing the measurement technologies in use to outlining the practical issues of sensor placement, trade-offs between sample size and sampling frequency, sensor placement, calibration, validation, and maintenance. In keeping with the "how to" theme, the authors provide several very informative case studies including the discussion of a soft sensor for dissolution monitoring. This chapter is one of the more valuable contributions of the book.

Given that discussion of models of the unit operations is already covered in the various unit operations chapters, the principal contribution of the Chapter 11 on process modeling is to explain and illustrate that for process studies a range of model types from mechanistic to data driven and hybrids of these two types may well be used. For process development purposes, the key is to be sure that the model captures the essential relations and then to use parameter fitting to minimize any model-plant mismatch. The chapter does address the issue of integrating such models into a complete process flowsheet and presents examples of use of such integrated models to investigate multiunit dynamics. Chapter 13 builds on this exposition to discuss the use of these models in process optimization studies executed using well-established optimization methods. The key contributions of this chapter are to review various types of data-driven models, to summarize procedures for generating the data for such models via efficient experimental designs, to outline the approaches to parameter estimation/model training, and to highlight the importance of validation of the resulting model. It is certainly good to be made aware that there are different types of models one could use; however, building robust process models of sufficient fidelity to enable meaningful optimization studies is not an easily transferred skill. Moreover, the successful execution of process optimization studies likewise requires thorough knowledge of the algorithms used, the likely causes of failures to converge, and the strategies of mitigation such as scaling. It is too much to expect that the authors could consolidate and convey that body of knowledge in a couple of chapters—these chapters can and do only open the door.

Chapter 14 addresses a critically important component of continuous process operation, namely, the design and validation of the distributed control system for the process. In this chapter, basic control architectures are reviewed, approaches to choosing input-output variable pairing described, and controller tuning/parameterization strategies summarized. These are practices well established in the process control community. The authors correctly emphasize the use of a model base strategy for control system design and closed loop performance evaluation before actual physical implementation. They also very effectively consolidate their experience in implementing a control system for the Rutgers tableting pilot plant into a useful case study. The case study includes succinct description of the necessary information flow, data management, and DCS system integration. The chapter does a good job of capturing the issues that need to be addressed in control system design and implementation and thus can well serve as a preview of the work that will need to be done with any new process application.

The last chapter summarizes the regulatory issues associated with a continuous manufacturing facility and makes useful observation about differences that do arise in the continuous case. It clarifies the meaning of the basic concepts, such as those of a batch, state of control, and steady state from a regulatory perspective. It describes important features of continuous operations relevant from a regulatory perspective, such as residence time distribution and role of real-time release testing. The authors highlight some very important new options that arise in continuous operation, such as the ability to reject a portion of a batch. To be sure, many of these topics can be found discussed in the draft ICH guidance on continuous manufacturing which was not available at the time of writing of this chapter but was endorsed in July 2021 and is close to being fully approved. Nonetheless, this chapter overall does provide a very succinct and useful summary of regulatory considerations, which should be taken into account when undertaking a continuous manufacturing—based project.

Over the past several years, a number of edited volumes covering the general topic of continuous pharmaceutical manufacturing have been published. Most make no claims to having covered the domain in any systematic manner—the extent and depth of coverage is left to the choice of the individual chapter contributors. It should be evident that this book differs from these in that it clearly was assembled with a "how to" goal and a logical plan to systematically cover all of the relevant components of manufacturing system design and operation. While it is not a textbook, by virtue of that plan it can nonetheless serve as very useful guide to students and professional seeking the "how to" knowledge. I expect that it will become to be recognized and appreciated as an important milestone in the development of this area of pharmaceutical manufacturing.

G.V. Reklaitis
West Lafayette, IN
October 27, 2021

Chapter 1

Introduction

Sarang Oka[1,2] and Fernando J. Muzzio[1]

[1]*Engineering Research Center for Structured Organic Particulate Systems (C-SOPS), Department of Chemical and Biochemical Engineering, Rutgers, The State University of New Jersey, Piscataway, NJ, United States;* [2]*Hovione, Drug Product Continuous Manufacturing, East Windsor, New Jersey*

1. Foreword—our journey in CM

After decades of near stagnation, pharmaceutical manufacturing is experiencing unprecedented innovation. In the last decade, the pharmaceutical industry and its technology and ingredient suppliers have embraced a worldwide transformation from traditional, inefficient batch methods to continuous manufacturing, which is an emerging technology that has been shown to greatly reduce both the time and the cost of developing and manufacturing new medicines, while enabling significant improvements in the quality of the final product and the reliability of the manufacturing process.

How did this happen? Like many good ideas, continuous solid dose pharmaceutical manufacturing has been thought by many people in the past. In fact, many of the unit operations used in batch manufacturing (mills, tablet presses, roller compactors, packaging equipment) are intrinsically continuous. The so-called "batch process" is not truly batch, rather it is actually a mishmash of intrinsically continuous and intrinsically batch processing steps, utilized in asynchronous sequence. So, it was only natural, and only a matter of time, until someone decided to remove the intrinsically batch steps and replace them with intrinsically continuous steps, to enable the whole process to be continuous and to enjoy the many advantages discussed in the next section and also throughout this book.

For the team involved in writing this book, efforts to achieve complete implementation of a continuous line going from powder raw materials to finished products started around 1998, during a visit by F.J. Muzzio to the GSK (then Glaxo) facility in Verona, Italy, to teach a week-long course on pharmaceutical manufacturing methods focused on a topic that commanded enormous attention at the time: powder mixing and blend homogeneity sampling. As part of many discussions with the excellent team of Italian GSK scientists regarding this topic, an idea emerged—if most blending problems

How to Design and Implement Powder-to-Tablet Continuous Manufacturing Systems
https://doi.org/10.1016/B978-0-12-813479-5.00008-2

were caused by the basic approach of using batch blenders, which are intrinsically difficult to sample and which promote blend segregation during discharge, why not avoid the problem altogether by creating a continuous feeding/mixing system that would be easy to characterize (as the stream discharged by the blender could be conveniently sampled), would not promote segregation, and would even be amenable to closed-loop process control?

Following these discussions, the Muzzio team at Rutgers University wrote many proposals to industry seeking funds to create this technology. Invariably, industry representatives rejected those proposals, indicating that they believed that regulators would never approve continuous processes. The argument, more or less, boiled down to "the regulatory framework requires a batch process." It took 5 years to change this mindset, but in 2003, during discussions of the CAMP consortium in New Brunswick, NJ, focusing on the PAT initiative, Dr. Janet Woodcock of the US Food and Drug Administration (FDA) indicated that efforts to implement continuous manufacturing methods would be welcome by the US regulatory authority. FDA documented this conceptual support in 2004 in the revised PAT guidance. Since then, progress has been rapid, fueled by unwavering support from the FDA, soon followed by European and Japanese regulators.

The rest is history. In 2004, the Rutgers team formed a continuous manufacturing consortium, integrated at the time by Merck, Pfizer, Apotex, and GEA. This small band of confederates focused primarily on feeding and blending, demonstrating, using rudimentary versions of the technology, that continuous feeding and mixing was indeed feasible, generating some of the first conference presentations and papers in this area [1−5]. Soon thereafter, in 2006, the Rutgers team, in partnership with Purdue University, NJIT, and the University of Puerto Rico, succeeded in attracting significant funding from the National Science Foundation (NSF) to establish the Engineering Research Center on Structured Organic Particulate Systems (C-SOPS). C-SOPS was entirely dedicated to the systematic application of engineering methods to pharmaceutical product and process design. In the 14 years since C-SOPS was established, more than 60 companies, as well as the FDA and the USP, would join C-SOPS, which focused on continuous manufacturing as its main research effort. This academic−industrial−regulatory partnership attracted research funding in excess of $100 million, published over 500 peer-reviewed papers, and enabled implementation of advanced manufacturing methods at numerous companies.

Rapid progress followed NSF funding. C-SOPS formed a mentor team that included more than 20 industrial representatives, who worked closely with the students, postdocs, and the occasional professor, to integrate a working system. By 2008, this team had integrated gravimetric twin screw feeders, a continuous tubular blender, and a tablet press; implemented hyperspectral NIR sensing; and closed the mechanical controls loop using both Siemens and Emerson control systems.

Shortly thereafter, representatives of JnJ approached the Rutgers and UPR teams and challenged them to reach a new milestone—to create a commercial system. This effort was the development of Prezista continuous manufacturing, described in Chapter 15. Many more collaborations followed, both with JnJ and with other companies, expanding from continuous direct compression (CDC) to continuous wet granulation (CWG), continuous roller compaction (CRC), and more recently, continuous API manufacturing.

Since then, continuous manufacturing has become a reality and is quickly becoming a major modality of manufacturing with active projects all over the world. Importantly, many other organizations have made very substantial contributions to the current state of the art. While JnJ was working closely with C-SOPS, many other companies, among them Vertex, Pfizer, Eli Lilly, GSK, Merck, and Novartis, were working on their own in-house efforts. Many of those efforts have begun to bear fruit, as we currently see, about a decade later, with rapid growth in both the number of approved continuous processes and the number of companies that are able to implement continuous methods to completion.

To be fair, C-SOPS was not the only academic/industrial coalition that contributed to this major change in manufacturing methods. Several others took place in parallel, or soon thereafter. A major effort funded mainly by Novartis at MIT undertook the development of end-to-end continuous manufacturing, integrating the process from the final steps of organic synthesis all the way to the finished product. Another major effort, the Research Center for Pharmaceutical Engineering (RCPE), emerged around 2009 in Austria, led by the Technical University of Graz. Among many accomplishments, the RCPE team took discrete element method (DEM) simulation of pharmaceutical processes to the next level and also demonstrated full integration of a holt-melt extrusion (HME) system. Shortly thereafter, another major effort, CMAC, emerged in the United Kingdom, focusing primarily on continuous API synthesis and crystallization. In the last few years, these efforts have been joined by more consortia, centered in Dublin, Ghent, Sheffield, and Kuopio and more recently in Japan and China.

At about the same time when C-SOPS was getting started, equipment suppliers begun to offer process components and eventually fully integrated systems. The first company to do so was GEA, which launched the ConsiGma system for commercial implementation, soon to be joined by the Excellence United consortium led by Glatt. GEA and Glatt were soon joined by LB Bohle and Powrex, more recently by Bosch and Fette, and soon, we anticipate, by many others.

Finally, funding agencies and regulatory agencies of the US government have played a really important role in enabling and fueling this progress, both by reassuring industry and by providing funding to academia to continue the effort. Our team is forever grateful to both NSF for the early funding and the FDA for its enduring support.

2. The many benefits of continuous solid dose manufacturing

As discussed throughout this book, continuous manufacturing methods enable modeling, sensing, and closed-loop real-time process control. This can lead to better understood processes, better product quality, increased process reliability, facilitate real-time release, enhanced product quality, lower manufacturing cost, increased yield, smaller process footprint, and many other self-evident advantages.

2.1 Improving product quality

Continuous manufacturing processes can enable superior product quality. Our collaborations with the FDA and industry, and our partnership with organizations like the United States Pharmacopoeia, help to ensure this outcome. There are three main reasons for this. First, the near-steady nature of the process enables all portions of material to be processed under equivalent conditions at a constant state of control. Second, because only a small amount of material is processed at any given time, quality attributes of every portion of the process stream can be rigorously monitored to assure quality. Any defective product units can be tracked and scrapped, while retaining only quality-compliant product units. Third, and again because the system is nearly steady and continuous, real-time monitoring, active control, and advanced optimization can be used to ensure that the process remains within operational specifications at all times. In addition, continuous manufacturing enables detailed and accurate computer modeling, assuring a much deeper scientific understanding. This improvement in quality can translate directly into health benefits because defective product may fail to provide its therapeutic benefit or, in extreme cases, cause harm to patients.

2.2 Faster product and process development

For solid dose products such as tablets and capsules, which comprise the great majority of drugs taken by patients, continuous manufacturing has been shown to greatly reduce both the time and cost of developing new medicines. A typical continuous manufacturing line for solid dose product reaches an operational state of control in a matter of minutes. Therefore, extensive studies examining alternative product formulations and multiple process conditions can be performed in just a few days, using only a small amount of raw materials. Moreover, because such development studies are performed using the same equipment that will be subsequently used for manufacturing, no scale-up studies of the process are needed, and process development is further accelerated. As mentioned, this ability to develop products and processes faster and with less waste can have a major impact on the profitability of both brand-based products (which are protected by patents with finite life) and generic products (where the first company to file an approvable application often accrues a larger share of profits). In our opinion, an even more important benefit is the ability of accelerating access to life-saving new medicines to patients that literally cannot wait, providing the strongest incentive for implementing technologies that enable rapid product and process development.

2.3 Faster responses to shortages and emergencies

By enabling faster product and process development, continuous manufacturing can allow manufacturers to develop products quickly to respond to emergencies, to address shortages, and to bring breakthrough therapies to market. The current state of knowledge often enables a skilled practitioner to create a formulation and a process for a given product in just a few weeks. Under emergency conditions, such processes need not be optimum, just adequate, which further enables rapid development. As mentioned, such processes can be developed at the full manufacturing scale and using only a small amount of material, which is often critical early in the life cycle of a product, or when quality issues are detected, because under such conditions, suitable raw materials can be scarce. Moreover, the intrinsically higher reliability of continuous processes should make them safer and easier to approve by regulatory authorities, further enabling rapid response during emergencies.

We believe that this ability to enable faster product development will become a major driver for the implementation of continuous systems in the post–COVID-19 world, not only for solid dose products but also for APIs, injectable products, vaccines, and other product forms. How such an initiative will come together remains to be seen, but the potential benefit is so large that in our opinion it is only a matter of time.

2.4 Potential for reducing drug prices

Continuous manufacturing can help reduce the cost of both prescription and over-the-counter (OTC) drugs in multiple ways. Some of the impact is direct: continuous manufacturing processes have smaller footprint, achieve higher yields, and require less direct labor than their batch counterparts, so they are able to directly impact the cost of making pharmaceutical products. Some of the impact is indirect: because continuous manufacturing processes also enable the manufacture of products with superior quality, and because they enable real-time quality control and, if desired, release, they reduce the cost of assuring product quality. While continuous manufacturing processes require upfront investments in both physical and human infrastructure, they can return this investment rapidly. Moreover, as mentioned, continuous manufacturing products and their required manufacturing processes can be developed faster than their batch-based counterparts. As a result, products developed and manufactured using continuous manufacturing technology can reach the marketplace faster, extending profitability periods for the companies making them. These factors could contribute to lower drug prices to the US consumer, if continuous manufacturing technologies could be adopted in the highly price-competitive generic and the OTC sectors of the pharmaceutical industry.

3. The engineering toolbox, applied to pharmaceutical manufacturing process design

While not entirely new from an engineering perspective, implementation of continuous manufacturing systems in the pharmaceutical industry brought renewed interest in various methodologies, including materials characterization and process modeling, and redefined their use in pharmaceutical process design. The higher level of complexity of continuous systems, emerging from the need to operate multiple processes simultaneously all at the same rate and interacting with each other, enhanced the need for in-depth process understanding and required the creation of new pathway for regulatory evaluation. Traditionally, except perhaps for API synthesis and purification, process engineering methods had encountered little use in pharmaceutical process design. Continuous manufacturing changed this almost overnight, as the need for process modeling and process control became immediately apparent. Responding to this need was an inherently multidisciplinary effort that required the creation of public—private partnerships to bring the technology forward. As mentioned, these partnerships emerged in multiple places in the United States and Europe and more recently in Japan and China.

By the mid-2010s, the question was whether all of this research activity would result in a major new manufacturing mode. Today, with six solid dose continuous manufacturing approvals in the United States as of early 2020 and a similar number in other countries, and dozens of ongoing filings with the US FDA both for dose and drug substance, this manufacturing approach has reached the full commercialization stage. We expect that the adoption process will continue to accelerate, until reaching full maturity.

As the rate of adoption increases, a concise resource on the basic principles of the technology, focusing on providing practical advice regarding implementation, appears to be needed. This book, *How to Design and Implement Continuous Manufacturing Systems for Solid Oral Dosage Pharmaceutical Products*, seeks to meet this need. Our aim is to present the engineering toolbox, as it is applied to continuous pharmaceutical manufacturing process design for solid dose products. Much of the material in the book comes from firsthand experience of the authors and their collaborators as "first adopters" of the technology. The book follows a systematic stepwise approach to the Design of an Integrated Continuous Manufacturing System. To be successful at implementing or at evaluating continuous manufacturing systems, the reader needs to gain an appreciation of the highly integrated nature of continuous manufacturing processes before getting started. One of the key aspects of advanced manufacturing and continuous processing is the ability to work with increased amounts of information and to relate process data to the variability of inputs. These concepts were examined in detail in a previous publication of the authors [6]. Although not collated in a single chapter in this book, the concepts have been revisited in the individual chapters as applicable.

Chapter 2 covers advances in material characterization and what to do with the data to enhance process understanding and increase the speed of process development in a material sparing manner. Chapters 3 through 9 are each devoted to a single unit operation which can collectively be coupled to produce integrated systems for CDC, CWG, or continuous dry granulation or CRC. In many cases, "extrusion" is used as a CWG technology. Till date, applications of this method are largely limited to extrusion granulation. Importantly, similar equipment could be used for continuous melt granulation or paste extrusion processes, and while we do anticipate that such systems will become part of the range of implemented systems in due time, they are not explored in detail in this book.

Chapters 10 through 13 present the key elements of the engineering toolbox as applied to advanced solid dose pharmaceutical manufacturing, including process monitoring, modeling and optimization, automation, and closed-loop process control. While full implementation of these toolbox components remains work in progress, in our opinion it is only a matter of time until they become standard practice, given their proven ability to make manufacturing approaches both more reliable and more efficient.

In addition to covering the technical aspects of process development, the book also touches upon associated, nontechnical considerations in bringing a CM drug product to the market. Chapter 14 is dedicated to understanding regulatory expectations. Aspects such as the evolving nature of engagement with regulatory bodies, difference between filing a batch and continuous processes, international harmonization, and role of academia have also been discussed in this chapter. Chapter 15 and Chapter 16 are industrial case studies. They are experiences penned by pharmaceutical organizations describing what it took to build a continuous manufacturing (CM) line, develop their first product, and have it approved by the US FDA. Chapter 15 is authored by a group from Janssen Pharmaceuticals and describes the story of converting an existing batch process to a continuous process for their product *Prezista*. Chapter 16 is authored by scientists from Vertex Pharmaceuticals and narrates the effort of bringing their first CM drug product, *Orkambi*, to the market. Lastly, Chapter 17 presents the editors' outlook on the technology— existing technology gaps, maturation of the technology and its implications, future role of academia and CMOs, role of regulatory bodies, future skill set and human resource needs, and more. The book is intended to serve as a guidebook for a relatively new adapter of CM technology. The idea is for the book to serve as a primer, walking the adopter through the steps of designing and building a CM line and developing a tableted product. It is hoped that experts in the field will find useful chapters on regulatory expectations, industrial experiences, and technology outlook, in addition to the design chapters.

References

[1] Portillo P, Ierapetritou M, Muzzio FJ. Development of Control Strategies for Blending Operations in Pharmaceutical Processes. In: Paper 414m, AIChE annual meeting, Austin, TX; November 2004.

[2] Portillo PM, Muzzio FJ, Ierapetritou MG. Modeling Granular Mixing Processes Utilizing a Hybrid DEM-Compartment Modeling Approach. In: AICHE annual meeting; November, 2005.

[3] Portillo PM, Ierapetritou M, Muzzio FJ. Characterization and Modeling of Continuous Convective Powder Mixing Processes. In: AICHE annual meeting; November, 2006.

[4] Portillo PM, Muzzio FJ, Ierapetritou MG. Characterizing powder mixing processes utilizing compartment models. Intl J of Pharmaceutics 2006;320:14−22.

[5] Portillo PM, Muzzio FJ, Ierapetritou MG. Modeling and designing powder mixing processes utilizing compartment modeling. Comp Aid Chem Eng 2006;21(C):1039.

[6] Oka S, Escotet-Espinoza SM, Singh R, Scicolone J, Hausner D, Ierapetritou M, Muzzio FJ. In: JKP K, et al., editors. Design of an integrated continuous manufacturing system. John Wiley & Sons Ltd; 2017.

Chapter 2

Characterization of material properties

Sonia M. Razavi[1], Sarang Oka[1,4], M. Sebastian Escotet-Espinoza[1,2], Yifan Wang[5], Tianyi Li[1], Mauricio Futran[3] and Fernando J. Muzzio[1]

[1]*Engineering Research Center for Structured Organic Particulate Systems (C-SOPS), Department of Chemical and Biochemical Engineering, Rutgers, The State University of New Jersey, Piscataway, NJ, United States;* [2]*Oral Formulation Sciences and Technology, Merck & Co., Inc., Rahway, NJ, United States;* [3]*Janssen Supply Chain, The Janssen Pharmaceutical Companies of Johnson and Johnson, Raritan, NJ, United States;* [4]*Hovione, Drug Product Continuous Manufacturing, East Windsor, NJ, United States;* [5]*US Food and Drug Administration, Center for Drug Evaluation and Research, Office of Pharmaceutical Quality, Silver Spring, MD, United States*

1. Introduction

A thorough understanding of raw materials and intermediate blends is crucial to the successful design and operation of a robust continuous process, wherein achieving and maintaining a steady operation is paramount [1−3]. The effect of raw material properties and material history on the unit operation performance, intermediate materials, and final products attributes is conceptually self-evident and has been confirmed in other areas such as ceramics [4−7], metallurgy [8−11], catalysts [12−14], and food [15−20].

It is thus clear that a thorough characterization of raw and intermediate material properties is essential for the development of a robust process. Due to the complex nature of bulk material properties, there are numerous measurement techniques, each one quantifying a specific aspect of behavior, and often performed under different conditions. Moreover, a theory of powder flow, rooted in first principles, does not currently exist. While powder behavior and its impact in a process is a multivariate phenomenon, the use of multivariate methods in process development of pharmaceuticals has only recently begun to be considered, after the PAT and QbD initiatives of the FDA circa 2003 [3,21−24].

Given the complex and multivariate nature of the problem, building comprehensive property databases of raw and intermediate properties of granular components and examining their effects on the process and the product is essential to understand how material properties affect the process

How to Design and Implement Powder-to-Tablet Continuous Manufacturing Systems
https://doi.org/10.1016/B978-0-12-813479-5.00002-1

9

performance. A material property library that complies with modern quality systems set forth within the Current Good Manufacturing Practices paradigm can be used in many applications. Such databases can serve as a repository of information describing effects of material properties on process performance, facilitate the standardization of testing methods and the specification of property value ranges for commercially available ingredients, support the identification of equivalent (i.e., interchangeable) materials based on their properties, enable the identification of surrogate materials for process development based on their properties, and facilitate the development of mathematical models relating material properties to unit operation performance. A schematic of the material property database applications is shown in Fig. 2.1.

Having standardized testing methods can help the process developers in establishing a framework for characterizing new materials, which serves as a starting point for their development efforts. Material property knowledge which is gathered during the manufacturing stage can be used to detect trends and evaluate current and future processes [26]. Furthermore, the collected measurements can be used to determine the concise number of material properties that need to be measured as well as to identify the specific measurements that can provide relevant information about the material.

The foundation of developing material property databases is a robust, reproducible, and standard method of measuring powder properties. There exist several commercially available powder characterization tools such as powder rheometers,[1] particle size analyzers, and segregation testers. In addition, there are other measurements for powder characterization that have recently gained attention such as powder cohesion, dilation, permeability, compressibility, shear sensitivity, hydrophobicity/wettability, and electrostatics. However, the latter methods have not yet been fully standardized,

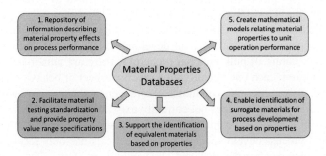

FIGURE 2.1 Potential applications of material property databases [25].

---

1. While, rigorously, a rheometer is a device that creates a rheometric flow with a uniform shear field, the term "rheometer" has been popularized for powder flow testing and is used here in that manner.

which prevents the creation of reproducible characterization metrics. Moreover, results obtained using different methods for shear cell measurements cannot be easily translated between instruments [27,28] and different particle size testers can give different results [29]. It is worth mentioning that the different results that are obtained by various measurement techniques are often highly collinear. This makes it possible to collapse the relevant information to a smaller number of degrees of freedom that can be sampled equivalently by different methods. Usually, at the early stages of library development, it is not clear which testing methods are more accurate or meaningful. Thus, all available methods of material property measurements should be considered in developing a comprehensive database. After the formation of the initial library, with the gathered information, one can evaluate which material properties are most relevant to a given process, the relative accuracy of the measurements, and the degree of co-linearity among measurements. This analysis can be performed using several multivariate analysis techniques, some of which will be discussed in this chapter.

In this chapter, a summary of the characterization techniques that are currently used in the Engineering Research Center for Structured Organic Particulate Systems (ERC-SOPS) is briefly described. This is followed by a description of how to use and analyze the material characterization data library. The chapter continues by providing different applications of material database: selecting the suitable surrogate or tracer in residence time distribution (RTD) studies, creating models for unit operation performance incorporating material properties, predicting the performance of a new powder using the existing knowledge, and choosing the desired powders/blends for different design of experiments (DOEs) based on their material properties.

2. Summary of the characterization techniques and their description

In this section we will discuss the commonly used standardized characterization techniques to build a material library database. Several measurement techniques described below are widely known and thus will not be described in detail. For a more comprehensive description of these methods, the reader is directed to references specified in the text. These techniques are not limited to what is provided in this chapter. Other techniques, such as specific surface area, dynamic vapor sorption, static image analysis, loss on drying, and dynamic avalanche angle, to name a few, can be found elsewhere [30].

2.1 Bulk density test

This property is estimated following the standard procedure outlined by US Pharmacopeia Chapter <616> [31]. A 250-mL graduate cylinder is gently filled with powder, preferably near the 150-mL level, and the powder mass is recorded. The initial aerated density (i.e., mass of powder divided by the occupied volume, $\rho_B = (M/V)$) is calculated using the mass and volume measurements.

2.2 Particle size distribution test

Several methods are used to measure the particle size distribution of granular materials. Laser diffraction, sieve analysis, and image processing of true or magnified particles are the most popular methods. The selection of the method is dictated by the size, shape, and sometimes cohesivity of the particles being measured. A thorough review of different particle size measurement techniques can be found in Ref. [29].

2.3 Powder flow measurements

Properties commonly used to describe flow behavior of powders can be measured using a shear cell. Properties such as major principle stress, unconfined yield stress, flow function, cohesion, angle of internal friction, and angle of wall friction can be measured using shear cell testers. In addition, some shear cell testers are provided with accessories, which allow the measurement of properties such as permeability, compressibility, powder flow stability, and variable flow energy tests. Moghtadernejad et al. [26] describe in detail the methods used to measure flow properties using shear cell testers and powder rheometers. As previously mentioned, there are several commercially available shear cells. Koynov et al. [27] examined the performance of the most widely used shear cells. Wang et al. [28] recently demonstrated that for many powders the data generated by shear cells at various consolidation stresses can be collapsed onto a single master curve.

2.4 Powder hydrophobicity/wettability measurements

Hydrophobicity/wettability is a measure of a material's affinity for water (or lack thereof). One of the primary quality characteristics of the final dosage form is its dissolution in aqueous media. The hydrophobicity of the blends comprising the dosage forms is therefore critical. Because blend properties are often a function of the raw material properties as well as the processing conditions, and because blends can become more hydrophobic during processing, the wettability of the raw materials and the effect of processing on blend hydrophobicity should be characterized.

The wettability of a powder can be measured using a modified Washburn technique. The uptake of water into a powder bed due to capillary action was first described by Washburn in 1921 [32]. The volume of water that permeates the powder bed increases linearly with the square root of time. A hydrophobic powder will resist the capillary action resulting in a slower rate of water uptake. The relationship of water uptake to time can also be expressed as a linear relationship between mass of water in the powder bed squared and time.

$$t = \frac{\eta}{C\rho^2 \gamma cos\theta} m^2$$

where t is the time, η is the liquid viscosity, C is a geometric factor comprising the effect of powder packing density and particle size, ρ is the liquid density, γ is the liquid surface tension, θ is the contact angle between the liquid and particles, and m is the mass of the liquid in the powder bed column. The slope of this line is related to the hydrophobicity of the material. Alternately, when comparing the hydrophobicity of one material to the other, one may also compute the contact angle with water by the same method. The geometric factor of the powder material, which is a property of the powder alone, can be computed by first performing the measurement with a liquid whose contact angle with the material is zero (for example, n-hexane or silicone oil). The contact angle with water can then be computed by repeating the test using water as a wetting medium and using the value of "C" obtained from the previous measurement. A thorough description of the Washburn can be found in Refs. [33,34,34a].

Powder hydrophobicity can also be measured using drop penetration tests. In this method, drop volume as a function of time is plotted and analyzed as the drop penetration profile. Silicone oil (Poly(dimethylsiloxane) or any other fully wetting solvent can be used as the reference liquid. Generated by a manual syringe, a droplet is ejected and deposited on the powder bed. Sequences of droplet penetration images are recorded using a CCD camera. Using image analysis, the penetrated volume is calculated assuming the shape of the droplets is axisymmetric. The same procedure is repeated using water or the liquid of interest. The contact angle can be calculated using dimensional analysis. A smaller cosine value is indicative of lower hydrophobicity. A thorough description of the drop penetration method can be found in Ref. [35].

2.5 Electrostatic measurement (impedance test)

Several pharmaceutical blends are electric insulators, meaning they can easily acquire and hold charge upon friction or contact with other materials. Particle charging is a complex phenomenon. While some of the fundamental causes of this phenomenon have been examined, a complete theory of powder electrostatics, much like powder flow, does not exist [36−38]. Various techniques have been proposed to characterize various aspects of electrostatic behavior [39,40]. A thorough review of the various measurement methods used to investigate powder electrostatics can be found in Ref. [41].

3. Developing a material property database

A comprehensive material characterization exercise generates large quantities of data. Data are generated not only from the benchtop testing of materials in powder property testers but also from the process analytical sensors during process development and normal operation. Availability of such extensive amount of data enables the application of statistical and data-driven tools to these databases. The application of these tools transforms material databases from just a collection of information into powerful predictive methodologies. They can reduce dimensionality of the measurements, informing the practitioner which tests capture the majority of the variance in the measurements. It also enables prediction of material behavior and process performance and consequently final product quality. This can facilitate future process development. There are ongoing efforts, both in industry and academia, aimed at developing extensive databases of material properties. A primary objective is the application of statistical and data-driven methods in order to identify material properties that will best characterize the performance of a certain unit operation. For example, there are statistically significant correlations that can predict the RTD and material holdup in blenders as a function of material properties and process parameters. Besides facilitating process development, such databases can also be used to select a surrogate material—an alternate, widely available, inexpensive material which has "similar enough" properties as that of an expensive, scarce material that is not available in bulk quantities for process development. Concentrated efforts to develop such databases have become a priority for organizations that have multiple products in their CM pipeline. Such databases would be equally valuable, if not more, for contract development and manufacturing companies (CDMOs) who are likely to examine, handle, and process a large number of different materials during process development and operation, given the nature of their business. The different applications of material data library are discussed in more depth in the following sections.

4. Multivariate analysis

Material property libraries can aid researchers significantly during development to reduce the amount of material needed for characterization. On the other hand, the library needs to be broad enough to encompass all the behaviors of interests. Multivariate analysis is a useful tool to evaluate material property libraries that are constantly growing in both dimensions and variability by combining partially collinear variables and describing them by fewer, so-called latent variables. The main purpose is to separate useful information from noise and detect patterns within the data, especially when there are no visually obvious natural groupings. Among several multivariate methods that can be applied to extract information from them, principal component analysis (PCA) and clustering analysis will be discussed in detail in this chapter. An example of using partial least squares regression (PLSR) to predict process performance will also be discussed.

4.1 Principal component analysis

The oldest and most common latent variable projection method is PCA. PCA uses a vector space transform to reduce the dimensionality of a large data set and replace it by a smaller number of independent linear combinations of its data variables [42,43]. Specifically, a model is used to represent a data set (X) in a reduced dimension (principal component [PC] space) such that the major axes of variability are identified [44]. The data set X can be decomposed, based on the equation below, into a set of scores (T) and loadings (P), while the remaining variability is modeled as random error (ε).

$$X = TP^T + \varepsilon$$

By retaining only a small number of eigenvectors, the original set of measurements is effectively mapped into a smaller set of new variables that are mutually orthogonal and each new variable is a linear combination of the original variables. The reduction of data dimensionality by projecting the data onto orthogonal coordinates is an effective approach to efficiently generating process models for development, as well as minimizing the number of variables (i.e., number of material properties being measured) [45]. Some of the major disadvantages of PCA are (1) ambiguous physical interpretation of the new variables, (2) large amounts of experimental data required, and (3) lack of robustness of the method in some instances (i.e., PCs can be miscalculated and yield different results). These disadvantages are often addressed by performing sequential PCA analysis, which in turn allow researchers to examine the model limitations.

The PCs, which initially match the number of dimensions, are obtained as the correlation matrix's eigenvectors of the original measurements. However, as the goal is to minimize the data set's dimensionality, not all eigenvectors need to be retained, given that they contain decreasing amounts of information. An important question is how to select the optimum number of eigenvectors such that each one captures a statistically significant fraction of the observed variability. Several approaches are proposed, including a cumulative variance criteria and a statistical significance method. The cumulative variance criterion selects a threshold to determine the number of PCs that will be retained after analysis. This method is one of the most widely used, as it can be adjusted based on the researchers target number of PCs needed. The more rigorous statistical method determines which PCs are statistically significant by means of the Bartlett test, which examines the fractional variance of each eigenvalue/eigenvector pair and calculates the corresponding P-value using the Chi-square distribution and degrees of freedom in the matrix [46].

To facilitate comprehension of applying PCA methods to a material database, an example is discussed below. Seven materials were characterized to build a model using PCA, with each material comprising of 30 flow indices. In the addition to the battery of shear cell tests performed to develop the database, additional tests such as compressibility test, permeability test, and dynamic flow test were also performed. Coarse γ-alumina, fine γ-alumina, fine zeolite, zeolite (Y-type CBV 100), Satintone (calcined kaolin SP-33), and lactose monohydrate were used as part of this exercise. Although the example presented here is not comprised of commonly used pharmaceutical materials, the principles of developing a model can be translated across materials.

Fig. 2.2 shows a 3D score plot with three PCs, which encompass information from a total of 210 measurements used to develop a database. The applications of PCA performed on material databases are discussed in Section 5.1.

A similar exercise has also been performed by Van Snick et al. [30], where a total of 55 different raw materials were characterized and described by more than 100 raw material property descriptors to develop an extensive material property database. PCA was applied to reveal similarities and dissimilarities between materials and to identify overarching properties.

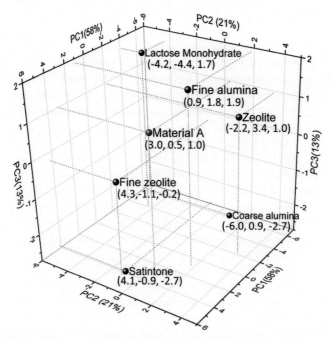

FIGURE 2.2 A cubic score plot is used to illustrate how different materials are distributed in the projected space. The coordinates of each material are shown as the scores of each principal component [28].

4.2 Clustering analysis

Another common multivariate method that can be applied to the material property library is clustering analysis. Clustering analysis is an unsupervised machine learning method used to classify any set of data into groups (i.e., clusters) with similar attributes without requiring a previously developed learning set [47,48]. The method consists of forming data element groups representing proximate collections of data elements based on a distance or dissimilarity function measured using different distance methods (e.g., Euclidean, Manhattan), wherein identical data pairs have zero distance or zero dissimilarity. This methodology proves to be very helpful when studying library sets, as it provides a means to establish relationships and partitions between the different data points. Thus, this methodology becomes of interest when considering materials in a property library, particularly when attempting to classify these materials based on their properties. Among available algorithms, hierarchical categorical clustering and nonhierarchical K-means clustering are the most commonly used methods. The reasons for their popularity include ease of implementation, speed of convergence, and adaptability to sparse data. Mathematically, categorical clustering determines the distance between points and, given a predefined number of groups, allocates the components in each group based on the maximum distances between groups. K-means clustering partitions all data points into predefined number of sets to minimize the within-cluster variance and, as the total variance of the data is constant, it also maximizes the between-cluster variance.

An example is presented to further elucidate the method. Details of the example, the materials characterized, the measurements performed, and the clustering method can be found in Ref. [25]. A total of 20 commonly used pharmaceutical materials were used to devise a material property library for analysis. Five active pharmaceutical ingredients (APIs) and 15 excipients composed the material data set. A total of 32 measurements were collected for each of the 20 materials, leading to a data set of 640 measurements. The clustering method evaluated distance using the squared Euclidean distance metric to utilize all available material properties (i.e., 32 measurements). The distance grouping algorithm to cluster materials was based on their proximity in Euclidian space. The goal was to perform the analysis using standard clustering methods, rather than using dimensionality reduction methods, to analyze the material property database. Fig. 2.3 shows the results from the hierarchical clustering using the Ward method. Fig. 2.3A shows the dendrogram resulting from the clustering, Fig. 2.3B shows the distance between clusters, and Fig. 2.3C represents the distance added by increasing the number of clusters.

Fig. 2.3A shows which materials share a closest neighbor based on their properties. The results indicate that acetaminophen and API 3 share the most commonality based on their material properties, which points to the use of

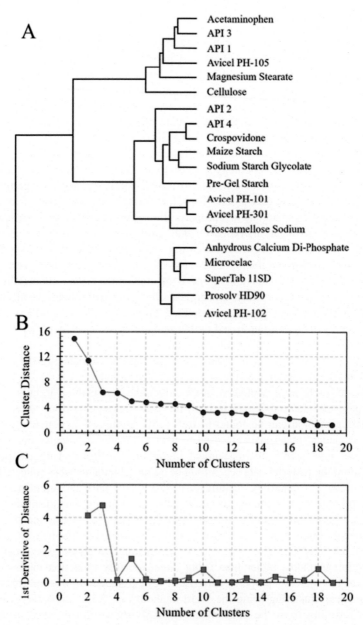

FIGURE 2.3 Hierarchical clustering results of material properties in a database. (A) Dendrogram, (B) distance between clusters as a function of the number of clusters, and (C) first derivative of distance between clusters.

acetaminophen (a common available pharmaceutical API) as a surrogate for API 3. Moreover, acetaminophen and API 3 are then closest similar to API 1, based on the clustering. The distance between the clusters is shown in Fig. 2.3B. Two groups decreased the cluster distance from 15.3 to 11.2. Nevertheless, two clusters lead to large groups of materials, which would not be representative of the vast differences captured by the material properties. Three clusters further decreased the distance between the groups to 6.4, which indicated that the selected cluster were displaying less within-group differences and more between-group differences. Increasing the number of clusters to four did not significantly decrease the distance within clusters (distance = 6.3), suggesting that the differences represented between four and three clusters are comparatively small. Moreover, Fig. 2.3C shows that the addition of information by adding clusters is significant until the third cluster and then it is close to zero when increased beyond a fourth cluster. Therefore, in order to express the most amount of differences in the tested data set in the fewest number of clusters and to maintain the number of materials in a group larger than the number of groups, three clusters were selected for the analysis in this case. Similar to a PCA method, which will be discussed in Section 5.1, clustering analysis can also be used to predict process performance of materials in various unit operations. For the sake of brevity, this will not be discussed here in detail. Escotet et al. showed how materials within the same clusters exhibit similar process performance [25].

5. Application of material property databases

The availability of a comprehensive material property database can be a powerful tool in the arsenal of a continuous manufacturing process development group. Material databases enable prediction of material behavior and process performance and consequently final product quality. Statistical methods can be applied to such databases in order to identify material properties that will best characterize the performance of a certain unit operation. As mentioned, besides facilitating process development, such databases can also be used to select surrogate material. Multivariate methods, when applied to material databases, also facilitate the selection of appropriate tracers for accurate characterization of RTD experiments. For the sake of brevity, only two examples will be discussed here: a PCA-based method to identify a surrogate material and a PLSR method to predict process performance.

5.1 Identifying similar materials as surrogates for process development

Material property information captured on a PCA score plot (or by other multivariate analysis techniques described above) can be applied to identify surrogate materials. The example presented here illustrates how a surrogate material can be used to predict the performance of a "similar" material. Specifically, the following three questions are attempted to be answered in the presented example: (i) How can we compare a given new material to previously characterized materials? (ii) For a material with given properties, can we predict its feeder performance? (iii) Furthermore, for a material with given properties, can we predict the optimal tooling that achieves a specified feeding performance?

A good indication of similarity between two materials is a measure of the distance between them in the re-projected material property space [49]. PCs are orthogonal to each other, and each is associated with the value that explains the proportion of variability in the data set [50]. Usually, only a few PCs are retained based on their statistical significance. The similarity scores between Material A and Material B can thus be calculated based on weighted Euclidean distance $(d_{w(a,b)})$:

$$d_w(a, b) = \sqrt{\sum_{i=1}^{n} w_i * (a_i - b_i)^2}$$

where n is the total number of principal components selected in the model, a_i is the score of Material A in the ith principal component, b_i is the score of Material B in the ith principal component, w_i is the weight of the ith principal component, namely the relative variability explained by the ith principal component:

$$0 < w_i < 1$$

The similarity score is based on the concept of distance between objects, which quantifies the similarity or dissimilarity between observations in the data set [49]. A relatively small value of d_w suggests similarity between materials. Thus, if a surrogate for a material "X" needs to be identified, this method suggests that a material closest to this material in the PC space will result in the most similar performance in a unit operation. This method of first identifying similarity between materials and then using unit operation performance of the most similar material to predict performance is called PCA-SS (PCA followed by similarity scores).

In this example, where the objective was to identify a material that behaves similar to Material A, Material A was characterized and plotted on the PC space, as shown in Fig. 2.2. The Euclidian distance of Material A from all other materials in the database was calculated and the results are presented in Fig. 2.4.

As can be observed, fine zeolite was closest to Material A in the PC space. Moreover, the identified similarity between Material A and fine zeolite using PCA-SS also translated to similarity in performance of the materials in a screw feeder. The relative standard deviation (RSD) and the relative deviation from the mean (RDM) were used to evaluate and compare the feeder's feeding

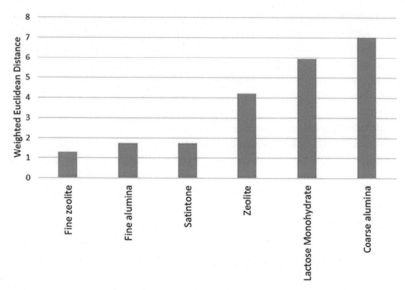

FIGURE 2.4 Weighted Euclidean distance between Material A and other existing materials was calculated. Based on the concept of distance between objects in the reduced dimensions, fine zeolite was identified to be the most similar material to Material A [51].

performance for each calibration materials (six materials originally plotted on the PC space). Gravimetric feeding tests were also performed with Material A to obtain Material A's RSD and RDM.

Based on the results observed in Fig. 2.5, fine zeolite is able to accurately predict the performance of Material A in a feeder.

The advantage of the PCA-SS approach is the ability to quickly identify similar materials considering all the available flow property measurements.

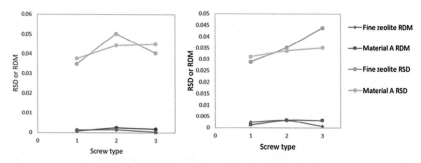

FIGURE 2.5 Comparison of the feeding performance between fine zeolite and Material A at (A) 50% of the initial feed factor for the fine concave screw and at (B) 80% of the initial feed factor for the fine concave screw. Screw type 1 corresponds to the fine concave screw, screw type 2 corresponds to the fine auger screw, and screw type 3 corresponds to the coarse concave screw [51]. *RDM*, relative deviation from the mean; *RSD*, relative standard deviation.

The predicted material is also included when developing the PCA model, which enables the model to fully explore the overall data structure and patterns. The prediction error can be reduced significantly by using similarity scores, instead of projecting the predicted material data to the existing PCA models. When the material available is limited, this approach is especially useful to identify surrogate material that can be used to quickly identify the design space and avoid failure modes during process development. Additionally, the selected surrogate material can be used to accelerate process scale-up and technology transfer. However, there are limitations to the PCA-SS approach. The approach is based on an assumption that materials with similar flow properties will perform similarly in identical physical configuration of the equipment. If the calibration materials do not cover a wide range of flow properties, the difficulty in finding similar materials may potentially increase prediction error.

5.2 Predicting process performance using material property databases

To consider the potential quantitative correlation between material properties and process performance, PLSR models can also be developed. The example discussed here leverages the same material property database to predict feeder performance using a PLSR method. Four-factor nonhierarchal PLS model were fitted to the material property data consisting of 6 calibration materials and 30 flow parameters described previously. The response variables were the RSD and RDM for each screw at feed rates corresponding to 80% of the initial feed factor. The cumulative percent variance explained by the fourfactors in all the models was above 95%. Fig. 2.6 suggests that good parity was observed between the predicted and actual y values (RSD or RDM).

The root mean square error of calibration (RMSEC) and root mean square error of cross-validation (RMSECV) are also shown. The regression coefficients for the RSD models are presented in Fig. 2.7. The regression coefficients indicate the contribution of each material flow property to the prediction models, with higher absolute values of the coefficients indicating larger contributions. A positive value of the coefficients indicates a positive correlation with the RSD, and a negative coefficient value indicates inverse correlation. Material flow properties with relatively small coefficients suggest that the contribution to the prediction is small. Fig. 2.7 also suggests that material properties contribute to feeder performance differently when different screws are selected.

The RSD and RDM of a material during the feeding process can be predicted based on the obtained regression coefficients and material flow properties. Fig. 2.8 shows the predicted results of Material A in comparison to the measured results. It can be seen that the developed PLSR models were able to predict RSD and RDM values that are reasonably similar to the experimental values.

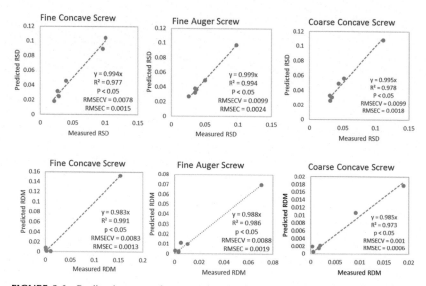

FIGURE 2.6 Predicted versus reference parity plot for feeding performance, represented by relative standard deviation (RSD) and the relative deviation from the mean (RDM), for three screws.

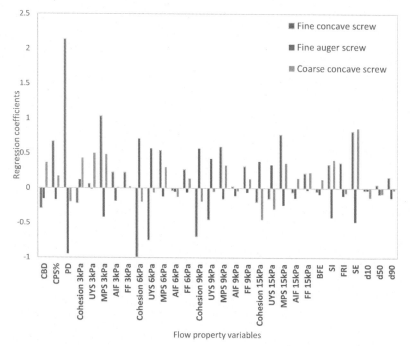

FIGURE 2.7 Regression coefficients for partial least squares regression models of relative standard deviation prediction.

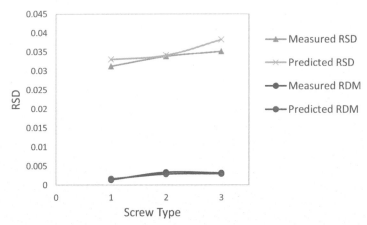

FIGURE 2.8 Predicted feeding results for Material A, in comparison with experimental results. Screw type 1 corresponds to the fine concave screw, and screw type 2 corresponds to the fine auger. *RDM*, relative deviation from the mean; *RSD*, relative standard deviation.

As previously mentioned, similar to PLSR methods, clustering analysis among other techniques can also be used to predict process performance by leveraging a database of material properties. Escotet and coworkers have shown the use of clustering analysis to predict process performance in loss-in-weight screw feeders and continuous blenders [25,52]. Fig. 2.9 attempts to

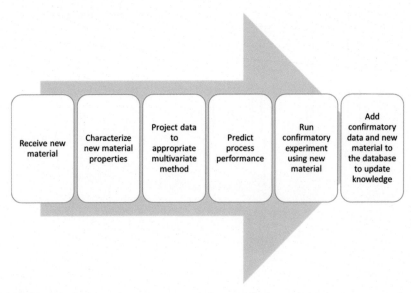

FIGURE 2.9 Workflow of multivariate methods as applied to material databases to predict process performance.

formalize a workflow using any appropriate multivariate method to predict performance of a material whose material properties are either known or can be measured.

In addition to the applications discussed above, material property databases have also been used to select appropriate tracers for performing RTD experiments. It is generally accepted that disparity in the properties of the tracer and bulk material results in improper characterization of the RTD in a continuous flow system. Multivariate methods discussed above can be applied to material databases in order to identify tracers that best mimic the bulk material (but are still chemically distinguishable) resulting in accurate characterization of RTD. Details of the work describing the above can be found in Refs. [53,54].

6. Conclusions

At the time of this writing, continuous manufacturing has been shown to be a feasible method for pharmaceutical manufacturing that is attracting rapidly growing interest. Moreover, for solid dose products, a consensus is rapidly emerging where direct compression is the preferred mode of implementation. However, the main limitation in implementing continuous direct compression is whether material properties of ingredients enable its success. Powder flow properties determine whether accurate and consistent feeding is possible at a desired rate, electrostatics determines the tendency to stick to equipment surfaces or to agglomerate, both of which can pose risks to blend and content uniformity, and API compactability determines the limit in the maximum amount of active ingredient in a formulation. Understanding and controlling these properties in fact determines the applicability range of direct compaction (and to some extent, of any continuous manufacturing method).

Given the interest in continuous direct compaction, "preprocessing" techniques aiming to provide ingredients with suitable properties are rapidly emerging. However, a central need in the development of such methods, important also to the efficient selection of formulation ingredients, is the ability to predict the target range in material properties required for process success. This is an active area of research, leading to a rapidly growing interest in the creation of predictive process models capable of predicting process performance directly from material property measurements.

Other key needs have been identified. As the relevant properties are largely well defined, the next obvious steps are (1) to standardize material characterization methods so that measurements from different laboratories can be shared and (2) to create material property databases where the most common materials of interests are characterized and their properties are known.

Meeting these goals is only a matter of time, and once they are achieved, manufacturing will be greatly facilitated. We enthusiastically anticipate updating the contents of this chapter along these lines in the near future.

References

[1] Hasa D, Jones W. Screening for new pharmaceutical solid forms using mechanochemistry: a practical guide. Adv Drug Del Rev 2017;117(Suppl. C):147−61.

[2] Sun CC. Microstructure of tablet—pharmaceutical significance, assessment, and engineering. Pharmaceut Res 2017;34(5):918−28.

[3] Yu LX. Pharmaceutical quality by design: product and process development, understanding, and control. Pharmaceut Res 2008;25(4):781−91.

[4] Becher PF. Microstructural design of toughened ceramics. J Am Ceramic Soc 1991;74(2):255−69.

[5] Meyers MA, Mishra A, Benson DJ. Mechanical properties of nanocrystalline materials. Prog Mat Sci 2006;51(4):427−556.

[6] Sigmund WM, Bell NS, Bergström L. Novel powder-processing methods for advanced ceramics. J Am Ceramic Soc 2000;83(7):1557−74.

[7] Zhilyaev AP, et al. Mechanical behavior and microstructure properties of titanium powder consolidated by high-pressure torsion. Mat Sci Eng 2017;688(Suppl. C):498−504.

[8] Amherd Hidalgo A, et al. Powder metallurgy strategies to improve properties and processing of titanium alloys: a review. Adv Eng Mat 2017;19(6). 1600743-n/a.

[9] Kallip K, et al. Microstructure and mechanical properties of near net shaped aluminium/alumina nanocomposites fabricated by powder metallurgy. J Alloys Compd 2017;714(Suppl. C):133−43.

[10] Khodabakhshi F, Simchi A. The role of microstructural features on the electrical resistivity and mechanical properties of powder metallurgy Al-SiC-Al$_2$O$_3$ nanocomposites. Mater Des 2017;130(Suppl. C):26−36.

[11] Shen J, et al. The formation of bimodal multilayered grain structure and its effect on the mechanical properties of powder metallurgy pure titanium. Mater Des 2017;116(Suppl. C):99−108.

[12] Bezemer GL, et al. Cobalt particle size effects in the Fischer−Tropsch reaction studied with carbon nanofiber supported catalysts. J Am Chem Soc 2006;128(12):3956−64.

[13] Min M-k, et al. Particle size and alloying effects of Pt-based alloy catalysts for fuel cell applications. Electrochim Acta 2000;45(25):4211−7.

[14] Xie HY, Geldart D. Fluidization of FCC powders in the bubble-free regime: effect of types of gases and temperature. Powder Technol 1995;82(3):269−77.

[15] Chegini GR, Ghobadian B. Effect of spray-drying conditions on physical properties of orange juice powder. Drying Technol 2005;23(3):657−68.

[16] Moreyra R, Peleg M. Effect of equilibrium water activity on the bulk properties of selected food powders. J Food Sci 1981;46(6):1918−22.

[17] Agudelo C, et al. Effect of process technology on the nutritional, functional, and physical quality of grapefruit powder. Food Sci Technol Int 2016;23(1):61−74.

[18] Fitzpatrick JJ, Barringer SA, Iqbal T. Flow property measurement of food powders and sensitivity of Jenike's hopper design methodology to the measured values. J Food Eng 2004;61(3):399−405.

[19] Kondor A, Hogan SA. Relationships between surface energy analysis and functional characteristics of dairy powders. Food Chem 2017;237(Suppl. C):1155−62.

[20] Mani S, Tabil LG, Sokhansanj S. Effects of compressive force, particle size and moisture content on mechanical properties of biomass pellets from grasses. Biomass Bioenergy 2006;30(7):648−54.

[21] Aksu B, De Beer T, Folestad S, Ketolainen J, Linden H, Lopes JA, de Matas M, Oostra W, Rantanen J, Weimer M. Strategic funding priorities in the pharmaceutical sciences allied to quality by design (QbD) and process analytical technology (PAT). Eur J Pharma Sci 2012;47:402–5.

[22] FDA. In: Rockville MD, editor. Guidance for Industry: Pat — a framework for innovative pharmaceutical development, manufacturing, and quality assurance, U.D.o.H.a.H. Services; 2004.

[23] Almaya A, et al. Control strategies for drug product continuous direct compression-state of control, product collection strategies, and startup/shutdown operations for the production of clinical trial materials and commercial products. J Pharm Sci 2017;106(4):930–43.

[24] ECA. Why did FDA change their guideline on process validation? GMP News; 2011 [cited 2013]. Available from: http://www.gmp-compliance.org/eca_news_2600.html.

[25] Escotet-Espinoza MS, Moghtadernejad S, Scicolone J, Wang Y, Pereira G, Schäfer E, Vigh T, Klingeleers D, Ierapetritou MG, Muzzio FJ. Using a material property Library to find surrogate Materials for pharmaceutical process development. Powder Technol 2018.

[26] Moghtadernejad S, Escotet-Espinoza MS, Oka S, Singh R, Liu Z, Román-Ospino AD, Li T, Razavi S, Panikar S, Scicolone J, Callegari G, Hausner D, Muzzio FJ. A Training on: continuous manufacturing (direct compaction) of solid dose pharmaceutical products. J Pharm Innov 2018;13(2):155–87.

[27] Koynov S, Glasser B, Muzzio F. Comparison of three rotational shear cell testers: powder flowability and bulk density. Powder Technol 2015;283(Suppl. C):103–12.

[28] Wang Y, et al. A method to analyze shear cell data of powders measured under different initial consolidation stresses. Powder Technol 2016;294(Suppl. C):105–12.

[29] Allen T. Particle size measurement, vol. 4. USA: Springer; 1990 [Dordrecht].

[30] Van Snick B, et al. A multivariate raw material property database to facilitate drug product development and enable in-silico design of pharmaceutical dry powder processes. Int J Pharm 2018;549(1–2):415–35.

[31] Pharmacopeia, T.U.S. <616> bulk density and tapped density of powders. In: Stage 6 harmonization. The United States Pharmacopeia: The United States Pharmacopeial Convention; 2012.

[32] Washburn EW. The dynamics of capillary flow. Phys Rev 1921;17(3):273.

[33] Llusa M, et al. Measuring the hydrophobicity of lubricated blends of pharmaceutical excipients. Powder Technol 2010;198(1):101–7.

[34] Oka S, et al. The effects of improper mixing and preferential wetting of active and excipient ingredients on content uniformity in high shear wet granulation. Powder Technol 2015;278:266–77.

[34a] Wang Y, Liu Z, Muzzio F, German D, Gerardo C. A drop penetration method to measure powder blend wettability. Int J Pharm 2018;538:112–8.

[35] Liu Z, et al. Capillary drop penetration method to characterize the liquid wetting of powders. Langmuir 2016;33(1):56–65.

[36] Harper W. The Volta effect as a cause of static electrification. Proc R Soc Lond A 1951;205(1080):83–103.

[37] Lowell J, Rose-Innes A. Contact electrification. Adv Phys 1980;29(6):947–1023.

[38] Jones T. Electromechanics of particles. New York: Cambridge University Press; 1995.

[39] Matsusaka S, Masuda H. Electrostatics of particles. Adv Powder Technol 2003;14(2):143–66.

[40] Rowley G. Quantifying electrostatic interactions in pharmaceutical solid systems. Int J Pharm 2001;227(1−2):47−55.

[41] Naik S, Mukherjee R, Chaudhuri B. Triboelectrification: A review of experimental and mechanistic modeling approaches with a special focus on pharmaceutical powders. Int J Pharma 2016;510(1):375−85.

[42] Smith LI. A tutorial on principal components analysis. 2002.

[43] Kroonenberg PM. Applied multiway data analysis, 702. John Wiley & Sons; 2008.

[44] Wold S, Esbensen K, Geladi P. Principal component analysis. Chemomet Intell Lab Sys 1987;2(1−3):37−52.

[45] ten Berge JM. Least squares optimization in multivariate analysis. Leiden University Leiden: DSWO Press; 1993.

[46] Snedecor GWC, William G. Statistical methods/george W. Snedecor and william G. cochran. 1989.

[47] Aggarwal CC, Reddy CK. Data clustering: algorithms and applications. CRC press; 2013.

[48] Gan G, Ma C, Wu J. Data clustering: theory, algorithms, and applications, vol. 20. Siam; 2007.

[49] Ferreira AP, et al. Use of similarity scoring in the development of oral solid dosage forms. Int J Pharm 2016;514(2):335−40.

[50] Geladi P, Kowalski BR. Partial least-squares regression: a tutorial. Anal Chimica Acta 1986;185:1−17.

[51] Wang Y, et al. Predicting feeder performance based on material flow properties. Powder Technol 2017;308:135−48.

[52] Escotet Espinoza M. Phenomenological and residence time distribution models for unit operations in a continuous pharmaceutical manufacturing process. Rutgers University-School of Graduate Studies; 2018.

[53] Escotet-Espinoza MS, et al. Effect of tracer material properties on the residence time distribution (RTD) of continuous powder blending operations. Part I of II: experimental evaluation. Powder Technol 2019;342:744−63.

[54] Escotet-Espinoza MS, et al. Effect of material properties on the residence time distribution (RTD) characterization of powder blending unit operations. Part II of II: application of models. Powder Technol 2018.

Chapter 3

Loss-in-weight feeding

Tianyi Li[1,3], Sarang Oka[2,3], James V. Scicolone[3] and Fernando J. Muzzio[3]
[1]Drug Product Development, J-Star Research, Cranbury, NJ, United States; [2]Hovione, Drug Product Continuous Manufacturing, East Windsor, NJ, United States; [3]Engineering Research Center for Structured Organic Particulate Systems (C-SOPS), Department of Chemical and Biochemical Engineering, Rutgers, The State University of New Jersey, Piscataway, NJ, United States

1. Introduction

Loss-in-weight (LIW) feeders are a vital component in powder-based manufacturing, as they are used to accurately meter a given weight of material per unit of time, continuously, and for long periods of time. For continuous manufacturing of solid dose pharmaceutical products, powder feeding is one of the most important steps of the process and is a precursor to all other unit operations. Feeding controls the weight ratios of all the constituents of the formulation, and thus has a high impact on the critical quality attributes of the final product.

While there are many different designs of LIW feeders, they operate under a similar principle. A hopper is affixed on top of a load cell, which constantly monitors the weight of the material within the hopper. Aided by the action of a bridge-breaking agitation system, powder in the hopper enters a set of conveying screws placed below the hopper. The rotating speed of the screws dictates the rate at which powder is delivered by the feeders. Feeders can be operated in either volumetric or gravimetric mode. The volumetric mode sets the screws to run at a constant speed (revolution per minute). Volumetric feeding is normally used for easy flowing materials that show no significant changes in density during the process, i.e., incompressible and monodispersed material. In the gravimetric mode, closed loop control of the dispensing rate is enabled by using the signal from the load cell to constantly monitor the mass of material over time. The rate of change in mass over time informs the feeder's controller of its instantaneous mass flow rate, which is compared against a set point. Deviation from the set point is corrected by implementing changes in screw speed by the controller. Gravimetric control (mode) is a closed loop control system, referred as LIW feeding, where the controller dynamically adjusts the screw speed to achieve a desired mass flow rate. The objective of a feeding operation is to minimize variability in feed rate and continuously dispense material at a rate equal to the desired set point.

How to Design and Implement Powder-to-Tablet Continuous Manufacturing Systems
https://doi.org/10.1016/B978-0-12-813479-5.00017-3
29

During the refill of the feeder, which is required to maintain uninterrupted operation, the feeder hopper is refilled with powder. During this time, as mass is being added to the feeder, the gravimetric controls cannot accurately obtain a mass flow rate because the addition of fresh material creates a perturbation to the load cell. During this time, the feeder control system will switch to volumetric mode, operating with a constant screw speed. During the switch in operation to volumetric mode, feeders can often have mass flow rates that deviate from the set point, typically delivering an excess of the ingredient for a period of up to 30 s.

Although ensuring accurate mass flowrate is a crucial objective of process development for feeders, a precursor to optimizing performance (minimizing federate deviations) is determining feeding capacity, and consequently choosing a hardware configuration of the feeder that is able to deliver the desired mass flow rate for each material of interest. The range in operation of LIW feeders, that is, maximum and minimum capacity, depends on the size of the feeder (scale and motor capacity and gearbox), screw configuration, and the properties of the material being fed. The volumetric capacity of a feeder depends on screw diameter, screw speed, and screw type for a given motor/gearbox. The higher feed rates are positively correlated to larger diameter screws and higher screw speeds. Feeders are available for single or twin-screw configurations. Single screw feeders are primarily used for freely flowing, noncohesive, materials, while twin screws can be used for both free flowing and poorly flowing materials. Screws come in different shapes (concave, auger, and spiral), and the choice of the screw depends on the type of material being fed. Concave and auger screws are recommended for a wide range of materials, while the spiral screws are recommended for free-flowing powders. The screw types are also available in two grades meant for higher (coarse) or lower (fine) throughput. As previously mentioned, feeders typically include an agitation mechanism designed to keep the powder flowing within the hopper and into the screws. The purpose of the agitation system is twofold. First, it is designed to prevent the powder from bridging over the screws, which would prevent fresh powder from entering the screw flight. Second, to send material of consistent density into the screw flight, which promotes a stable screw speed and an accurate feed rate.

The last variable in determining feeder capacity is the material itself. In general, a higher density material will have a higher range in throughput than a low-density material. Besides the powder density, properties including particle size, compressibility, electrostatics, and flowability (e.g., cohesion) can be dominant descriptors of "feedability." Cohesive materials that tend to agglomerate, or to entrap air within the powder bed, will have narrower feed rate ranges, for a given set up, than easy flowing materials that pack well. Irregularity in bed density, due to entrapped air, can result in unstable flow rate and in consequent variability. Another common challenge with powder feeding is electrostatic effects, such as charge buildup, which can lead to material sticking to the equipment at the exit of feeders, or to any other exposed metal surface, resulting in compromised feeding performance.

Another important characteristic of a feeder is the feed factor. To dispense material out of the feeder, screws are rotated at a given angular speed (ω) to displace a certain volume of material per unit time ($V_{displacement}$). The volume displaced per screw revolution is known as the screw sweeping volume (V_{screw}) and is calculated from the annular dimensions of the screw. When the volumetric displacement ($V_{displacement}$) is considered for a material with a given density value (ρ), the material's mass flow rate (m) can by calculated using the equation $\dot{m} = \rho V_{displacement}$. However, it is important to note that powder feeding operations typically do not have a constant value of material density. due to the compressible nature of the materials. Furthermore, the fill efficiency (ε) of material in the V_{screw} is highly dependent on the ability of the material to go from the feeder hopper into the screws at the bottom of the unit. Thus, powder feeding operations often refer the material density at the screws (ρ_{screw}) by defining it as $\rho_{screws} \approx \rho \varepsilon$. However, since the density at the screws cannot be measured inside of the unit, a feed factor (ff) value, shown in the equation below, is used to denote the mass of material being dispensed per screw revolution. The feed factor has units of mass (dispensed) per revolution.

$$ff = \rho_{screw} * V_{screw}$$

Equipment vendors have provided alternative definitions for the feed factor, one of the most common being the maximum feeding capacity of the feeder for that configuration and material, that is the mass flowrate of the feeder when motor is running at its maximum speed. Both definitions will be used throughout this chapter and the specific definition being used will be stated as necessary.

The chapter is divided into four major sections. This introduction is followed by a section on characterizing LIW feeders along with information about selection of tooling, determination of feeder refill schedule, and characterization of residence time distribution. Section 3 discusses the effects of material properties on LIW feeding and how material databases can be used to inform process development. Section 4 touches on feeder unit operation modeling and Section 5 covers concluding comments.

2. Characterization of loss-in-weight feeders

LIW feeders have improved the ability to control powder feed rate and minimize flow variability caused by bulk density changes associated with the emptying of the feeding hoppers [1,2]. This is helpful once a feeding system is set up. Unfortunately, the selection and setup process for a feeding system is typically based on experience and empirical knowledge that is not readily available to the general user. For LIW feeders, most of the existing knowledge regarding either (i) the effect of powder properties on flow rate variability or

(ii) the effect of feeder design and operation on powder properties, resides with the equipment manufacturers. There has been some published work on improving feeder performance, for instance by using various devices at the discharge [3], or vibratory hopper agitation [4], but actual specification and sizing information is lacking. Selection of feeder tooling (screw, discharge screen, etc.) has been performed typically using trial and error methods. This section will first illustrate how to characterize a LIW feeder's feeding performance and then introduce the concept of development of the ideal design space of LIW feeders based on material properties.

This section focuses on the development of a method for the characterization of LIW feeders that can be used to aid in the proper selection of feeder tooling for a given powder at a given feed rate. The method includes the experimental setup and procedure for collecting feeding data and the data filtering and data analysis methods that are used to obtain useful values for comparison. The experimental procedure is a multistep process that involves running the feeder in both volumetric and gravimetric modes. Volumetric studies are performed to determine capacity, followed by gravimetric studies, which are used to determine overall performance. The performance data for each condition is analyzed using relative standard deviation (RSD) calculations and analysis of variance (ANOVA) methods. The method is illustrated with the help of an example of a Coperion K-Tron KT35 LIW feeder where its performance is evaluated for its operational range using three pharmaceutical grade powders. This is followed by a discussion on the development of ideal design space for LIW feeders based on material properties. This work was validated with a K-Tron KT20 LIW feeder with multiple pharmaceutical and catalyst ingredients. Lastly, characterization of the feed rate deviation caused by refill of a feeder hopper is discussed, with the ultimate objective of optimizing refill scheduling.

2.1 Gravimetric feeding performance

As mentioned, the characterization method of a LIW feeder is illustrated in this chapter using an example of three pharmaceutical grade material tested on a K-Tron KT35 feeder [1].

2.1.1 Materials and equipment

The materials used in the study are listed in Table 3.1. These pharmaceutical powders were chosen to test a range of cohesiveness and flowability in the feeder characterization experiments.

A Schenck Accurate AccPro II was used as a "catch scale" for characterization of the performance of the LIW feeder. A catch scale is needed because the internal load cells used in gravimetric LIW feeders use different filtering algorithms to pretreat the gravimetric signal, which may not allow for

TABLE 3.1 Materials used to characterize Loss-in-weight feeders. Three materials were used for the study.

Material	Flow index	Dilation	Density	Average particle size (μm)	Vendor
316 Fast Flo Lactose	27.8	10	0.58	100	Foremost
Avicel PH 102	38.0	15	0.30	100	FMC
Ceolus KG-802	49.2	22	0.21	50	Asahi-Kasei

accurate performance comparison between different feeders and between different feeding trials. The AccPro II scale was chosen as a catch scale in this study as it was large enough to handle the typical feed rates of the K-Tron KT35 feeder, but still has a high resolution that can resolve the small variations associated with feeding powders at relatively low throughput.

The K-Tron KT35 twin screw LIW feeder is designed to handle a large range of pharmaceutical powders, including those with very poor flowability, which are often lumpy and tend to form bridges. The design consists of a modular twin-shaft feeder mounted on a sanitary weigh bridge. There are a variety of feeding screws and discharge screens, which allows one to feed a large range of bulk powder materials. Fig. 3.1 shows the K-Tron KT35 feeder

FIGURE 3.1 (left) K-Tron KT35 feeder with Schenck Accurate AccPro II catch scale. (Middle) K-Tron KT35 feeder tooling. From left to the right, fine concave screw (FCS), coarse concave screw (CCS), fine auger screw (FAS), and coarse auger screw (CAS). (Right) There were two screen used in the study: fine square screen (FSqS—top image) and coarse square screen (CSqS—bottom image). The feeder can also be run without a screen (NoS).

with the Schenck Accurate AccPro II catch scale (left) and displays a representative sample of feeder tooling, that is, feeder screws and screens (middle and right, respectively) for the KT35 feeder. At the bottom of the feed hopper is a bowl containing a horizontal agitator that helps fill the flights of the feed screws. The agitation speed is set at 17% of the feed screw speed. The gearbox controlling the screws is a type B with a gear ratio of 6.7368:1 combined with a motor with a maximum speed of 2000RPM. At 100% of the motor speed, the screw rate is 297 RPM (327 RPM @ 110% is also achievable by overspeeding).

2.1.2 Methodology

Determining the performance of a powder feeder includes an experimental setup to collect feed stream data, filter noise, and analyze results. Of the many benefits of a method for characterizing LIW powder feeders, the most significant is a means of determining differences in feeding performance that can be used to optimize the feeder and tooling selection. Quantified feeding performance also provides general users of feeding equipment an additional tool to validate that the feeder is performing according to the feeder controller displays. The gravimetric control of the LIW feeder involves a significant amount of noise filtering, and as a result, the process variables displayed by the feeder controller often appear more consistent. In addition, a poor or erroneous calibration of the LIW feeder's load cell will cause the controller to display a feed rate that is different from the actual value.

2.1.3 Experimental setup

In the case studied below, characterization experiments were performed using the Schenck AccPro II scale to record the mass of powder dispensed by the feeder every 0.1 s. A cylindrical bucket, 9″ in diameter and 9″ in height, was used to collect the samples. Fig. 3.2 shows a graphical representation of the experimental setup used for monitoring feed rate and determining steady state performance. The feeder was placed on a sturdy lab bench. The catch bucket and scale were placed on a separate lower stand with the bottom of the bucket at 10″ in below the outlet of the feeder. When a bucket becomes full, it was quickly replaced by an empty bucket. Due to the sensitivity of the load cells in the equipment, careful consideration was given to isolating and minimizing outside disturbances on the feeders and catch scale. In determining equipment placement and filtering methods, the various general considerations listed in the work by Erdem have been taken into account [5]. Most importantly, a curtain was placed around the setup to minimize effects from air currents.

In addition to catch scale data, the feeder process values were recorded, including screw drive speed and hopper fill level. The data from each feeder were compared to the data obtained from the catch scale. Testing proceeded with first determining the volumetric capacity of the feeder operating at

Loss-in-Weight Feeder

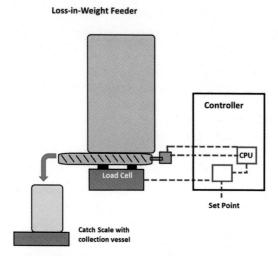

FIGURE 3.2 Loss-in-weight feeder characterization setup for monitoring feed rate and determining steady state performance. The catch scale is used to collect gain-in-weight data from the outlet of the feeder.

various volumetric speeds (without engaging the feeder gravimetric control system). Following this, gravimetric performance was characterized by monitoring the feed rate from the feeder for more than 30 min. A time longer than 30 min was chosen so that there would be sufficient data for meaningful statistical comparison between runs. As volumetric capacity testing only requires an estimated average feed rate, these tests can be short, only requiring achieving a steady state (or as close to it as possible) for a given set of experimental parameters (powder, tooling, screw speed, hopper fill level).

2.1.4 General volumetric test run procedure

The general procedure for the volumetric capacity experiments was as follows:

a. Calibrate the catch scale.
b. Fill the feeder to 100% of the maximum hopper fill level.
c. Run tests with volumetric set points at 10%, 20%, 50%, 80%, and 90% of the control magnitude or screw speed.

Volumetric mode is not the typical operation of the feeder and as such, the equipment may not have settings to select the speed of the screw manually. The K-Tron feeder used in this experiment required manually setting the initial feed factor to 100 kg/h. After manually setting this value, the volumetric set points could be entered directly with the default units of kg/h signifying % screw speed. The initial feed factor, in this case, is the control value that refers to the capacity of the feeder at 100% of the control magnitude.

2.1.5 General gravimetric test run procedure

The general procedure for the gravimetric characterization experiments was as follows:

  i. Calibrate the feeder and catch scale.

  ii. Fill the feeder to 100% of the maximum fill level.

  iii. Find the maximum feed rate for each experimental combination (powder, screw type, screens, agitation rate, and agitation depth), and use this for the initial feed factor controller value.

  iv. Run tests with set points at 20%, 50%, and 80% of the maximum controllable speed with initial fill level at 100% of the maximum fill level.

The maximum feed rate in step iii comes from volumetric testing performed in the previous test. The maximum controllable feed rate for each feeder can also be determined as the result of the built-in auto feed factor calibration program of the feeder's controller, rather than running volumetric capacity tests. This returns the value of the initial feed factor, which is the estimated feed rate at 100% of the screw speed that is used to control the feeder. An issue with using the built-in auto feed factor calibration program of the feeder is that if the relationship between the average feed rate and volumetric screw speed is not linear, the estimated feed factor could have some error, as it assumes a linear relation to extrapolate the value. Although a strongly nonlinear relationship is not very common for free-flowing powders, it becomes more common with powders that are cohesive and are unable to consistently fill the flights of the screw at higher rotation rates. The control system is always updating the averaged feed factor data, when in gravemetric mode, to accomodate any changes to the feedfactor due to volumetric fill.

The feed factor is used primarily for a volumetric reference point for the feeder and may be used whenever the feeder may need to be run in volumetric mode. A feeder running in gravimetric mode will occasionally switch to volumetric mode in instances where gravimetric control is impossible, such as during refilling of the feed hopper or when the feeder load cell experiences external perturbance.

The initial calibration of the feeder and catch scale load cells is of utmost importance, because if either of them is miscalibrated then the values collected from these load cells would be meaningless. Miscalibration of the feeder load cell has an additional implication since the feeder uses this signal for mass flow rate control. If this is incorrect, the feeder will misinterpret changes in weight, thus controlling to a different value than setpoint. This is a common mistake made when the wrong units are used for a check weight. This can be quite confusing to an operator that enters a desired setpoint of 5 kg/h which is then displayed on the controls of the feeder, yet the actual feed rate being fed is 5 lbs/hr (or 2.27 kg/h). Unless checked with a correctly calibrated catch scale, or until the calibration is rechecked with a check weight, it may go unnoticed until problems are discovered downstream.

The initial filling of the feeder is important as there is often a substantial change in the screw filling at lower fill levels. To avoid this issue altogether, it is recommended to fill the feeder close to maximum for testing, thereby ensuring that the minimum operation level is exceeded. Most feeding manufacturers state that this minimum is ~20% hopper fill level, but this is dependent on powder properties and may vary.

2.1.6 Analysis and filtering

The data collected from a catch scale are gain-in-weight information that can be used similarly to how the controller in a LIW feeder extracts useful values of feed rate from the LIW signal of the feeder's built-in load cells. To analyze the data, the mass dispensed every 1 s is used to calculate the fed material mass for the interval. From these data, the average feed rate $\left(\dot{m}_i\right)$ can be calculated for each 1 s (Δt) interval:

$$\dot{m}_i = \frac{\Delta m_i}{\Delta t}$$

From all the mass flowrates at each interval, a distribution can be determined; and from this distribution, the standard deviation (SD) (σ) and RSD of the mass flow rate can be calculated:

$$\sigma = \sqrt{\frac{\sum_{i=1}^{n}\left(\dot{m}_i - \overline{\dot{m}}\right)}{n-1}}$$

$$RSD = \frac{\sigma}{\overline{\dot{m}}}$$

where $\overline{\dot{m}}$ is the arithmetic mean mass feed rate of the distribution and n is the number of samples in the distribution.

During the analysis, in order to eliminate disturbances in the feed rate data caused by refilling, machine startup/shutdown, etc., the data are filtered by rigorously removing disturbances from the original data set. As an initial rough filtering method, 3 s of data are removed before and after each disturbance (total of 6 s in addition to the perceived duration of the disturbance), as this allows adequate time for the equipment to settle after a disturbance. Disturbances can be detected in the data set by setting appropriate bounds to the acceptable data. Since the feeder is under gravimetric control, the feed rate should not be deviating more than 10% from the setpoint, as a general rule of thumb. This is a modest set of bounds, as the feeder controls the feed rate much more tightly than this criterion. Thus, these bounds will detect significant physical disturbances to the catch scale, such as bucket changeovers.

Fig. 3.3 shows a sample set of unfiltered catch scale data with a disturbance at ~100 s into the set, when a bucket became full and was replaced with a new

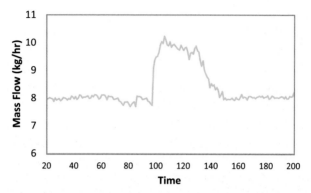

FIGURE 3.3 Sample 1 s interval catch scale data before any applied filter with a catch bucket change at ~100 s.

empty one. Fig. 3.4 shows the data after filtering. After filtering out disturbances, the leftover data are representative of the quasi-steady behavior of the
feeder. The "steady state" distribution of the feed rate data is obtained with the
feeder operating in gravimetric LIW mode. The example presented in Fig. 3.4
shows feed rate data as a function of time and respective distribution, which,
for properly filtered data, approaches a Gaussian curve. There are two
important parameters for performance that can be gathered from steady feed
rate signal data: spread of the data (SD), ideally as narrow as possible; and
deviation of the average from the setpoint, ideally zero.

By comparing the distribution of the original unfiltered data with the
filtered data, it is possible to further optimize the filtering procedure. The
initial filtering of an extra 6 s in addition to each disturbance results in a
roughly filtered data set that has an average feed rate that is an initial estimate
of the quasi-steady behavior and a SD that is relatively close to the true value.
The ideal filter uses the mean and SD of the filtered data, so an iterative

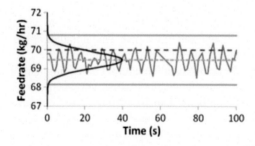

FIGURE 3.4 Sample filtered 1 s interval catch scale data (Blue) with its normal fitted distribution
(Red). Also marked with a *horizontal line* is the mean value (Light Blue), the setpoint (Purple), and
the ±3σ (Green).

procedure with this initial estimate is needed. The bounds used for each iteration are three SDs about the average. After each iterative pass, increasingly better estimates of the mean and SD are calculated. After several filtering passes, the iterative filter self-tunes and results in a final average and SD that is unchanging with additional filter passes and is representative of the data without the outlying datapoints caused by disturbances.

2.1.7 Results and discussion

A complete parametric set of characterization runs was performed for the K-Tron KT35 twin screw feeder. This included every combination of three screen conditions [no screen (NoS), coarse square screen (CSqS), and fine square screen (FSqS)], four paired sets of screws (coarse concave, fine concave, coarse auger, and fine auger), and the three powders at the three feed rate setpoints to perform the characterization exercise. Both Fast Flo lactose and Avicel 102 were fed successfully through their entire tested range of speed setpoints for all combinations of feeder tooling. Ceolus, on the other hand, did not run for either of the auger twin screws with the FSqS due to a motor overload. To test for reproducibility of results of the data, all of the runs with Avicel 102 were repeated.

ANOVA is a statistical tool that can determine the significance of differences in the gravimetric performance data based on the potential sources of change to the performance such as: screw, discharge screen, and screw speed and was applied to the entire dataset. A sample ANOVA for the Avicel 102 in the K-Tron KT35 is shown in Table 3.2. SD as a function of average federate is shown in Fig. 3.7 and RSD as a function of feed rate in Fig. 3.8.

The ANOVA shows that for the K-Tron KT35 feeding Avicel 102, the speed is the most influential source for change in feeder performance and has a statistically significant effect ($F > F_{critical}$ or $P < \alpha$). The screw type is also found to be statistically significant. Following this is the screen, which was not shown as a statistically significant variable for the dataset collected. The effects on RSD examined in this case may not apply universally to all powders or to all feeding equipment, as this is just descriptive of the data that was analyzed in this ANOVA. With some powders, the screen may become a very significant source of performance change.

Fig. 3.5 shows the RSD plotted against the feed rate. A plot of RSD as a function of feed rate can be used to select the best available feeder tooling for an application. For the case of Avicel 102 in the K-Tron, KT35 shown in Fig. 3.8, this plot can be used to choose the best-performing tooling for any desired feed rate in the tested range. This can be done by selecting the set of tooling that has the lowest RSD at any feed rate. As the coarse auger screws (CASs) with NoS has the highest RSD values, it would only be used for high rates that the other screws cannot achieve. The fine concave screw (FCS) with NoS should be used for lower feed rates as the RSD values for all other

TABLE 3.2 Analysis of variance of Avicel 102 feeder characterization data with interactions ($n = 2$ and $\alpha = 0.05$).

Source	df	SS	MS	F	P	Fcrit	
Screw	3	4.14E-05	1.38E-05	3.794	0.018	2.866	Significant
Screen	2	2.34E-05	1.17E-05	3.213	0.052	3.259	
Speed	2	2.99E-04	1.50E-04	41.18	4.96E-10	3.259	Significant
Screw*Screen	6	5.11E-05	8.52E-06	2.343	0.052	2.364	
Screw*Speed	6	5.48E-05	9.13E-06	2.512	0.039	2.364	
Screen*Speed	4	2.26E-05	5.65E-06	1.554	0.208	2.033	
Screw*Screen*Speed	12	6.63E-05	5.52E-06	1.519	0.162	2.033	
Error	36	1.31E-04	3.64E-06				
Total	71	6.90E-04					

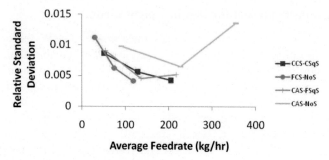

FIGURE 3.5 Relative standard deviation (RSD) plotted as a function of average feed rate of the KT35 feeder characterization data for Avicel 102.

combinations are higher at feed rates less than ~ 100 kg/h. For the intermediate speeds between ~ 100 and 200 kg/h either of the other options, coarse concave screw (CCS) with the CSqS or the CAS with the FSqS would be appropriate, as the RSDs are very similar for both.

In brief, a method for characterization of LIW feeders was demonstrated using a case study. In this method, a catch scale was used to monitor the feed rate of material dispensed from the LIW feeder. The feeder was monitored as it ran in two different modes: volumetric (constant screw speed) and gravimetric (variable screw speed based on feedback control).

Volumetric capacity trials were used to determine gravimetric setpoints for the gravimetric trials. Gravimetric testing was performed for the K-Tron KT35 with setpoints that would result in screw speeds that fall within the manufacturer's recommended range of 20%–80%. By postprocess filtering and analysis, it was possible to fit the data to a normal distribution that allows the performance to be quantified by two values: average feed rate and SD of feed rate. This allows the performance to be compared between the different gravimetric trials with different feed tooling and powders.

ANOVA of the feeder characterization data was used to determine significance of feeder variables (feeder tooling, powder, and speed) on feeder performance. The significance of screw, screen, and speed may vary with powder. For instance, a free-flowing powder may not have as significant a screen effect as a very cohesive powder that may be prone to forming clumps that may be broken up by a discharge screen.

The method presented here, based on using a catch scale, greatly improves feeder tooling selection. With the characterization method described, a database of feeder performance and powder properties can generate a predictive model such that feed tooling can be selected based on desired feed rate and measured powder properties rather than trial and error. The use of such predictive models to predict feeding performance is discussed in the next section and in Chapter 2 of this book.

2.2 Ideal design space for loss-in-weight feeders

As demonstrated in the case study described above, a LIW feeder has its optimal operational range for a given powder-screw combination. Understanding how the feeder and the material behave is vital to the manufacturing process. Powder properties, such as density and cohesion, can cause large variability in the flow rate of ingredients fed from powder feeders. It is important to understand the operational ranges or the capacity of the feeder, i.e., percent motor power, which correlates directly with screw speed. Knowing an ideal range of operation and correlating powder properties to process performance can lead to accurate determination of the operational (design) space for a feeder, and to faster optimization of feeder performance, that results in savings in effort and material.

As an illustrative example to determine the ideal design space of a K-Tron KT20 C-gearbox LIW feeder, the process of analyzing the flow of multiple material was performed. Again, four screws were used in this analysis: coarse auger, fine auger, coarse concave, and fine concave. Multiple feed rates were studied with all screws and utilizing material that varied greatly in bulk properties, i.e. particle size, flowability, and bulk density. The results show an optimal motor range, which could be represented as percent drive command or screw speed, for the feeder. Another value obtained from these studies was the feed factor (mentioned previously as the maximum flow rate for a given material), for each feeder and screw combination. Based on powder material properties, a correlation between material packing and feed factor was determined. By predicting the maximum flow of a material from material characterization, and an optimal drive command range, which relates to feed factor (flow at 100% drive command), the optimal mass flow ranges of each screw was determined. This correlation can be used to select a feeder and a screw type based only on information of the properties of material of interest. Since material characterization requires less time, effort and material compared to performing full characterization on a feeder, the method can result in faster development times with minimal use of expensive, scarce, material, particularly active ingredients.

2.2.1 Materials and experimental setup

Many different material were used in this work, with various particle sizes, flowabilities, and densities, to obtain a broad understanding of feeder behavior. Both powders and granules of API and excipients were used. Multiple PH grades of Avicel (FMC Corporation) were used, including 101, 102, 105, 301, and 200, which encompasses the full range of available density and particle sizes. Additional excipients included Lactose Monohydrate 310 NF (Kerry Inc.), croscarmellose sodium (FMC Corporation), magnesium stearate (FMC Corporation), Prosolv HD 90, HD200, and 50 (JRS Pharma), and Ceolus KG-802 (Asahi Kasei Corporation). The APIs used in this study were powder grade acetaminophen/paracetamol (APAP, Mallinckrodt Pharmaceuticals) and a granulated API (provided by Janssen Ortho LLC-Gurabo). Additional materials included inorganic powders typically used in ceramic applications, including coarse grade alumina (Albermarle, Amsterdam, the Netherlands), and Molybdenum Oxide (Albermarle, Amsterdam, the Netherlands). These ingredients were chosen due to their higher densities, which enabled the determination of the relationship between feed factor and material properties over a wider range. A shear cell test using the Freeman FT4 powder rheometer (Freeman Technologies Ltd., Worcestershire, UK) was performed to obtain the conditioned bulk density (cBD) for each powder. Details of the test procedure to obtain cBD of a powder can be found in Ref. [6].

A K-Tron KT20 twin screw LIW feeder with a type C gearbox (with a gear reduction ratio of 12.9:1) was used in this study. The KT20 consists of three main parts: a volumetric feeder, a weighing platform (load cell), and a gravimetric controller. The volumetric feeder is mounted on top of the weighing platform with a 10-L hopper containing a horizontal agitator, which helps break powder bridges in order to allow the powder to flow into the channel of the feeder screws. The agitation speed is set to 17% of the screw speed; therefore, the speed of the agitator changes with the changes in screw speed. The KT20 with a type C gearbox used in this study has a maximum processing screw speed of 154 RPM and a maximum motor speed of 2000 RPM. (170 RPM @ 110% is also achievable by over-speeding).

2.2.2 Methods

The following procedure was used to characterize the feeder:

i. Calibrate the feeder's scale platform, and the catch scale, by using standard weights.
ii. Fill the feeder with a 100% hopper level.
iii. Perform the feeder and material calibration three times and get the average of the initial feed factor, which is considered as the maximum feed rate with the combination of certain material and type of screws.
iv. Calculate 10%, 30%, 50%, 70%, 90%, and 110% of the initial feed factor and make these values the target set points.
v. Run the feeder under gravimetric mode for 20-30 min at all the target set points separately with the same initial fill level of 80% of the hopper by refilling after each setpoint.

The maximum controllable feed rate was achieved by initial feeder calibration rather than running volumetric capacity tests as discussed previously. This approach returns the value of the initial feed factor in Kg/h, which as mentioned is the estimated feed rate at 100% of the screw speed. From the initial feed factor, the desired feed rates to test are 10%, 30%, 50%, 70%, 90%, and 110% of the drive command. The purpose is to determine the feeding range of materials with different flow properties under gravimetric control. It should be noted that the actual recorded drive command was not a one-to-one match with the experimental design. Specifically, while the designed experiments targeted drive command values of 10%, 30%, 50%, 70%, 90%, and 110%, the dynamic nature of the feed factor does not allow for the drive command to stay constant throughout the run, nor is it possible to operate precisely at the desired drive command value since the feeder controls for federate rather than drive command. For example, at high screw speeds, screw filling may be less than 100%, resulting in a lower calculated feed factor, and therefore a higher drive command. Thus, the average of the real drive command was used as the tested value instead of the designed one.

For each material with each type of screws, each setpoint was run from an initial fill level of 80%. The final hopper level at the end of the test differed from each setpoint by approximately 40% to 70%. Each experiment was run for 20-30 min to obtain enough data for analysis. The reason for running each setpoint for 20 min (instead of running a full hopper of material) is that the feeder's performance is affected by the emptying of the hopper since the feeder factor decreases as the feeder empties.

The feeder control box was connected to a laptop, which was used to record multiple feeding performance parameters. Values recorded include time, setpoint, mass flow (the instantaneous feed rate calculated by the feeder's controlling system), initial feed factor (kg/h), average feed factor (kg/h), screw speed, drive command (the percentage of instantaneous screw speed compared to the maximum screw speed), net weight, and perturbance value. In addition to recording the feeder parameters from the LIW control box of KT20, a gain-in-weight catch scale was also used to monitor the discharge performance. As before a collection bucket was placed on an Ohaus laboratory scale, underneath the outlet of the feeder, to collect the material as it was discharged. The mass was recorded every 1 s by the catch scale and recorded by the same laptop that was recording the feeder parameters.

In summary, to determine the operational range of gravimetric feeding, each combination of process parameters, material, and tooling was fed at multiple set-points, selected from 10%, 30%, 50%, 70%, 90%, and 110% of the initial feed factor. The feeder is able to reach a maximum screw speed with a drive command of 110% determined its intrinsic design. The feeder was filled to 80% of the feeder hopper capacity and was run for 20 min. The catch scale's data were recorded to calculate the RSD and the relative deviation from the mean (RDM) of the actual feeding setting. The combined results are plotted in Fig. 3.6 as a function of the percent drive command of the motor for Prosolv HD 90, Prosolv 50, Prosolv HD 200, Avicel PH 101, Avicel PH 102, Avicel PH 105, Avicel PH 200, Avicel PH 301, powder APAP, magnesium stearate, and lactose monohydrate 310. These results show that the RSD is highest at low drive command levels. Above about 20% drive command, the RSD is lowest and very similar, for all values up to 90% drive command, and

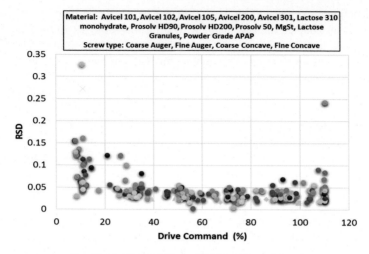

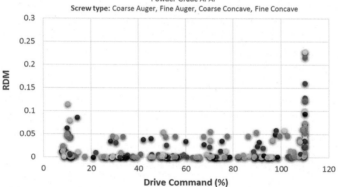

FIGURE 3.6   Combined RSD (top) and RDM (bottom) versus drive command (%) of lactose monohydrate 310, Prosolv SMCC, Avicel 101, Avicel 102, Avicel 105, Avicel 200, Avicel 301, magnesium stearate, lactose granules, and powder grade APAP.

then it increases again. The reason for the high RSD at drive command higher than 90% is that the feed factor decreases due to the emptying of the hopper, which causes insufficient screw filling at high screw speeds. A similar trend was found for the RDM. High deviations were observed below 10% drive command, and above 100%. The largest deviation from the mean occurs at the highest drive command, for the same reason as for the RSD; the powder is simply unable to flow into the screw channel at sufficient rates. The maximum acceptable limit of the feed rate variability, as quantified by the RSD, is dependent on several factors, including the feeding application. For example, acceptable variability on an API is different than acceptable variability on a functional excipient, which is different from acceptable variability on a nonfunctional excipient. This issue will not be discussed here. For the purposes of this chapter, the acceptable RSD limit was set to 5% of the mass flow rate, when characterized using 1 s intervals. The maximum limit of RDM also depends on the application, for example, the width of composition intervals is different for product and is not discussed further in this chapter.

Based on the information from these two figures, the optimal range in operation of the K-Tron KT20 C-gearbox feeder is between 20% and 90% of the drive command for CCS, FCS, and CAS and 40% to 90% for FAS. This range is not an absolute range of operation because it is possible to obtain low RSD and low RDM below 20% and above 90% drive command for certain materials, but that likelihood is much lower outside the optimal range.

An ideal operational space for a gravimetric feeder would be defined by the range of feed mass flow rates in which the feeder will perform accurate gravimetric feeding at the required target mass flow rate, with acceptable RSD and acceptable RDM to the setpoint with certain tooling selections for a certain material. On the other hand, acceptable gravimetric feeding performance is not guaranteed if a LIW feeder is not operated within its ideal operational design space (between 20%, 40% and 90%). Evidently, if feeding performance can be correlated to specific material properties, a large amount of material and effort would be saved, compared to determination by exhaustive experimentation.

In order to identify a correlation, the feed factor was analyzed against material properties. Most of the properties, such as particle size, flow function coefficient, permeability, and tapped density, did not result in an identifiable trend. Correlation was identified once the cBD was plotted versus the feed factor. Fig. 3.7 shows a linear relationship between feed factor and the cBD of the material. The cBD is obtained from the compressibility test of the Freeman FT4. The cBD is the mass of material that fits into a 10 mL volume after a

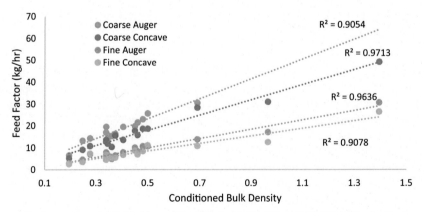

FIGURE 3.7 The relationship between feed factor and material conditioned bulk density, for the various material listed, over a wide bulk density range, and for the four screw types.

helical blade passes through the powder bed. While cBD shows a linear relationship to feed factor, it should be noted that none of the other density properties from the compressibility test, from 1 to 15 kPa, showed a comparable correlation. The aerated and tapped densities showed a linear treand with feed factor; however, the R-squared fitting for the trendlines were not as good for the cBCBackspaceD. Therefore, by analyzing 10 mL of material, the maximum flow rate of a particular material and screw combination could be approximated.

Based on the results from feeder's operational range, which identified a drive command between 20%, 40% and 90%, and the feed factor's correlation to cBD, an effective design space for the different screw types can be determined. The example of FCSs is shown in Fig. 3.8. It should be noted that the case study discussed here is specific for the K-Tron KT-20 C-gearbox feeder. It can be assumed that the trends for feeders fitted with an A or a B gearbox would be similar, as the operational parameters and control system are the same.

By knowing the target throughput and cBD of a material, a proper screw and feeder type can be identified that is able to deliver the desired throughput for that material. If it is determined that the throughput is too low or too high for the identified screws, the recommendation would be to switch feeder types, for example, from a KT-20 C-gearbox, studied here, to a KT-20 B-gearbox (for higher throughput), or an MT-16 (for lower throughput).

In summary, feeder characterization was generally performed using a trial by error approach to determine optimal feeder and screw selection, resulting in delays in product and process development. The method discussed above enabled the determination of a LIW feeder ideal design space by correlating the ideal operational range of the feeder with materials of significantly

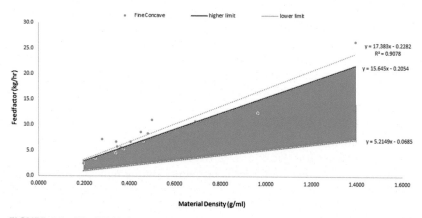

FIGURE 3.8 The KT-20 feeder design space (flow rate range for a given density) for fine concave screws (FCSs).

different bulk properties. In this work, 10 materials including both APIs and different grades of excipients with a wide range of particle sizes, densities, and flow properties were tested to understand and develop the design space of a K-Tron KT20 LIW feeder, fitted with a C-gear box. First, by feeding all these materials at different drive commands of 10%, 30%, 50%, 70%, 90%, and 110%, the optimal range of operation was located by calculating the RSD and RDM with an acceptable value of 0.05 for both metrics. Values of 20%, 40% and 90% were determined as the lower and upper limit of drive command for optimal operational range of KT20 C-gear box. This observation was confirmed for all four feeding screws, i.e., CASs, fine auger screws, CCSs, and FCSs. Based on this work, it was determined that there was a good linear correlation between density and feed factor. With this information, a K-Tron KT-20 feeder's feeding capacity, for different types of screws, can be predicted by the cBD of the material. Specifically, this means that by characterizing as little as 10 mL of material sufficient information can be obtained to determine the feeder's feed factor, and consequently a feeder's ideal design space.

2.3 Feed rate deviation caused by hopper refill

As previously discussed, a gravimetric feeder does not monitor and control the feed rate during hopper refill, which often results in deviations from set point. This refill problem is a known challenge, and manufacturers of feeder equipment have developed several methods to attempt to address it, including: refill modes that have a variable screw speed during refill [7], redundant refill [8], and/or feeder systems that try to bypass the refill [9].

All of the approaches mentioned above may work to reduce or eliminate the issue. However, these patented techniques do not always completely eliminate the problem and often involve purchasing extra equipment. The work discussed in this section constitutes a method for quantifying the effects of feeder refill and provides an approach for optimizing its refilling schedule. By using a gain-in-weight catch scale that collects and weighs material as it is fed, deviations from the feed rate setpoint can be monitored during hopper refill. It has been observed that size (range) of refill has a significant impact on feeder consistency and performance, and it is the use of this information that is primarily leveraged for the method discussed in this section.

2.3.1 Operation during hopper refill

During continuous operation as the hopper empties, powder in the hopper needs to be replenished (See Fig. 3.9). To maintain continuity of operation, the hopper is refilled while the feeder is operating. During refill the feeder must switch to nongravimetric operation where screw speed is instead controlled volumetrically. The reason for the switch in operation mode is because the

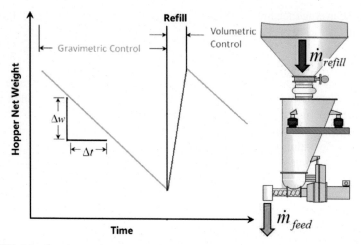

FIGURE 3.9 Loss-in-weight operating principle depicting the loss-in-weight feeding cycle created by periodic hopper refill.

change in weight with time during refill includes the mass flow of material refilling the feeder, $\dot{m}_{refill}$, in addition to the mass being fed from the feeder, $\dot{m}_{feed}$. The instantaneous rate of change of mass as observed by the feeder's load cells during refill is represented by the following equation:

$$\frac{\Delta w_{feeder}}{\Delta t} = \dot{m}_{refill} - \dot{m}_{feed}$$

This indicates that there are two unknown mass flows, whereas there is only one measurement. The net hopper weight measured by the feeder's internal load cell confounds the two feed streams. To distinguish the feed rate for each of these varying streams would require more information about one of the streams (i.e., by using a LIW feeder to refill the system of interest, which is not current practice). Hence, the feeder operates in volumetric mode, which can lead to feed rate inconsistencies, in particular, as the feeder is refilled and flight filling varies due to an increase in the weight of the powder above the screws.

2.3.2 Quantifying deviation

The variability of the feed rate due to refill can be characterized in multiple ways, including the magnitude of the maximum deviation, the time feed rate is away from setpoint, and the total deviation of material mass flow rate in excess (or in defect) of set point. The best value for quantifying the effect of deviation depends on the application and subsequent unit operations.

If the sensitivity of a unit operation is constrained by a maximum feed rate, the maximum deviation may be the best quantifying method to use. The maximum deviation is calculated by subtracting the peak of the deviation from the average feed rate (See Fig. 3.10), and is represented by the following equation:

$$\Delta\dot{m} = \dot{m}_{max} - \overline{m}$$

where $\dot{m}_{max}$ is the maximum feed rate (at the peak) and $\overline{m}$ and is the mean feed rate calculated from steady state feeding. This value is the quickest and easiest method for comparing multiple refill deviations or for detecting out-of-specification material. Because it has no time dependence, its quantity only represents a deviation in feed rate, but it does not contain information on the length of the time of the deviation.

In relevant cases, the blending unit operation is affected by the duration of the deviation, and therefore the time the feed rate is out of specification will be useful for quantifying the actual magnitude of the deviation. This method uses an upper bound for detection that should be equal to the value that represents an out of specification feed rate. In this case, it was defined to be at three SDs above the mean feed rate. As previously discussed, the SD is determined by the steady state feeding signal. The duration of the perturbation is defined as the period that begins when the feed rate first exceeds the boundary and ends when the feed rate returns within range (See Fig. 3.10), and is represented by the following equation:

$$\Delta t_{OOS} = t_{OOS,\ final} - t_{OOS,\ initial}$$

where $t_{oos,\ initial}$ and $t_{oos,\ final}$ are the initial and final times where the flow rate is out of specification. The time the feed rate is out of specification can still produce acceptable product, because short pulses may not have a large enough

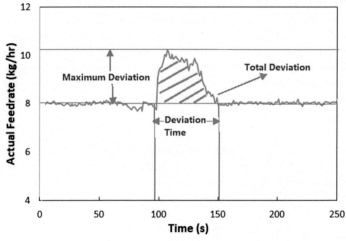

FIGURE 3.10 Methods for quantifying the deviation from setpoint: magnitude of the maximum deviation, the time that the feed rate is out of specification, and the total deviation/powder fed in excess.

effect to drive product to failure in a robust system. However, if the deviation occurs for an extended period of time, it may cause disruption or product failure.

The total amount of excess powder fed during a perturbation is also a useful parameter, because it captures both the magnitude and duration of the deviation. This quantity, total deviation, is calculated by determining the area between the feed rate target and the feed rate profile (See Fig. 3.10) for out of specification material, and can be represented with the following equation:

$$m_{\text{excess}} = \int_{t_{\text{OOS, initial}}}^{t_{\text{OOS, final}}} \left[\dot{m}_{\text{feed}}(t) - \left(\overline{\dot{m}} + 3\sigma \right) \right] dt$$

where $\dot{m}_{\text{feed}}(t)$ is the feed rate data, $\overline{\dot{m}}$ is the mean feed rate calculated from steady state feeding, σ is the SD, $t_{\text{oos, initial}}$ is the initial time of out of specification material, and $t_{\text{oos, final}}$ is the final time of out of specification material.

2.3.3 Effect of refill level

The effect of refill level, which is the fill volume at which the hopper is refilled back to its starting fill level, was investigated with multiple powders. Results are shown in Fig. 3.11. When the hopper is filled more frequently with a smaller amount of material, the deviation is reduced until it becomes nearly

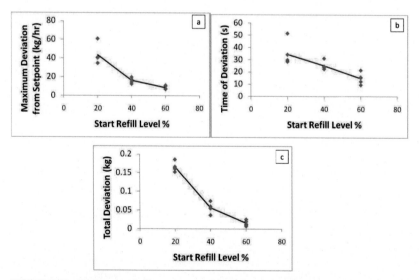

FIGURE 3.11 Performance indicators extracted from the catch scale feed rate data for series of five repeated manual refills of the K-Tron KT35 feeding Grillo Pharma8 at three refill levels. (A) The maximum deviation from setpoint, (B) total time of deviation, and (C) total deviation (total amount of excess powder delivered per refill) all decrease as the refill is performed at higher hopper fill levels.

undetectable at the highest refill initiation level of 60%, filling the hopper up to 80% fill. The maximum deviation (Fig. 3.11A) and total amount fed in excess (Fig. 3.11C) reduced sharply when the refill initiation level changed from 20% to 40%, in each case returning the hopper fill level to 80%. The difference in maximum deviation and total amount fed in excess continued to decrease, but not as much, when the refill level changed from 40% to 60% initial fill level. The time of deviation (Fig. 3.11B) showed a more gradual reduction across all low hopper fill levels.

2.3.4 Investigation of refill method

Another critical factor that determines the behavior of a feeder during refills is the rate at which its hopper is refilled. The results discussed in this section revealed that the modality of the refilling operation (not just its magnitude) can have an important role on deviation reduction. To further investigate the effect of refill method, the hopper replenishment of a K-Tron KT20 was tested. High-rate refilling with the automatic vacuum refill system was compared to rate controlled refilling, utilizing the K-Tron KT35 as an automatic refill system. The powder used for testing was Fast Flo Lactose, and the feed rate setpoint was 20 kg/h.

The feed rate results for both refill methods are shown in Fig. 3.12. The refills from the automatic vacuum refill system were quick and delivered the powder with a high impulse. As shown in Fig. 3.12A, this refill modality resulted in deviations even when the feeder was refilled very frequently (at 60% fill level). The deviations during refill can be clearly observed in the plot at ~550 and ~1100 s. When using the K-Tron KT35 as the refill device (See Fig. 3.12B), the two refills were also performed at ~550 and ~1100s, but no discernable deviations were observed in the feed rate data. The lower refill rate of the K-Tron KT35 resulted in lower compressive forces that did not cause significant density changes of the powder in the hopper. The high-rate

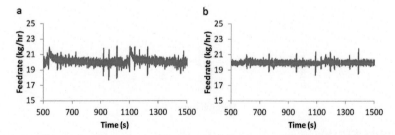

FIGURE 3.12   Feed rate data during 60%−80% hopper refills of the K-Tron KT20 using (A) an automatic vacuum refill system and (B) the K-Tron KT35 setup as a volumetric screw refill system.

automatic vacuum refill system, which dumps a sudden charge of powder, caused significant compressive force, which resulted in feed rate deviations.

To recapitulate, continuous processing requires a steady, consistent, and nonstop feed stream of powder. The use of gravimetric LIW feeders with limited hopper size inevitably means that the feed hoppers will need to be refilled often, leading to potentially large deviations from feed rate setpoint. When designing the refill apparatus and scheduling the hopper refill, it is necessary to mitigate and minimize such deviations. By quantifying the size of the deviations, it is possible to compare multiple feeder/refill system configurations and refilling schedules in order to select the optimal setup.

By refilling the feeders more frequently with less material, the magnitude of perturbations are reduced and the accuracy of the feeders are improved. However, this results in the feeder functioning in volumetric mode a longer fraction of the time. In volumetric mode, the feeder is essentially blind to changes in screw filling or powder density. In order to minimize this potential problem, it is best to select a refilling regime that minimizes both the magnitude of deviations as well as the number of refills needed. The size of refills should be based on reducing the deviations from feed rate setpoint to a level that is acceptable to the unit operations downstream of the feeder. This requires an understanding of the downstream process, its mixing capabilities, and the acceptable level of variability of a given ingredient in the finished product. This has been discussed in detail in Chapter 4 of this book.

Alternatively, in cases where even small levels of variability are unacceptable (i.e., API levels in products with narrow therapeutic windows), the refilling system should be designed to gently refill the feeder. This requires more control over the refilling stream, which also demands a specialized refilling apparatus. In the extreme, this might require using a larger feeder that would need to be refilled less often. In the case shown in this study, where a volumetric feeder is used to control the refill rate, this will also require an additional refilling system to refill the volumetric feeder. While filling an intermitten volumetric refill feeder is a simplier concept, this leads to a higher initial equipment cost, higher overhead space requirements, etc. However, where increased control over fluctuations is necessary, this may be the preferred option.

Understanding how a feeder behaves with different materials, tooling and workable range are essential points to achieve good feeding performance. In this section, methods for characterization of LIW feeder's gravimetric feeding performance, a method to identify an ideal design space for LIW feeders and a method for characterization of the feed rate deviation caused by hopper refills were discussed in detail. Combining this with knowledge of properties of materials, a great amount of material and time can be saved during the process development of continuous processes.

3. Effect of material flow properties on loss-in-weight feeding

LIW feeder performance depends not only on equipment design, feeding capacity, and its intrinsic control systems but also on properties of the material being fed. In general, a specific feeder cannot handle every conceivable material, even when everything is perfectly set up to achieve its best feeding performance. On the one hand, cohesive powders can adhere to the surface of the screws or can bridge across the feeder hopper, causing a phenomenon called "rat holing," or they can form a thick cohesive layer at the bottom of the hopper that affects the filling of powder into the screw flights. On the other hand, free-flowing powders usually have higher densities, and thus requires more energy and more torque for the feeder to convey these materials. They can also "flush" through the feeder, causing pulsating flow rates. A poorly paired powder-feeder combination typically results in complete failure of feeding and/or high variability in feed rate. Therefore, understanding the effect of material flow properties on the performance of LIW feeders, including regular gravimetric feeding and the deviations caused by hopper refills is a critical element of continuous process development.

There has been a substantial body work performed to understand the role of material properties on feeding performance. This literature will not be discussed in detail here. Interested readers can refer to Refs. [10–12] for a general review. A powerful application of large material database in combination with statistical methods to predict feeder performance is discussed in Ref. [13]. The use of material property databases to predict feeder performance was discussed in detail in Chapter 2. Readers are referred to Chapter 2 for discussions on development of material databases, coupling process performance of feeders to materials databases, and predicting LIW feeding performance using statistical methods. However, as previously discussed, the behavior feeders during refill is crucial to understanding their overall performance. The use of material property databases and statistical methods to predict feeder deviations during refill is thus discussed here.

In summary, deviation in feed rate during hopper refill has been a common but critical issue for LIW feeders. As discussed in previous sections, during the refill process, feeders run in volumetric mode, and this results in momentary variability in the behavior of the federate. As discussed, previous research has investigated the effect of the refill level on the feed rate deviation. Enormous differences were observed between materials with different flow properties. A poorly designed refill strategy has been known to produce deviations that exceed the acceptance thresholds of product composition during commercial production. Understanding the effect of materials properties on feed rate deviation caused by hopper refill are thus valuable to the design and success of a continuous manufacturing process.

Multivariate analysis methods can be applied to establish correlations between material properties and the deviations caused by hopper refills. A Principal Component Analysis - Sum of Squares (PCA-SS) method was developed [14] based on the analysis of the weighted Euclidean distance between the location of each material (as defined by its properties) in the principal component (PC) space, followed by the development of a Principal Least Squares Regression (PLSR) predictive model to predict maximum deviation during a refill, deviation time, and percentage of total deviation. Fig. 3.13 shows PLSR prediction results versus measured values of (A) maximum deviation, (B) deviation time, (C) % total deviation of all the rest materials in the predictive library. All of the three plots show very good correlations between the predicted values and the measured values, which means there is sufficient variability in the calibration thereby implying that a feeder's feed rate deviation during hopper refill is highly correlated to properties of the material being examined. Detailed information of this study, and the associated methods can be found in Ref. [14].

A feeder's feeding performance not only depend on feeding condition and the control system but it is also highly correlated to the properties of the material being fed. Studies have shown that the gravimetric feeding performance of a LIW feeder and the feed rate deviation caused by hopper refills can be correlated and predicted by the application of statistical methods in conjunction with material flow property databases. Moreover, materials with similar flow properties, as defined by the weighted Euclidean distance in material property PC space, exhibit similar feeding performance. This approach is especially powerful when the amount of a given new material is limited, expensive, or dangerous since an appropriately chosen surrogate that is statistically *similar* to the material of interest can be used to perform experimental work during process development. Moreover, using data-driven models such as PLSR, the amount of time and material needed for process development can be significantly reduced thus lowering the cost for development.

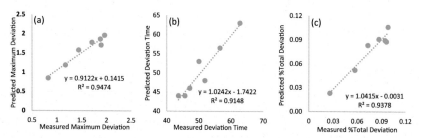

FIGURE 3.13   Predicted values versus measured values of (A) maximum deviation, (B) deviation time, (C) % total deviation of all the rest materials in the predictive library.

4. Modeling of loss-in-weight feeders

LIW feeders have been modeled by a variety of approaches over the past decade. The use of purely statistical approaches, in combination with material properties characterization, to predict feeder performance has been discussed in this chapter and in Chapter 2 of this book. A similar approach that also combines the configuration of the feeder was proposed by Tahir et al. [15]. Metamodeling techniques such as Kriging and response-surface methods have also been used to model feeders [16]. More recently, semimechanistic or hybrid approaches have also developed [17]. Attempts have also been made to use mechanistic methods such as discrete element modeling [18]. The challenges associated with modeling any solids process also apply to modeling of LIW feeders. Furthermore, mechanistic or semimechanistic approaches are challenged with not having access to proprietary control algorithms of feeder manufacturers that control the feed screw to ensure feeding at target set point. The steady stream of peer-reviewed research on this topic suggests that a winning modeling approach is yet to be identified, although the authors of this textbook believe that the use of a statistical approach in combination with material properties and feeding performance database discussed in this book have proved to be the most powerful.

5. Conclusions

This chapter brings together different concepts related to LIW feeding in an attempt to provide a unifying framework for process development of feeding processes. Characterizing feeding performance, an indispensable element of processes development that enables comparison between design choices, was discussed in detail. This was followed by a discussion on the definition of feeding design spaces using material characteristics. Research suggests that a single material descriptor, the bulk density of the powder, is a strong predictor of the feed factor of the material. Discussions related to identifying the optimal screw combination that results in the best feeding performance were excluded since they were covered in Chapter 2 of this book. Characterizing refill noise, and the effect of material properties on refill noise was discussed, along with an approach to predict refill noise using statistical performance databases. The chapter concludes with a brief discussion on modeling LIW feeding.

References

[1] Engisch WE. Loss-in-weight feeding in continuous powder manufacturing. Rutgers University - Graduate School - New Brunswick; 2014.
[2] Hopkins M. Loss in weight feeder systems. Meas Control October 2006;39(8):237—40. https://doi.org/10.1177/002029400603900801.

[3] Kehlenbeck V, Sommer K. Possibilities to improve the short-term dosing constancy of volumetric feeders. Powder Technol November 2003;138(1):51−6. https://doi.org/10.1016/j.powtec.2003.08.040.

[4] Weinekötter R, Reh L. Continuous mixing of fine particles. Part Part Syst Char 1995;12(1):46−53. https://doi.org/10.1002/ppsc.19950120108.

[5] Erdem U. A guide to the specification and procurement of industrial process weighing systems. Meas Control February 2003;36(1):25−9. https://doi.org/10.1177/002029400303600105.

[6] Moghtadernejad S, et al. A training on: continuous manufacturing (direct compaction) of solid dose pharmaceutical products. J Pharm Innov June 2018;13(2):155−87. https://doi.org/10.1007/s12247-018-9313-5.

[7] Wilson DH, Loe JM. Apparatus and method for improving the accuracy of a loss-in-weight feeding system. US4635819A. January 13, 1987.

[8] Aalto P, Björklund J-P. Loss-in-weight feeder control. US6446836B1. September 10, 2002.

[9] Wilson DH, Bullivant KW. Loss-in-weight gravimetric feeder. US4579252A. April 01, 1986.

[10] Wang Y, Li T, Muzzio FJ, Glasser BJ. Predicting feeder performance based on material flow properties. Powder Technol February 2017;308:135−48. https://doi.org/10.1016/j.powtec.2016.12.010.

[11] Engisch WE, Muzzio FJ. Loss-in-weight feeding trials case study: pharmaceutical formulation. J Pharm Innov March 2015;10(1):56−75. https://doi.org/10.1007/s12247-014-9206-1.

[12] Escotet-Espinoza MS, et al. Improving feedability of highly adhesive active pharmaceutical ingredients by silication. J Pharm Innov May 2020. https://doi.org/10.1007/s12247-020-09448-y.

[13] Wang Y. Using multivariate analysis for pharmaceutical drug product development. Rutgers University - Graduate School - New Brunswick; 2016.

[14] Li T. Predictive performance of loss-in-weight feeders for continuous powder-based manufacturing. Rutgers University - School of Graduate Studies; 2020.

[15] Tahir F, et al. Development of feed factor prediction models for loss-in-weight powder feeders. Powder Technol March 2020;364:1025−38. https://doi.org/10.1016/j.powtec.2019.09.071.

[16] Jia Z, Davis E, Muzzio OJ, Ierapetritou MG. Process design, optimization, automation, and control predictive modeling for pharmaceutical processes using kriging and response surface. 2009.

[17] Bascone D, Galvanin F, Shah N, Garcia-Munoz S. A hybrid mechanistic-empirical approach to the modelling of twin screw feeders for continuous tablet manufacturing. Ind Eng Chem Res March 2020;59(14):6650−61. https://doi.org/10.1021/acs.iecr.0c00420.

[18] Bhalode P, Ierapetritou M. Discrete element modeling for continuous powder feeding operation: calibration and system analysis. Int J Pharm July 2020;585:119427. https://doi.org/10.1016/j.ijpharm.2020.119427.

Chapter 4

Continuous powder mixing and lubrication

Sarang Oka[1] and Fernando J. Muzzio[2]

[1]*Hovione, Drug Product Continuous Manufacturing, East Windsor, NJ, United States;*
[2]*Engineering Research Center for Structured Organic Particulate Systems (C-SOPS), Department of Chemical and Biochemical Engineering, Rutgers, The State University of New Jersey, Piscataway, NJ, United States*

1. Fundamentals of powder mixing

Powder mixing is a ubiquitous unit operation in particulate processes, where manufacturing product with uniform composition is a key requirement. This is more so in the case of mixing of pharmaceutical ingredients, where the quality constraints for blend and final product homogeneity are strict and are needed to ensure correct drug content in finished dosage forms. Excipients and active pharmaceutical ingredients (APIs), both in powder and granular form, are mixed before being further processed and compressed into tablets or filled into capsules—and the homogeneity of the final tableted product is directly affected by the homogeneity of the powder mixture entering the tableting machine.

Discussions on mixing in this chapter will be restricted to dry mixing of solid ingredients, although unit operations such as feeding, granulation, milling, compression, and coating encompass varying degrees of solid–solid and fluid–solid mixing.

The chapter aims to serve a guide for the design and operation of continuous blending systems. It is not meant to be a reference or a primer on the fundamentals and physics of powder mixing. Several books and hundreds of publications have been devoted to this topic. Detailed discussions can be found in several sources including Ref. [1]. With this in consideration, this section only introduces the basic concepts and fundamental terms related to solids mixing, especially those which are relevant to continuous mixing and referred to in the chapter. Details on the classification of powder mixtures, followed by the general mathematical framework of classifying them are provided. Ideas and challenges in sampling are introduced, and the section is concluded with an introduction to powder mixing mechanisms.

How to Design and Implement Powder-to-Tablet Continuous Manufacturing Systems
https://doi.org/10.1016/B978-0-12-813479-5.00013-6

1.1 Type of mixtures

Although the broad definitions of the types of mixtures are generally agreed upon, there is a lack of uniformity and consistency in definitions associated with classifying mixtures. The chapter does not dive into the details and arguments that currently persist in classifying powder mixtures, keeping in mind the broad objectives of the chapter. Only the basic concepts and the most widely accepted classification are provided.

1.1.1 Perfect mixture

A perfect mixture of two types of particles is defined as one in which any sample randomly taken from the mixture will contain the same ratio of both particles as the ratio present in the mixture taken as a whole. Fig. 4.1 (top-left) shows a perfect mixture consisting of black and white particles. A perfect mixture is an idealized state, and such mixtures are rarely found in nature or industrial operations.

1.1.2 Random mixture

A random mixture is a state of mixedness where both kinds of particles in a two-component mixture have a chance of being sampled that is equal to their average composition. Fig. 4.1 (top-right) shows a random mixture of equal number of black and white tiles. Each tile has an equal chance of being black or white. Said another way, in a perfect mixture, there is no correlation between sample location and sample composition. Further mixing will not result in an increase in homogeneity of the system. It is the highest degree of homogeneity that can be achieved in a real system (except for "ordered mixtures," see below). Random mixtures are obtained with noninteracting particles with almost identical properties. For particles with differing properties, which is more often the case during mixing of pharmaceutical ingredients, it is extremely challenging to achieve a completely random mixture. A random mixture cannot be achieved in the presence of significant interparticle forces or when the material has a non-zero tendency to segregate.

1.1.3 Ordered mixtures

For a system of particles with discernible cohesion, agglomerates may form where certain particles (guests) may attach to a larger, unlike particle (host). For the ordered mixtures depicted in Fig. 4.1 (middle-left) and Fig. 4.1 (middle-right), there must exist a force of attraction between the guests and the host. In the ideal case, as shown in Fig. 4.1 (middle-left), an identical number of guests coat each host particle, resulting in a higher degree of homogeneity than the random mixture. However, usually, the number of guests on each host particle varies, resulting in a distributed number of guest particles coating the host particles, along with the presence of free guests and/or free hosts (Fig. 4.1 (middle-right)). These factors usually result in mixture homogeneity that is lower (i.e., a higher degree of compositional variability) than the homogeneity of the random mixture.

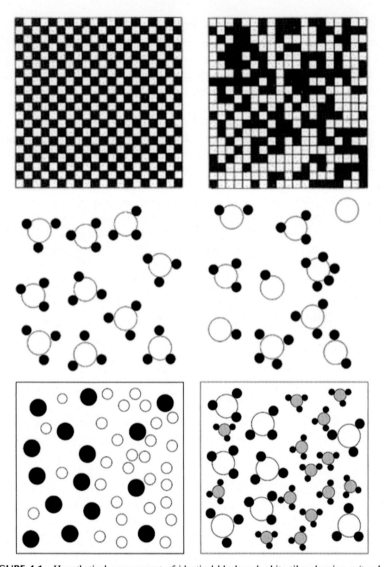

FIGURE 4.1 Hypothetical arrangement of identical black and white tiles showing a (top-left) perfect mixture and (top-right) a random mixture. Particles exhibiting cohesion from agglomerates where the large particles act as hosts and smaller particles as guests. These can form (middle-left) ideal ordered mixtures where each host is coated by an identical number of guests or (middle-right) a more realistic form of ordered mixtures. Segregation can occur for free-flowing materials (bottom-left) and even (bottom-right) cohesive materials can exhibit segregation due to disparity in properties, such as size, density, and shape.

1.1.4 Textured (segregated) mixtures

Textured mixtures are formed when properties of one particle species cause that component to concentrate in specific regions of the mixture. The extent of segregation depends on the type of agitation applied to the mixture. Such behavior is more commonly exhibited by free-flowing materials. Cohesion is generally shown to inhibit mixture segregation, as individual particles find it difficult to move independently of the bulk mixture.

The bottom two images in Fig. 4.1 show two types of segregated mixtures, one for free-flowing materials (Fig. 4.1 (bottom-left)) and one for cohesive particles (Fig. 4.1 (bottom-right)). The pattern in Fig. 4.1 (bottom-left) is shown by free-flowing materials where particles with different size, density, or shape exhibit "differential mobility." The larger the disparity in these properties, typically, the more pronounced is the segregation tendency [2]. For more cohesive systems, like that shown in Fig. 4.1 (bottom-right), a more homogenized state can develop. Segregation of ordered units with different sized carrier particles can also occur, along with segregation of ordered units of the free guests.

1.2 Quantifying mixtures

In order to characterize the homogeneity of granular mixtures, their degree of "mixedness" must be quantified. The metric used for the quantification often depends on the type of mixture and nature of the ingredients. Several metrics involve counting the total number of particles present. For fine powders, this is impossible for any realistic sampling method seeking to characterize a macroscopic blend, where the number of particles, even in a small sample, become so large that it makes the metric unsuitable. Such intricacies must be given due consideration when selecting a quantification metric. Also very important is the scale of examination (i.e., the size of the individual sample used to characterize composition variability). For most measures of homogeneity, the measured degree of homogeneity depends on the scale of examination. Because very small samples will invariably reveal a higher degree of compositional variability than larger samples, it is important to select a scale of examination that is relevant to the application of interest.

One of the most useful measures of mixedness is the intensity of segregation. This is normalized variance of concentration measurements. The intensity of segregation, I, is defined as

$$I = \frac{\sigma^2 - \sigma_r^2}{\sigma_0^2 - \sigma_r^2}$$

where σ^2 is the variance of the sampled data, σ_r^2 is the variance that would be exhibited by samples of the same size obtained from a random mixture, and σ_0^2 is the variance of an initial, fully segregated state, again based on samples of the same size. In practice, estimating σ_r^2 can be challenging. Users of this metric often assume that the blend composition distribution is Gaussian (as would be the case for a random mixture of equal-size ingredients). However, it is unclear whether mixing of granular ingredients ever results in a true Gaussian distribution of compositions. Most blends exhibit at least some degree of segregation tendency. The observed "almost Gaussian" behavior of some blends is likely to be due, at least in part, to the cumulative effects of composition variability, sampling variability, and analytical errors.

However, the assumption that the blend composition distribution is Gaussian pervades blend characterization, both implicitly and explicitly. A Gaussian distribution implies that the probability of deviations from the blend average is known, which enables the experimenter to make assessments about the likelihood of certain types of composition failure. However, the Gaussian assumption also implies that certain samples in the blend (or tablets, when measuring uniformity in tablets) will be outside the desired therapeutic range. Moreover, the assumption of a Gaussian distribution also promotes a tendency by manufacturers to extract the smallest possible samples from the blend, as (1) the shape of the distribution is assumed to be known and therefore does not need to be characterized, and (2) the absolute probability of encountering at least one out-of-specification sample increases with the total number of samples analyzed.

The most common metric in pharmaceutical manufacturing to estimate the variability in the amount of API in powder mixtures, or tablets, is the coefficient of variation of the composition of unit dose samples or, as it is or more commonly referred to in pharmaceutical manufacturing, the relative standard deviation (RSD), which is defined as

$$\mathrm{RSD} = CoV = \frac{\sigma}{M}$$

where σ is the standard deviation, and M is the estimated (i.e., "measured") mean concentration of API of the samples. The RSD is used for multiple purposes in manufacturing of solid oral dosage forms, including to quantify variability in API concentration in blend samples, in total dose of the API in the finished product, and variations in drug dissolution and bioavailability. In general, regulations mandate a ceiling value of 5% in the RSD of the API concentration for acceptable blend uniformity and 6% for finished products (although for some products narrower criteria have been registered). Although seldom calculated, the confidence interval of the RSD estimate is also an important metric. The confidence interval depends on the number of samples used to compute the RSD and also on the actual degree of variation of the blend. The confidence interval of the RSD can be computed theoretically or by certain simulation methods, although neither is very practical. Gao et al. [3] developed an accessible method to estimate the confidence intervals of the RSD estimate. The main conclusion from this study is that under normal conditions typically relying on 10 to 30 samples, a realistic interval of confidence (e.g., 95%) for the RSD is usually quite broad. Under such conditions, the "measured" RSD value is just a coarse estimate of blend variability. Manufacturers are cautioned to always be conservative in making decisions based on estimates of the RSD exceeding values of about 3%, in particular, if the estimates are based in less than at least 100 samples.

1.3 Sampling

In order to evaluate the state of homogeneity of a mixture, in many cases a representative set of samples must be extracted and analyzed. The size of the sample, the location of sampling, the sampling tool, and the number of samples are important considerations when sampling. Theoretical discussions on this topic can be found in Refs. [4,5].

1.3.1 Sample size

The term "sample size" often leads to confusion between the amount of material contained in individual samples and the number of samples extracted to perform the characterization. In this book, by "sample size" we refer to the amount of material contained in a single sample. The size of an individual sample needs to be carefully selected to provide an accurate representation of the quality of the mixture with regard to the final application. Ideally, the size of the sample should be equal to the scale of the scrutiny of the mixture, where the scale of scrutiny is defined as scale at which homogeneity needs to be characterized. In manufacture of solid oral dose forms, the size of the sample (in grams) must be similar to the size of the tablet (in grams). Mathematically, it is fairly straightforward to show that for a random mixture, the sample variance is inversely proportional to the sample size. In general, a larger scale of scrutiny masks heterogeneity within system. A smaller scale of scrutiny, on the other hand, tends to amplify nonhomogeneity. Hence, it is critical that the sample size be approximately equal to the scale at which homogeneity assessment is needed. In the most common situation, the goal is to determine whether different portions of the blend contain similar amounts of active ingredient and will result in variability in drug content in the finished product units. The correct sample size to estimate this type of variability is the size of the final product.

This must be distinguished from situations where the goal is to estimate the variability within a single tablet. This is often needed to examine the presence of agglomerates, or in some cases to determine whether the blend is segregating as it flows into the tablet press dies. For such situations, the scale of examination must be at least two orders of magnitude smaller than the tablet weight.

1.3.2 Number of samples

In the absence of sampling bias, the value of composition variance estimated from the samples approaches the true mixture variance as the number of samples becomes very large (following chi-square statistics if the underlying composition distribution is Gaussian). It is thus recommended that a large number (often referred to as Large N) of representative samples be accurately extracted in order to minimize the sampling contribution to the estimate of the mixture variance. By the same argument, using a Large N also improves confidence interval in estimating the RSD. Confidence intervals can be computed by the method of Gao et al. [3] as previously discussed. A large number of samples also help detect agglomeration in blends—agglomerates display out-of-trend super potency values where they appear as deviations from normal distribution in the high tail. A large number of samples (more than 50, preferably more than 100) are typically necessary to diagnose segregation. Large N also helps detect sampling biases.

1.3.3 Sampling in batch versus continuous blenders

In batch blenders, samples are usually extracted using a device called "sampling thief." Although widely used, the tool suffers from several drawbacks, which have been extensively documented [5] and include generating sampling bias when sampling mixtures with ingredients with diverse properties, disturbing the existing powder bed, and dependence of the measurement on the operating personnel. Batch blenders are thus extremely challenging to sample accurately. Poor suitability for PAT, problems with the use of sampling thieves, lack of clarity in regulation, and limitations in capacity of the analytical lab often dictate the number of samples that are extracted from blends in batch systems. This number is often small and rarely more than 25. The application of PAT tools for online monitoring of blend uniformity in batch systems, while capable of generating a much larger number of measurements, also presents its challenges [6].

Continuous blenders, on the other hand, are more amenable to sampling. The exit end of the continuous blender discharges a continuous stream of powder through its exit mouth, which is typically between 3 and 6 inches in diameter, and it is easy to sample. For accurate and representative sampling of continuous powder streams, it is recommended to follow the two "golden rules of sampling" as proposed by Allen [7].

1. Sample a moving stream
2. The whole of the stream should be taken for many short increments of time, as opposed to part of the stream being sampled for the whole of the time.

These rules are much more easily followed for a continuous mixer than for a batch one. Continuous manufacturing systems are also more amenable for implementation of PAT, which enables sensing of a large number of "samples." A near-infrared sensor mounted on the outlet of the blender can sense and analyze the moving powder continuously. Care must be taken to ensure that portion of the stream that the spectrometer senses is representative of the entire powder mixture. Moreover, sensing and averaging frequencies must be adjusted such that the amount of powder sensed per measurement follows the recommendations for the scale of scrutiny discussed above. Intricacies of sampling using PAT systems are discussed in Chapter 10 of the book. With proper precautions to avoid redundancy and overaveraging, each data point sensed by the PAT tool can be considered equivalent to a single independent sample. The use of PAT tools thus lends itself to sensing a large number of samples automatically and with ease, enabling implementation of Large N methods and potentially enabling much more accurate estimates of the blend RSD.

Physical samples can also be extracted following Allen's golden rules. Several tools are available, which can extract a sample of desired size from a moving powder stream. For the sake of brevity, these tools will not be discussed here.

1.4 Mixing mechanisms

The literature discuses several mixing phenomena, which can be largely summarized into three major mechanisms as defined by J.C Williams [8], namely, convection, dispersion, and shear.

Convective mixing is the result of bulk movement of particles in space, and in most blending applications, it is responsible for the majority of the reduction in variance of the mixture. Convective mixing effects are easier to scale up and the mixing process that ensues is very rapid. However, they are limited by segregated flow structures and rarely result in the creation of completely random mixtures.

Dispersive mixing, on the other hand, is analogous to diffusion in fluids (except that a dispersion coefficient plays the role of the diffusion coefficient). It is driven by small-scale random motion of particles, typically when high particle mobility exists. Like diffusion, dispersive mixing is orders of magnitude slower than convective mixing but results in more complete mixing at the particle scale. However, because the dispersion coefficient itself can be scale-dependent, dispersive mixing effects are more challenging to scale up than convective mixing.

Shear-based mixing is caused by the presence of velocity gradients within powder streams. Particle "lumps" traverse across shearing planes, resulting in deagglomeration and mixing. Shear-based mixing occurs at particulate level resulting in change in the properties of the particulates. Shear-based mixing has been found to be extremely challenging to scale up.

Most mixing operations will involve all three of the above mechanisms occurring simultaneously across time and length scales. Efficient and successful mixing of granular ingredients needs a reasonable understanding of these mechanisms and their role in the mixing device of choice. Relationships between the mixing mechanisms and devices guide processing decisions such as degree of fill, loading pattern, rotation speed of the vessel or mixing aids, and design parameters.

Although the mixing mechanisms discussed above encompass almost the entirety of relevant phenomena, mixing mechanisms are also often classified based on the scale of mixing as macromixing and micromixing. As the names suggest, macromixing is mixing at the bulk particle scale and is largely caused by convective mixing, while micromixing is mixing at the single particle level. The latter is also associated with a change in a property of the individual materials being mixed (such as a change in the wettability due to the particle-level dispersion of a hydrophobic ingredient or decreases in blend cohesion due to the coating of particles by smaller constituents).

Definitions and terms described above will be often used throughout this chapter. The above discussion is also meant to provide the reader with an introduction to the fundamentals of powder mixing and references to find more detailed discussions. A more thorough description of the topics discussed in Section 1 can be found in Chapter 15 of the Handbook of Industrial Mixing [1].

2. Modes of powder mixing

2.1 Batch powder blenders

Powder mixing can be performed in two modes of operation—batch and continuous. In the context of pharmaceutical manufacturing, batch blenders exemplify the batch manufacturing paradigm. A typical batch blender is a large vessel, which can be sealed shut. The vessel could be shaped like a "V," could be a cylinder with a cone on either end, could be a cube with a conical bottom, or can have any one of the myriad available shapes. Powder ingredients are carefully loaded in the vessel, often in a horizontal layered pattern, and the vessel is sealed shut and then rotated until the process is deemed to have reached its "end point." The vessels can be as large as 15,000 L, processing over 5000 kg of material, depending on material density, although typical sizes of batch blenders used in pharmaceutical applications range from about 1 qt to about 150 cu ft. Ingredients are tumbled for a few hundred revolutions during which they get progressively mixed. Mixing occurs over time—the homogeneity of the blend increasing with increasing revolutions of the vessel (time). Processing in time is a defining characteristic of batch-based operations. Once blending is complete, the contents of the blender are discharged into one or more intermediate storage vessels for further processing or progressively emptied onto the piping connecting to a tablet press. Batch mixers, once properly characterized, can provide excellent and reliable mixing performance, in particular when they are equipped with high-speed internal devices (intensifier bars) and used to mix cohesive powders. However, they are often poorly suited for blending ingredients with high segregation potentials [2]. Scaleup of batch mixers is challenging; they have a large equipment footprint, are ill-suited for meaningful PAT measurements that rely on a single measurement point, and do not enable real-time process control of the mixing operation. Commercial-scale batch mixers are also known to be energy-intensive—several hundred kilograms of material is tumbled over large distances resulting in high frictional forces between particles. The energy dissipated during the mixing operation in large-scale devices has been known to generate changes in the material properties of the ingredients.

2.2 Continuous powder blenders

Continuous blenders operate very differently than batch ones. Convection is largely forced by the action of impellers that move the powder along the blender. Unmixed ingredients continuously enter at one end of the blender, while a mixed stream of powder ingredients exits through the other end. Mixing occurs in space, as opposed to time; the homogeneity of the powder stream increases as one traverses closer to the exit end of the blender.

These blenders are available in a variety of configurations and are gentler (lower shear rates) on the powder than batch blenders. They have much smaller equipment footprints, are shown to adequately blend segregating ingredients, and are well suited for online monitoring and real-time control. A review of the theory and practice of continuous mixing can be found in Ref. [9], while a broad review of different continuous blenders can be found in Ref. [10]. Besides the blenders discussed in aforementioned articles, newer designs include the continuous tumble blender by Velazquez and coworkers [11] and the PCMM design by GEA.

The most popular class of continuous blenders for pharmaceutical mixing applications is the tubular blender. Discussions about design, characterization, optimization, and modeling of continuous mixers in the remainder of this chapter will thus be confined to this class of blenders. These discussions, however, are easily translatable to other blender types.

Tubular blenders are characterized by a cylindrical tubular section with a diameter ranging from 3 to 8 inches and an axial length between 6 inches and 4 ft. Fitted along the axial centerline of the tube is a motor-driven impeller (agitator). The impeller has a number of blades distributed along its length. The speed of the impeller, type of blades, number of blades, and their orientation vary from blender to blender, all of which have implications on the performance of the blender. Most tubular blenders accommodate a design

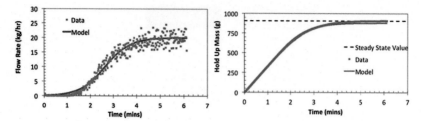

FIGURE 4.2 Dynamic behavior of a continuous tubular blender during blender startup. (Left) Mass flow rate out of the blender during startup shows slow initial change followed by a rapid increase and an asymptotic approach to the final steady state value. (Right) The blender holdup rises almost linearly before tapering off at its steady state value. The blue dots/line indicate actual process data while the red dots/line indicate fitted models.

adjustment which allows for the manipulation of the blender holdup, most commonly through the adjustment of the exit weir/gate of the blender [12]. Others allow for a change in their angle of inclination of the tube, which has the same effect [13,14].

A continuous blender exhibits two states of operation—a dynamic state and a steady state. Consider a tubular blender to be completely empty before operation has begun. Once the operation begins, powder begins to enter the blender, and the blender begins to fill up with powder. At early times, very little powder exits the blender. Fig. 4.2 (left) shows the exit mass flow rate out of the blender as a function of time, while Fig. 4.2 (right) shows the mass of the powder within the blender over time. As the blending tube begins to fill up, the exit flow rate from the blender begins to rise quickly and then asymptotically reaches the value that equals the input mass flow rate into the blender. The state of operation where mass flow rate into the blender equals the mass flow rate out is defined as "steady state" operation. The mass of the material inside the blending tube during the steady state operation remains constant (shown by the horizontal line in Fig. 4.2 (right)) and is defined as the mass holdup of the blender at those conditions. Operation during which the mass holdup within the blender and the exit mass flow rate out of the blender change over time is defined as the blenders' dynamic state of operation. Blender startup and shutdown are characterized by dynamic operation. Change in process parameters, such as impeller speed and input mass flow rate, results in a transient dynamic state, before the blender reacts to the change and reaches its new steady state. Unless otherwise mentioned, all discussions about the blender will assume that the blender is operating at steady state.

3. Mixing in continuous tubular blenders

Powder mixing during steady state operation in continuous tubular blenders can be categorized into two mixing modes: axial and cross-sectional (called

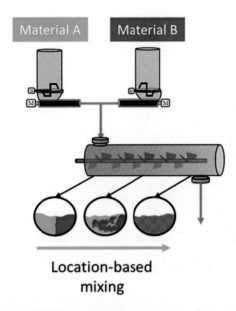

FIGURE 4.3 Illustration of radial mixing phenomenon in a continuous tubular blender along the blender length. Each radial cross section in the blender exhibits the same arrangement of components over time but subsequent cross sections do not.

"radial" for short). Consider two powders, A and B (Fig. 4.3), being fed into the blender. For the sake of argument, consider them to be completely unmixed at the entrance of the blender, as shown in Fig. 4.3, which illustrates a radial cross section at the mouth of the tube. The blades of the impeller lift the powder settled at the bottom of the tube and tumble it over, leading to mixing in the radial direction. In an ideal case, the powder components at the exit of the blender will be completely mixed. At steady state, radial mixing can be considered to be largely time-independent. Each radial cross section in the blender exhibits the same arrangement of components over time but subsequent cross sections do not. A blender in which radial mixing is the only mode of mixing is said to be operating under complete plug flow conditions. Radial mixing is responsible for the majority of the reduction of incoming variance; in other words, radial mixing performs the majority of the homogenization operation in a continuous blender.

It is, however, close to impossible to set up a blender to operate under complete plug flow. When the blades of the impeller lift and tumble the powder in the radial direction, parts of the material are also pushed forward and backward depending on the orientation of the blades. The degree to which the blade configuration encourages backward versus forward movement determines the extent of axial mixing in the blender. It is also undesirable to operate the blender under complete plug flow. As discussed in Chapter 3, the granular nature of all powders makes accurate and steady feeding challenging. All feeding operations have a degree of variability associated with them. If the blender is configured for complete plug flow, the noise in the mass flow rate from the feeders will pass through the system unfiltered and will cause content variations in the final product.

3.1 Residence time distribution in continuous powders blenders

In order to have a comprehensive understanding of the noise filtering ability of continuous blenders, it is critical to understand the concept of *residence time distribution (RTD)* in continuous blenders (the concept and the underlying mathematical framework of RTDs are applicable and common for all continuous flow systems). It is easy to visualize that when powder enters the blending tube at a steady state rate, some particles experience greater convection in the forward axial direction, while some particles experience an unusual amount of backward pushing or become trapped for some time in a "dead" volume within the blender. Thus, while most particles stay in the blender for a time close to a certain mean value, defined as the "mean residence time," some particles might exit the tube very quickly, while other particles spend an unusually long time inside the tube. There thus exists a distribution of residence times for a representative group of particles entering a blender at steady state. This is called its RTD.

The RTD of the blender can be measured by introducing an instantaneous pulse of a tracer material in the system and measuring the concentration of the tracer in the outlet stream as a function of time. A pulse of tracer is meant to embody a representative group of particles of interest entering the blender and it is assumed to mimic the motion of those particles within the blender [15−17].

As seen in Fig. 4.4, the pulse, initially sharp, broadens as it traverses the blender length. Some particles exit the blender quickly while some take unusually long to exit the system. Most spend an amount of time close to a mean value, which can be calculated by computing the mean of the distribution or by

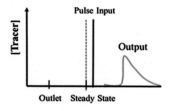

FIGURE 4.4   Inlet and outlet profile of a tracer introduced as a pulse to measure the residence time distribution of the blender. The sharp pulse is broadened due to convection in the axial direction, as the tracer traverses the blender tube.

the dividing the mass holdup of the blender by the operating mass flow rate. In the absence of dead zones, these two ways of calculating the mean residence time should give similar results.

The RTD function $E(t)$ is defined as

$$E(t) = \frac{C(t)}{\int_0^\infty C(t) \cdot d(t)}$$

The mean residence time or the average time that a particle spends within the blender is given by

$$\tau = \int_0^\infty t \cdot E(t) \cdot dt$$

Lastly, the mean centered variance (MCV) or the width of the RTD profile and a measure of its noise filtering ability is given by

$$\sigma_\tau^2 = \frac{\int_0^\infty (t - \tau)^2 \cdot E(t) \cdot dt}{\tau^2}$$

Alternately, the tracer could also be introduced as a step function and the response of the system can be measured as a function of time. The details of the method and the accompanying equations are not discussed here.

As will be discussed shortly, it is possible to tune the RTD profile of a blender by changing certain processing and design parameters. The wider the RTD profile, or the larger the MCV of the profile, the greater is the noise filtering ability of the blender. Consider the example described by Engisch et al. [18]. Figs. 4.5 and 4.6 show the same blender tuned to have a narrow RTD profile (Fig. 4.5A) and a wide RTD profile (Fig. 4.6A), respectively. In case of the former, high-frequency noise passes through the system almost unfiltered—the bimodal sine wave only shifts in the timescale, but the shape of the profile is nearly identical before and after the blender (Fig. 4.5B). In case of Fig. 4.6B, on the other hand, where the blender is tuned to have a much wider distribution, the incoming noise is substantially dampened. These results can also be visualized in frequency domain plots in Fig. 4.5D and Fig. 4.6D. For the filtering ability plots in Figs. 4.5C and 4.6C, a value of 1 indicates that fluctuations will pass through, and a value of 0 indicates that the fluctuation has been spread and therefore reduced in magnitude. Fig. 4.5C shows the filtering ability for the narrow distribution, which will not filter out most fluctuations with frequencies less than 0.15 Hz. In contrast, Fig. 4.6C shows the filtering ability for the broad distribution, which filters most fluctuations above 0.05 Hz.

The intrinsic noise associated with feeding during steady operation is not the only source of noise from the feeders. Feeders, operating continuously for extended periods of time, need to be periodically refilled. As discussed in

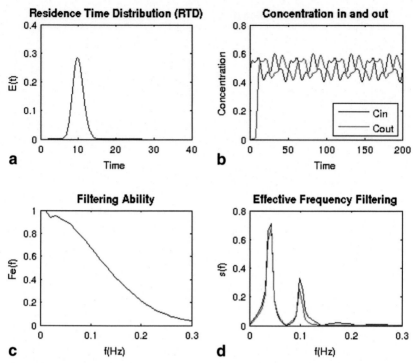

FIGURE 4.5 Simulated results for a bimodal sine wave feed stream being fed to a blender with a narrow residence time distribution (RTD) (in comparison to Fig. 4.6). (A) RTD. (B) Concentration profiles for the inlet and outlet of the blender. (C) Calculated filtering ability of the blender as a function of frequency. (D) Frequency domain of inlet and outlet streams.

Chapter 3, refill of a feeder results in the feeder discharging a sharp pulse of material (high-amplitude, high-frequency noise). If the blender does not incorporate sufficient axial mixing, this noise component can pass through the system inadequately filtered, resulting in an inhomogeneous blend, and maybe even in nonuniform product units. Using another example from Engisch et al. [18], Fig. 4.7A shows the concentration profile of the material of interest over time after each unit operation in a continuous manufacturing process train. A disturbance (pulse of magnitude 0.25 g) introduced in the feed stream passes unfiltered through the mill. This perturbation is dampened by the blender to below a maximum acceptable value (horizontal purple line). Note that the standard deviation of the RTD (nondimensionless square root of the MCV), which is also a measure of noise filtering ability, has a value of 12 s in this case. The blender with this value of the standard deviation of the RTD is, however, unable to dampen a perturbation of larger magnitude (pulse of magnitude 1 g), as exhibited by Fig. 4.7B.

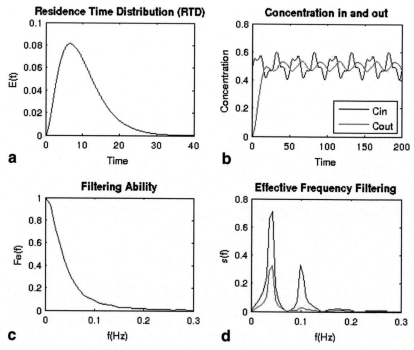

FIGURE 4.6 Simulated results for a bimodal sine wave feed stream being fed to a blender with a broad residence time distribution (RTD) (in comparison to Fig. 4.5). (A) RTD. (B) Concentration profiles for the inlet and outlet of the blender. (C) Calculated filtering ability of the blender as a function of frequency. (D) Frequency domain of inlet and outlet streams.

However, if the RTD is widened to a standard deviation value of 24.9 s by changing the blade pattern of the blender, it successfully dampens the larger pulse (Fig. 4.8). The RTD of the blender must be wide enough such that it is able to dampen the largest source of the incoming noise from unit operation upstream. In order words, the blender must incorporate sufficient axial mixing to dampen incoming noise of relevant magnitude.

To summarize the above discussion, a combination of radial and axial mixing converts heterogeneous powder streams into homogeneous powder mixtures in a continuous blender. Radial mixing eliminates the majority of the variance in incoming powder streams while the axial mixing eliminates incoming temporal variance along the axial length of the blender, serving as a high-frequency noise filter.

The impeller speed, number of blades, blade orientation, and angle of the exit weir are all design and process variables which impact the degree of radial and axial mixing and consequently final blend homogeneity.

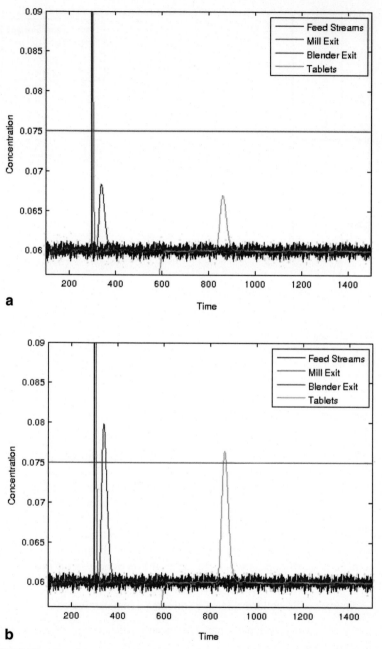

FIGURE 4.7 Simulation results showing the active pharmaceutical ingredient (API) concentration profile for the various unit operations and their response to a pulse of API added to the entrance to the mill. The blender has a mean residence time of 41.6 s and a standard deviation of 12 s. The sizes of the pulse are (A) 0.25 g and (B) 1 g.

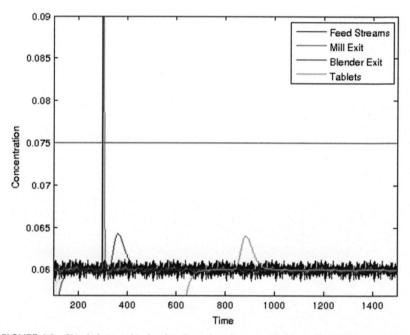

FIGURE 4.8 Simulation results showing the active pharmaceutical ingredient (API) concentration profile for the various unit operations and their response to a 1 g pulse of API added to the entrance to the mill. The blender has a mean residence time of 71.7 s and a standard deviation of 24.9 s.

3.2 Choosing an appropriate mixer configuration

The primary objective of characterizing a blender is to converge on a combination of process and design variables which ensure thorough mixing of the powder ingredients of interest. Vanarase and Muzzio [12,19] found that for free-flowing ingredients, the degree of mixedness of the powder stream exiting the blender bears a direct correlation to the number of impeller passes that the material experiences inside the blender. A higher number of impeller passes resulted in a higher degree of homogeneity of the exit powder stream. The objective function of the design problem can thus be distilled to maximizing the number of impeller passes that the material experiences in the process. The number of blade passes is given by

$$\text{Number of impeller passes} = \tau \ (\text{minute}) * \text{Impeller speed (rpm)}$$

where τ is the mean residence time. The mean residence time of the material can be expressed as follows:

$$\tau = \frac{\text{Holdup}(g)}{\text{Flowrate}(g/\text{minute})}$$

The equation for the number of impeller passes can thus be rewritten as

$$\text{Number of impeller passes} = \frac{\text{Holdup}\,(g)}{\text{Flowrate}\,(g/\text{minute})} * \text{Impeller speed}\,(\text{rpm})$$

It is important to note that the material holdup inside the blender is a function of the impeller speed. The number of impeller passes is thus not a product of two independent variables (the total mass flow rate is fixed for a given process and is usually not a variable in process design). For many blenders, the material holdup is inversely proportional to the impeller speed (Fig. 4.9C). The highest number of impeller passes thus occurs at intermediate rotation rates, where the product of the material holdup and the impeller speed go through a maximum. This can be observed in Fig. 4.9D. Vanarase and

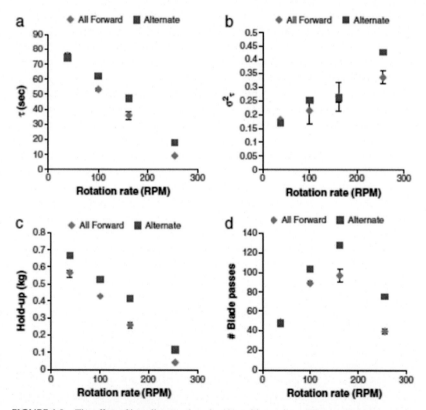

FIGURE 4.9   The effect of impeller speed on the (A) residence time, (B) mean centered variance (MCV), (C) mass holdup, and (D) number of impeller passes in a continuous tubular blender. The mean residence time and the mass holdup decrease with increasing impeller speeds, while the mean MCV increases. The maximum number of impeller passes is experienced by the powder at intermate rotation speeds.

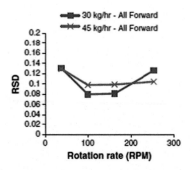

FIGURE 4.10 The mixing performance of a continuous tubular blender at 30 kg/h and 40 kg/h as a function of impeller speed of the blender. The best mixing performance, measured as the relative standard deviation (RSD) of mixed powder stream, is observed at intermediate impeller speeds.

coworkers [12] observed and validated this phenomenon for a Gericke GCM-250, and results presented in Fig. 4.9 are adapted from their work. Fig. 4.9A shows the relationship between mean residence time and impeller speed. Similar observations were made for an inclined blender similar in basic design to the blender commercialized by GEA.

As can be observed from Fig. 4.10, the best performance is obtained at intermediate impeller rotation rates for an all-forward blade configuration in the Gericke GCM-250 blender [12]. Similar observations were made for other blade configurations.

It is clear that the number of blade passes increases with increasing holdup, all other parameters held constant. The material holdup inside the blender can be modified independently by performing certain design alterations, specifically to the exit of the blender tube, as previously mentioned. For example, in a Glatt GCG-70, the angle of the outlet weir can be modified to increase material holdup, a higher angle of exit resulting in a higher holdup. In case of some blenders, the angle of inclination of the entire tube can be changed to modify holdup. Portillo et al. [13,14] studied two such blenders. The material holdup and consequently the number of impeller passes increased at high degrees of inclination, resulting in superior mixing performance. Similarly, the degree of openness of the exit gate in case of the Gericke GCM-250 can be adjusted to alter the material holdup inside the blender. Fig. 4.11 shows the effect of the angle of the weir in a Gericke GCM-250. It was observed that the highest blender holdup was observed when the degree of inclination of the exit weir was equal to the angle of the inclination of the powder bed.

A final means to modify material holdup is changing the blade configuration. Blades attached on the impeller can sometimes be changed to push the material forward or backward to varying extents based on their angle of orientation. The greater the number of blades pushing the material backward, the greater is the material holdup. Fig. 4.9C shows the effect of the angle of

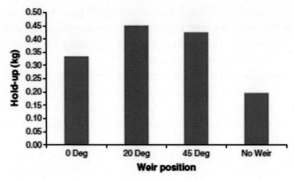

FIGURE 4.11 Effect of weir angle on the blender holdup in Gericke GCM-250 continuous powder blender. *Figure adapted from Vanarase AU, Muzzio FJ. Effect of operating conditions and design parameters in a continuous powder mixer. Powder Technol 2011;208(1):26–36. doi: 10.1016/j.powtec.2010.11.038.*

orientation of the blades on the mass holdup in the blender, all other parameters held constant. As can be observed, an alternate arrangement of blades, where every other blade is oriented 45 degrees backward, results in a higher mass holdup and a higher value of mean residence time compared to a configuration where all the blades are aligned 45 degrees forward. Consequently, the material in the blender also experiences more impeller passes (Fig. 4.9D).

As discussed in Section 3.1, another advantage of orienting blades backward is increased backmixing in the blender and consequently increased noise filtering ability of the blender. Fig. 4.9B shows the effect of the blade configuration on the MCV of the blender—the alternate backward–forward blade arrangement resulting in greater backmixing compared to the all-forward configuration. The greater the number of blades pushing the material backward, the greater is the MCV.

Intuition thus suggests that to maximize the blender holdup, and consequently the number of impeller passes, the practitioner should align all blades backward and then have an exit gate configuration which maximizes mass holdup. Orienting blade backward also has the added advantage of increasing the noise filtering ability of the blender. However, an excessive holdup in the blender often leads to reduction in processing capacity of the blender, sluggish response of the overall line to control actions, increased waste, and even worse, can result in choking of the blender and line stoppage.

An algorithm to maximize the number of the impeller passes, given the above constraint of choking and reduction in capacity, can thus be written as follows.

1. Choose an exit weir which maximizes material holdup.
2. Choose a blade configuration that maximizes material holdup.
3. Operate at an intermediate impeller rotation speed where the *impeller passes* versus *impeller speed* goes through its maximum.

Selections from Steps 1 and 2 should not result in reduction in capacity of the blender or blender choking/clogging. The blender should be able to process the desired mass flow rate with sufficient operating room to incorporate fluctuations in incoming mass flow rate.

There also exists a correlation between the impeller speed and the MCV. The MCV, and consequently the noise filtering ability of the blender, increases with increase in impeller speed. In case that the impeller speed selected from Step 3 does not result in sufficient noise filtering, it is suggested to increase the impeller speed.

As can be seen in Fig. 4.9B, an increase in the impeller speed, for both blade configurations, increases backmixing in the blender. Increase in the impeller speed also decreases blender holdup and increases operating capacity, giving the opportunity to further increase the blender holdup and backmixing by aligning more blades backward. If the operating space of the blender is well mapped out, then the blender design problem can be set up as an optimization problem with a two constraints.

Maximize (number of impeller passes)

Subject to i) the blender being able to filter out the largest source of incoming noise ii) with no loss of operating capacity or blender choking.

The largest source of incoming noise for the optimization approach stated above refers to the expected and characterized noise. Blenders cannot be expected to handle low probability events, for example, long-term feeder drifts or low-frequency noise accompanied with a high amplitude.

As previously discussed, in addition to the intrinsic feeder noise, feeder refill can result also in a large pulse of material being discharged by the feeder over a relatively short period of time. A feeder refill is typically the largest amplitude incoming noise source. If the optimization problem does not converge, then it is likely that the amplitude of the noise from the refill is too large for the blender to successively filter it. In such a case, it is suggested that the feeder be refilled more frequently and at higher levels of powder in the hopper, thus reducing the magnitude of feeder perturbations. The relationship between the refill quantity, frequencies, and resultant noise has been discussed in Chapter 3.

Solving the optimization problem posed above is an elegant method to solve the blender design problem and results in a robust and accurate result. However, in order to solve the problem, it is necessary that the design space of the blender is well mapped out. The effect of the impeller speed, weir angles, and blade configurations on the mass holdup, residence time, and residence time distribution must be well understood. Although the optimization approach is a more tedious method to solve the design problem, the method enables a higher degree of process understanding. This is in alignment with principles of Quality by Design. Moreover, given the critical role played by mixing processes on the evolution of regulatory expectations and requirements for process development, it is also increasingly likely that such thorough characterization studies will be expected by regulatory agencies in the future. Even if that was not the case, companies would greatly benefit from a thorough understanding of mixer performance and the interactions between feeders and blenders, as the interplay between these two pieces of equipment is responsible for ensuring blend homogeneity and finished product content uniformity.

4. Lubrication in continuous tubular blenders

This section discusses the mixing of lubricants in continuous systems. Significant attention is paid to lubricant mixing in batch systems due to risk of overlubrication. Mixing lubricants in continuous blenders brings its own challenges. The section begins with a discussion on the role of lubricants in pharmaceutical solid oral dosage forms, followed by measuring lubrication effects and then lubricant mixing in continuous blenders. The section concludes with comparison between lubricant mixing in batch systems versus continuous systems.

4.1 Role of lubricant

Most powder and granulation blends, prior to tablet compression or capsule filling, are mixed with a lubricant. Moreover, blends are often lubricated also prior to roller compaction. Lubricants such as magnesium stearate are added to the formulation to reduce sticking to metal equipment such as tablet press punches and dies and roller compactor rolls. Lubricants decrease the force of friction at the interface between the tablet surface and the die walls, facilitating ejection and reducing wear on the punches and dies. Lubricants also decrease interparticle friction, improve the packing efficiency of the blends, and improve their flow properties [20–22].

Lubricants, however, can also affect the processability and performance of the formulation. Magnesium stearate, generally present as small friable agglomerates, is known to coat larger, less shear-sensitive excipients and active ingredients [23]. This reduces the bonding ability of particles and consequently results in the loss of tabletability of the formulation. Magnesium stearate is also hydrophobic in nature. The coating of active ingredient particles with magnesium stearate thus decreases their wettability, resulting in a slower rate of dissolution for the active ingredient. Sometimes this effect is used deliberately to slow down the rate of wet granulation or the rate of tablet dissolution. In most cases, however, this is an unintended phenomenon that due to its shear-sensitive nature is also scale-dependent. Many scaleup efforts have failed due to lack of understanding of the effects of shear on MgSt-induced hydrophobicity.

The extent of coating of the lubricant and, consequently, the blend hydrophobicity is directly proportional to two factors—the total amount of lubricant and the total amount of strain the blend experiences when being mixed with the lubricant. The amount of lubricant is typically fixed between 0.25% and 1% (w/w) in most formulations. The amount of strain thus becomes a key design parameter in lubricant mixing.

4.2 Measuring lubricity

This phenomenon of coating of the blend particles, and the resultant change in hydrophobicity of the blend, allows a practitioner to quantify lubrication effects. The lubricity of the blend can be measured by measuring the hydrophobicity of the blend using the Washburn method [24,25] or by measuring the contact angle of the blend with water using the drop penetration method [26–28]. The Washburn method measures the rate of rise of water through a column of the blend due to capillary action. The more hydrophobic the blend, the longer it takes for the fluid to rise and thus greater is its hydrophobicity. Details of the method can be found in Ref. [24]. The method also allows to measure the contact angle of the blend with water. The drop penetration method, on the other hand, measures the time it takes for a drop of water to completely penetrate a conditioned powder bed due to capillary action. A longer penetration time indicates a larger contact angle and greater hydrophobicity.

4.3 Lubricant mixing in continuous blenders

As mentioned, the total amount of strain exerted by a continuous blender on the blend can be quantified by the total number of impeller passes experienced by the blend. The greater the number of impeller passes, the greater the strain and consequently the greater the associated lubrication effects [29,30]. It is interesting to note that the extent of lubrication only depends on the total amount of strain experienced by the blend, and under typical processing conditions, it is largely independent of the rate of strain. This means that in a continuous system, the lubricity of the final blend only depends on the total number of impeller passes it experiences in the blender and not on the impeller speed of the blender.

Fig. 4.12 shows the effect of total number of impeller passes on the blend hydrophobicity. Theta is the contact angle of the blend with water. The lower the $\cos\theta$ value, the greater the blend hydrophobicity. As expected, the hydrophobicity of the blend increases with increasing total impeller revolutions it experiences. However, blend hydrophobicity does not depend on the rate at which strain was applied, that is, the operating impeller speed of the blender. Similar observations were made for a batch system.

Once the relationship shown in Fig. 4.12 is established, designing a blender and selecting operating parameters to prevent overlubrication becomes accessible.

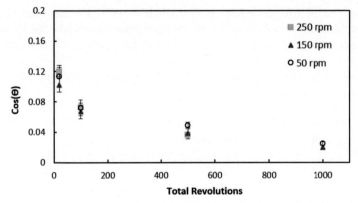

FIGURE 4.12 Effect of total strain and strain rate experienced by the blend on its hydrophobicity upon mixing with magnesium stearate. Hydrophobicity (lubrication) of the final only depends on the total amount of strain experienced by the blend and is independent of the rate of strain.

On the other end of the design problem, opposite to overmixing and overlubrication is ensuring sufficient mixing of the lubricant. Lubricants, due to their small particle size, are present as lumps. Uniform mixing involves deagglomeration of these lumps and uniform distribution of lubricant through the blend. The lack of uniform lubrication can result in serious processing difficulties. Poorly lubricated parts of the blend result in previously mentioned tableting issues, while the overmixed part of the blend may cause the previously discussed overlubrication problems. Interestingly, overall homogeneity of the lubricant in the powder blend is also a function of the total strain. Selecting operating parameters for lubrication is thus a balancing act between ensuring sufficient mixing and preventing overlubrication.

4.4 Lubricant mixing in continuous versus batch systems

It is generally accepted that powder in a continuous tubular blender experiences a more uniform degree of strain compared to a batch system. The ratio of dead volume in the system (i.e., volume not swept by the rotating blades) to the total blender volume is much smaller than it is in a batch blender (where the entire volume of the blender is considered free). In continuous blenders, a small amount of material is more uniformly subjected to the convective action of the blades, resulting in a more uniform nature of strain energy compared to what is found in batch systems, where there is little control on the path traversed by the blend particles, and where particles with dissimilar properties traverse different paths resulting in a nonuniform degree of strain experienced by them. In addition, depending on batch blender geometry, powder near the ends of the axis of rotation can experience slow motion. Risk of nonuniform mixing of the lubricant is thus considered much lower in a continuous blender compared to a batch system.

The risk of overlubrication is also much lower in a continuous blender compared to a batch system. Close attention must be paid to the extent of lubricant mixing in batch systems to ensure that the lubricant is not subjected to excessive strain levels during the blending operation. Typically, lubricants are added in a second, short stage, after the mixing of other ingredients, to minimize the strain experienced by the lubricant-containing blend. While the bulk ingredients are usually blended for several hundred revolutions in the blender before the lubricant is added to the system, the lubricant is generally mixed for a few tens of revolutions.

The material in a continuous blender spends a much shorter amount of time under shear as compared to a batch system. The material spends between a few seconds and a few minutes in the blender and experiences anywhere between 50 and a few hundred impeller revolutions. Moreover, the total mass of material in a continuous blender, at any time, is several orders of magnitude smaller than in a manufacturing-scale batch system. A smaller mass results in lower consolidation and frictional stresses. The blend is thus subjected to much lower degrees of strain. This eliminates the need for two-staged blending in continuous systems, unlike batch manufacturing. There are several commercially available continuous manufacturing systems which comprise of two-stage addition of the ingredients. Typically, this is achieved by having two blenders in series, where the lubricant is added in the second blender. The blade arrangement in the second blender is such that it minimizes backmixing and number of blade passes, thereby minimizing the degree of strain experienced by the material. However, many applications have shown that addition of lubricant to the first blender fitted with a "high shear" blade configuration would not result in overlubrication. In a direct compression system, this would eliminate the need for a second blender, reducing capital investment and system size and enabling the system to respond faster to automated process control actions. However, in processes where the blend undergoes granulation or where the blend passes through a traditional high shear environment (such as a mill), the presence of a lubricant in the blend should be avoided to prevent processing challenges and overlubrication. Two-staged addition with two blenders might be necessary in such cases—mixing and processing of blends/granulation in the first stage, followed by addition and mixing of lubricant in the second stage, just prior to compression.

5. Role of delumping in continuous powder mixing

As mentioned, the continuous tubular blenders discussed in the chapter so far are generally known to impart moderate amounts of shear on the powder. The shear action they provide has been found to be insufficient in situations that necessitate the application of high shear, such as during blending of cohesive

APIs which often tend to form agglomerates. Achieving blend homogeneity is usually not a challenge when the concentration of the API in the formulation is high. However, for potent drugs, where the APIs are usually present in the formula in small concentrations (typically less than 5% (w/w)), the challenge of achieving blend uniformity is magnified. The tendency of the API to form agglomerates, coupled with its low concentration in the formulation, creates a substantial powder mixing challenge—even the presence of a single agglomerate in a sample can result in the blend failing the typical homogeneity criterion. Likewise, a single agglomerate in a tablet can cause the product to fail content uniformity.

For these reasons, in batch manufacturing, powder ingredients are delumped in a sifter or a Comil prior to blending. Delumping also plays an important role in continuous manufacturing. Continuous tubular blenders, similar to batch blenders, are not very effective deagglomeration units. It is thus recommended that a Comil or a sifter, both intrinsically continuous devices, be integrated in the process. The absence of such a delumping device can diminish the flexibility of a continuous process. However, delumping must be implemented with additional care, as it has also been shown that for some APIs, high shear devices can induce electrostatic charging, which can lead to additional processing issues.

Vanarase et al. [19] examined the role of delumping in continuous processing/mixing. Figs. 4.13 and 4.14 demonstrate the benefits of combining a continuous Gericke blender and a Quadro 197-S Comil when mixing a binary powder system of micronized acetaminophen (APAP) and microcrystalline cellulose (Avicel PH-200). It is clear that, on its own, the Comil is a more effective mixer than the blender for blending this formulation, which contains 10% (w/w) of the highly cohesive APAP. Although operated at optimum

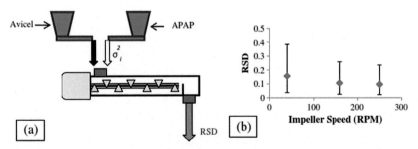

FIGURE 4.13 (A) Schematics of the experimental setup for mixing in a Gericke continuous mixer. (B) The effect of impeller speed on blend uniformity measured as the relative standard deviation (RSD). *Figures adapted from Vanarase AU, Osorio JG, Muzzio FJ. Effects of powder flow properties and shear environment on the performance of continuous mixing of pharmaceutical powders. Powder Technol 2013;246:63–72. doi: 10.1016/j.powtec.2013.05.002.*

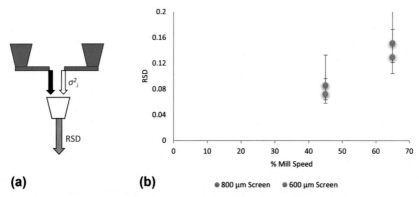

(a)　　　　　　　**(b)**　　　　　● 800 μm Screen　　● 600 μm Screen

FIGURE 4.14　(A) Schematics of the experimental setup for mixing in a Comil. (B) The effect of impeller speed and screen size on blend uniformity measured as the relative standard deviation (RSD). *Figures adapted from Vanarase AU, Osorio JG, Muzzio FJ. Effects of powder flow properties and shear environment on the performance of continuous mixing of pharmaceutical powders. Powder Technol 2013;246:63–72. doi: 10.1016/j.powtec.2013.05.002.*

mixing conditions (macromixing), the blender is ineffective at delumping and breaking up agglomerates. It is a poor micromixer. The Comil, on other hand, was found to be excellent at breaking up agglomerates and thus a very effective micromixer, and its overall mixing performance is found to be superior to the tubular blender for this formulation.

Either device, on its own, was found to be insufficient to achieve the desired blend uniformity. The two devices were then integrated in series and results are shown in Figs. 4.15 and 4.16. Fig. 4.15 shows the two powder

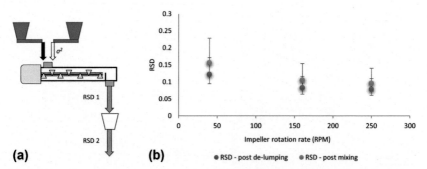

(a)　　　　　　　**(b)**　　　　　● RSD - post de-lumping　　● RSD - post mixing

FIGURE 4.15　(A) Schematics of the experimental setup for integrated low and high shear mixing (low shear mixing first). (B) Mixing performance after low and high shear mixing. The mixing performance is measured as the relative standard deviation (RSD). *Figure adapted from Vanarase AU, Osorio JG, Muzzio FJ. Effects of powder flow properties and shear environment on the performance of continuous mixing of pharmaceutical powders. Powder Technol 2013;246:63–72. doi: 10.1016/j.powtec.2013.05.002.*

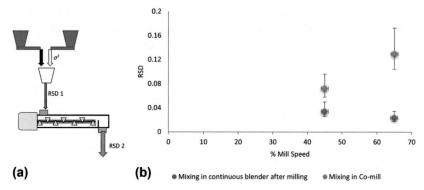

(a)   **(b)**   ● Mixing in continuous blender after milling   ● Mixing in Co-mill

FIGURE 4.16   (A) Schematics of the experimental setup for integrated high and low shear mixing (high shear mixing first). (B) Mixing performance after high and low shear mixing. The mixing performance is measured as the relative standard deviation (RSD). *Figure adapted from Vanarase AU, Osorio JG, Muzzio FJ. Effects of powder flow properties and shear environment on the performance of continuous mixing of pharmaceutical powders. Powder Technol 2013;246:63−72. doi: 10.1016/j.powtec.2013.05.002.*

ingredients first passing through the blender (macromixing) followed by the Comil (micromixing). The order of the blending units was then reversed and Fig. 4.16 shows a setup where the powder ingredients are micromixed in the Comil followed by macromixing in the blender. As expected, and in general, the two-stage blending was always found to be superior than single-stage blending in both setups.

It is interesting to note that the final blend homogeneity was *not* agnostic to the order of mixing. Micromixing and delumping (mixing in the Comil) prior to macromixing (mixing the blender) resulted in a superior performance compared to the reversed order. The blender, if present before the Comil, macromixes the powder ingredients, including agglomerates of the API. However, it fails to break up the agglomerates. The agglomerates in the blend breakup in the next blending stage, when it passes through the Comil. But the mill is a poor macromixer. As a result, the delumped agglomerates are not mixed with the rest of the powder and contribute to the nonuniformity of the blend in this case. In the reverse scenario shown in Fig. 4.16, the primary action of delumping results in breaking up of agglomerates. The continuous blending in the next step ensures macromixing of the delumped API, resulting in a superior blending performance.

In summary, the delumping step plays an important role in continuous blending and continuous processing, in general, and especially when dealing with potent, cohesive APIs that exhibit a tendency to form agglomerates. It provides flexibility to a continuous processing line and empowers it to handle a wider range of powder formulations. A delumping/sifting unit on its own, however, does not have sufficient mixing ability and must be followed by a convective macromixer. The order of mixing is also important—delumping or micromixing must be performed prior to blending or macromixing. The reverse order does not ensure sufficient dispersion of delumped agglomerates and results in suboptimal mixing.

6. Other topics

6.1 Modeling in continuous blenders

Modeling of continuous tubular blenders has been discussed in Chapter 11 and thus will not be discussed here. Interested readers can refer to Chapter 11. A thorough review of modeling in continuous blenders can also be found in Refs. [9,31−33].

6.2 Blending of segregating ingredients in continuous blenders

A generally accepted hypothesis in continuous manufacturing is that continuous processing is more forgiving than batch processing for manufacturing segregating formulations. It enables manufacturing by direct compression of those formulations that would be otherwise granulated in the batch paradigm due to segregation concerns. Segregation typically takes place when a large amount of powder is discharged onto an empty volume (as is the case in batch mixing when discharging the blend into a large hopper). In a properly designed continuous manufacturing system, such discharges do not occur, denying the blend the opportunity to segregate. However, the premise also assumes that the blending step of a continuous direct compression process satisfactorily mixes the segregating formulation, a nonobvious assumption given that batch blenders are known to be poor mixers of segregating formulations.

Oka et al. [2] compared batch and continuous blenders in terms of their ability to mix segregating ingredients and showed that continuous blenders are superior than their batch counterparts in this regard. The convective action of the blades in a continuous tubular blender ensures that the pathlines traversed by the powder ingredients are weak functions of their properties, namely, particle size and density. The pathlines are dictated by the action of blades, ensuring homogenization. On the other hand, in batch systems, the particles are free to follow their path during the blending motion. Dissimilar ingredients take different paths, resulting in segregation of ingredients rather than mixing. Fig. 4.17 illustrates the mixing performance of three highly segregating mixtures in a batch and in a continuous blender. It is clear that the mixing performance in the continuous blender is superior than the batch system.

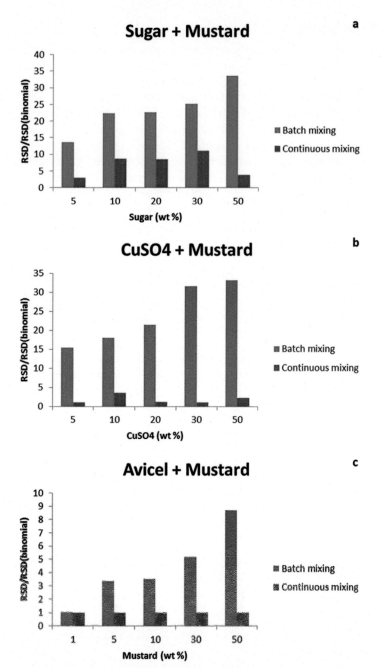

FIGURE 4.17 The relative standard deviation (RSD) of the samples normalized by the RSD of a binomial mixture as a weight function of the minor ingredient. Comparison between batch (blue) and continuous (red) blenders for (A) sugar and mustard, (B) $CuSO_4$ and mustard, and (C) Avicel PH101 and mustard.

7. Conclusions

This chapter provides a brief review of continuous blending of pharmaceutical ingredients, with reference to many relevant published literature reports.

In general, continuous blenders available in the market are able to provide effective macromixing in a robust and reliable manner. The main factor potentially affecting their performance is the presence of large composition perturbations during feeder refill. When properly instrumented, those perturbations can be detected and managed. Cohesive ingredients, even when present in small percentages, can be effectively homogenized by a combination of a mill and a blender. Moreover, if the system is properly designed, segregation poses a much lesser concern than in batch systems.

It is hoped that the reader will find the material useful. However, given the high level of activity in this area and its relative novelty, significant advances in the field are anticipated in the next few years.

References

[1] Muzzio FJ, et al. Solids mixing. In: Paul EL, Atiemo-Obeng VA, Kresta SM, editors. Handbook of industrial mixing. Hoboken, NJ, USA: John Wiley & Sons, Inc.; 2003. p. 887–985.

[2] Oka S, Sahay A, Meng W, Muzzio F. Diminished segregation in continuous powder mixing. Powder Technol March 2017;309:79–88. https://doi.org/10.1016/j.powtec.2016.11.038.

[3] Gao Y, Ierapetritou MG, Muzzio FJ. Determination of the confidence interval of the relative standard deviation using convolution. J Pharm Innov June 2013;8(2):72–82. https://doi.org/10.1007/s12247-012-9144-8.

[4] Muzzio FJ, et al. Sampling and characterization of pharmaceutical powders and granular blends. Int J Pharm January 2003;250(1):51–64. https://doi.org/10.1016/s0378-5173(02)00481-7.

[5] Muzzio FJ, Robinson P, Wightman C, Dean B. Sampling practices in powder blending. Int J Pharm September 1997;155(2):153–78. https://doi.org/10.1016/S0378-5173(97)04865-5.

[6] Sen M, et al. Analyzing the mixing dynamics of an industrial batch bin blender via discrete element modeling method. Processes June 2017;5(2):22. https://doi.org/10.3390/pr5020022.

[7] Powder Sampling and Particle Size Determination. 1st ed. https://www.elsevier.com/books/powder-sampling-and-particle-size-determination/allen/978-0-444-51564-3 [Accessed May 31, 2020].

[8] Williams JC. The mixing of dry powders. Powder Technol September 1968;2(1):13–20. https://doi.org/10.1016/0032-5910(68)80028-2.

[9] Pernenkil L, Cooney CL. A review on the continuous blending of powders. Chem Eng Sci January 2006;61(2):720–42. https://doi.org/10.1016/j.ces.2005.06.016.

[10] Oka S, et al. Continuous powder blenders for pharmaceutical applications. Pharma Manuf 2013. Accessed May 31, 2020, https://www.pharmamanufacturing.com/articles/2013/1308-continuous-powder-blenders/.

[11] Florian M, Velázquez C, Méndez R. New continuous tumble mixer characterization. Powder Technol April 2014;256:188–95. https://doi.org/10.1016/j.powtec.2014.02.023.

[12] Vanarase AU, Muzzio FJ. Effect of operating conditions and design parameters in a continuous powder mixer. Powder Technol March 2011;208(1):26–36. https://doi.org/10.1016/j.powtec.2010.11.038.

[13] Portillo PM, Ierapetritou MG, Muzzio FJ. Characterization of continuous convective powder mixing processes. Powder Technol March 2008;182(3):368–78. https://doi.org/10.1016/j.powtec.2007.06.024.

[14] Portillo PM, Ierapetritou MG, Muzzio FJ. Effects of rotation rate, mixing angle, and cohesion in two continuous powder mixers—a statistical approach. Powder Technol September 2009;194(3):217–27. https://doi.org/10.1016/j.powtec.2009.04.010.

[15] Escotet-Espinoza MS, et al. Effect of material properties on the residence time distribution (RTD) characterization of powder blending unit operations. Part II of II: application of models. Powder Technol February 2019;344:525–44. https://doi.org/10.1016/j.powtec.2018.12.051.

[16] Escotet-Espinoza MS, et al. Effect of tracer material properties on the residence time distribution (RTD) of continuous powder blending operations. Part I of II: experimental evaluation. Powder Technol January 2019;342:744–63. https://doi.org/10.1016/j.powtec.2018.10.040.

[17] Moghtadernejad S, et al. A training on: continuous manufacturing (direct compaction) of solid dose pharmaceutical products. J Pharm Innov June 2018;13(2):155–87. https://doi.org/10.1007/s12247-018-9313-5.

[18] Engisch W, Muzzio F. Using residence time distributions (RTDs) to address the traceability of raw materials in continuous pharmaceutical manufacturing. J Pharm Innov March 2016;11(1):64–81. https://doi.org/10.1007/s12247-015-9238-1.

[19] Vanarase AU, Osorio JG, Muzzio FJ. Effects of powder flow properties and shear environment on the performance of continuous mixing of pharmaceutical powders. Powder Technol September 2013;246:63–72. https://doi.org/10.1016/j.powtec.2013.05.002.

[20] Pingali KC, Mendez R. Physicochemical behavior of pharmaceutical particles and distribution of additives in tablets due to process shear and lubricant composition. Powder Technol December 2014;268:1–8. https://doi.org/10.1016/j.powtec.2014.07.049.

[21] Pingali K, Mendez R, Lewis D, Michniak-Kohn B, Cuitino A, Muzzio F. "Mixing order of glidant and lubricant − influence on powder and tablet properties. Int J Pharm May 2011;409(0):269–77. https://doi.org/10.1016/j.ijpharm.2011.02.032.

[22] Pingali K, Mendez R, Lewis D, Michniak-Kohn B, Cuitiño A, Muzzio F. Evaluation of strain-induced hydrophobicity of pharmaceutical blends and its effect on drug release rate under multiple compression conditions. Drug Dev Ind Pharm April 2011;37(4):428–35. https://doi.org/10.3109/03639045.2010.521160.

[23] Pingali KC, Mendez R. Nanosmearing due to process shear − influence on powder and tablet properties. Adv Powder Technol May 2014;25(3):952–9. https://doi.org/10.1016/j.apt.2014.01.016.

[24] Oka S, et al. The effects of improper mixing and preferential wetting of active and excipient ingredients on content uniformity in high shear wet granulation. Powder Technol July 2015;278:266–77. https://doi.org/10.1016/j.powtec.2015.03.018.

[25] Oka S, et al. Analysis of the origins of content non-uniformity in high-shear wet granulation. Int J Pharm August 2017;528(1):578–85. https://doi.org/10.1016/j.ijpharm.2017.06.034.

[26] Liu Z, Wang Y, Muzzio FJ, Callegari G, Drazer G. Capillary drop penetration method to characterize the liquid wetting of powders. Langmuir January 2017;33(1):56–65. https://doi.org/10.1021/acs.langmuir.6b03589.

[27] Wang Y, Liu Z, Muzzio F, Drazer G, Callegari G. A drop penetration method to measure powder blend wettability. Int J Pharm March 2018;538(1):112—8. https://doi.org/10.1016/j.ijpharm.2017.12.034.

[28] Gao T, et al. Granule formation and structure from single drop impact on heterogeneous powder beds. Int J Pharm December 2018;552(1):56—66. https://doi.org/10.1016/j.ijpharm.2018.09.036.

[29] Moghtadernejad S, Escotet-Espinoza MS, Liu Z, Schäfer E, Muzzio F. Mixing cell: a device to mimic extent of lubrication and shear in continuous tubular blenders. AAPS Pharm Sci Tech July 2019;20(7):262. https://doi.org/10.1208/s12249-019-1473-1.

[30] Oka S. et al., Lubrication in continuous tubular powder blenders. http://www.pharmtech.com/lubrication-continuous-tubular-powder-blenders [Accessed May 31, 2020].

[31] Boukouvala F, Dubey A, Vanarase A, Ramachandran R, Muzzio FJ, Ierapetritou M. "Computational approaches for studying the granular dynamics of continuous blending processes, 2 — population balance and data-based methods. Macromol Mater Eng 2012;297(1):9—19. https://doi.org/10.1002/mame.201100054.

[32] Dubey A, Sarkar A, Ierapetritou M, Wassgren CR, Muzzio FJ. Computational approaches for studying the granular dynamics of continuous blending processes, 1 — DEM based methods. Macromol Mater Eng 2011;296(3—4):290—307. https://doi.org/10.1002/mame.201000389.

[33] Berthiaux H, Marikh K, Mizonov V, Ponomarev D, Barantzeva E. Modeling continuous powder mixing by means of the theory of Markov chains. Part Sci Technol October 2004;22(4):379—89. https://doi.org/10.1080/02726350490516037.

Chapter 5

Continuous dry granulation

Nirupaplava Metta[1], Bereket Yohannes[2], Lalith Kotamarthy[3], Rohit Ramachandran[3], Rodolfo J. Romañach[4], Alberto M. Cuitiño[5]

[1]*Applied Global Services, Applied Materials Inc., Santa Clara, CA, United States;* [2]*Drug Product Development, Bristol-Myers Squibb Company, New Brunswick, NJ, United Staes;* [3]*Department of Chemical and Biochemical Engineering, Rutgers, The State University of New Jersey, Piscataway, NJ, United States;* [4]*Department of Chemistry, University of Puerto Rico-Mayagüez campus, Mayagüez, PR, United States;* [5]*Department of Mechanical and Aerospace Engineering, Rutgers, The State University of New Jersey, Piscataway, NJ, United States*

1. Introduction

The ultimate goal of a granulation process is to produce granules with a desired bulk density, shape, and size distribution to improve uniformity, flowability, compressibility, and to reduce the size and density segregation of ingredients ([1−6]). Granulation can be wet, in which an appropriate liquid binder is used to form agglomerates of particles, or dry, where particles are directly compacted and then milled to form granules of desired size. Dry granulation is especially appropriate when the blend contains ingredients that are chemically or physically sensitive to moisture, solvents, or temperature [7]. Dry granulation usually includes two major components: the compaction of the powder blend to form slugs or ribbons and the milling of the slugs or ribbons to form the granules. Slugging is an older process where the powders are tableted prior to milling. Roller compaction is a modern, economical, intrinsically continuous method used in dry granulation of powders. Typically, the powder is compacted into a ribbon by passing through the gap between two rolls, followed by milling of the ribbons using one of the several types of mills. In continuous manufacturing, dry granulation is preceded by continuous gravimetric feeding and blending of the ingredients and is followed by tablet blending of extragranular ingredients and subsequently by compaction. In this chapter, we present the parameters that are essential for the roller compaction process, methods of characterizing ribbons and granules, and prediction and quality control mechanisms that can be used in continuous manufacturing of tablets.

How to Design and Implement Powder-to-Tablet Continuous Manufacturing Systems
https://doi.org/10.1016/B978-0-12-813479-5.00014-8

2. Roller compaction

In all roller compaction processes, the blended powder ingredients are compacted to form ribbons by two counter-rotating rollers (Fig. 5.1A). The blend is fed through a feed hopper, sometimes assisted by screw feeders, and the friction between the blend and the two rotating rollers pushes the blend through a narrow gap between the two rollers, where the ribbon is formed due to the compaction pressure. The roller compaction process has three regions known as the slip, nip, and release regions (Fig. 5.1B). In the slip region, the blend slips on the roller surface, which occurs because the blend is moving at a lower speed than the roller surface. Within this zone the powder undergoes a rearrangement and deaeration with a low level of shear and compressive stress applied. The slipping of powder blend ends when the powder blend enters the nip region, which is also known as "no slip" region [8]. In the nip region, the powder blend flows as a solid with a speed equal to that of the roll surface. The transition from the slip region to the nip region starts at the nip angle α (Fig. 5.1B). As the powder moves toward the gap, where the distance between the rollers is minimum, the compressive stress applied on the blend increases significantly and a ribbon is formed. Beyond the gap, the distance between the rolls increases again, and the ribbon expands due to elastic recovery.

The characteristics of the ribbons depend on the powder properties and on the design and operating conditions including the feed rate, screw speed, the

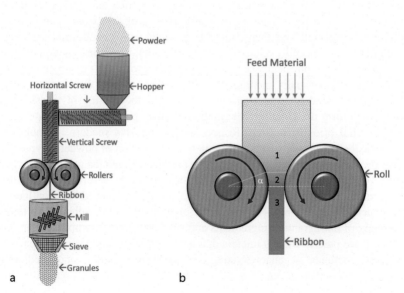

FIGURE 5.1   (A) Schematics of roller compactor setup and (B) roller compaction regions: (1) slip, (2) nip, and (3) release regions.

gap between the rollers, the speed of the rollers, the surface and diameter of the rollers, the compaction pressure applied by the rollers, the deaeration conditions, and to some extent, the sealing system that is used to prevent escape of powder at the sides of the rollers [9]. Roller compactors can be gap-controlled or pressure-controlled [4]. In gap-controlled roller compactors, the gap between the rollers is fixed and the thickness of the ribbons is determined by the width of this gap; compaction pressure fluctuates as a function of instantaneous powder flow rate. On the other hand, in pressure-controlled roller compactors, the compaction pressure is kept constant and the gap width between the rollers may fluctuate depending on the amount of blend passing between the rollers at a given point in time. The compaction pressure is the maximum at the minimum gap distance, which is along a straight line connecting the centers of the rollers. However, the compaction of the blend begins when the blend forms a strong contact with the rollers in the nip region. The nip angle depends on the particle properties of the preblend (such as size, density, cohesion, and plasticity), speed of rollers, gap between rollers, and surface texture of the rollers [6,10,11]. Nip angle values ranging between 5 and 30° have been reported in the literature for different materials and operating conditions [11,12]. Larger nip angles ensure a longer compaction step and better ribbon and granule quality.

Roller compactors are ideally suited for continuous manufacturing. They are intrinsically continuous and highly controllable. As the blend passes through the nipping region, a continuous sheet of powder compact is formed. The sheet is broken into pieces due to its own weight, and the ribbon is subsequently milled to produce desired size granules. The properties of the milled granules are highly dependent on the attributes of the ribbon and hence on all the factors that affect the properties of the ribbon.

The relative density (porosity) of the ribbon after compaction is the most important control parameter to maintain consistency of granulation and finished product properties. The relative density of the ribbons significantly affects the final particle size distribution (PSD). For example, larger relative density can result in larger particle size and mechanically weaker tablets [13]. To maintain constant relative density across several types of roller compactors, the gap between rollers for gap-controlled roller compactors, the compaction force per unit length of the rollers for pressure-controlled roller compactors, the peripheral speed of rollers, and the ratio of screw speed to the roller speed should be kept constant. Further adjustment of these parameters may be implemented as necessary after measuring the relative density of the ribbons. In addition, keeping the milling conditions (i.e., mill impeller speed, coarse and fine screens) constant during milling operation is important to control the size distribution of the granules.

Design and operating conditions, such as the position/orientation of rolls, speed of roll rotation, gap width, applied pressure, feeding system, and roller properties (diameter, width and surface texture), may differ in continuous manufacturing lines, where it is important to assure that the critical attributes of the granules produced in a dry granulation are retained independent of the roller compaction setup [13]. Some of the most critical attributes are the final granule size distribution, the mechanical properties of the granules, and content uniformity of the granulated blend across granule sizes. In the following sections, the factors that affect these attributes and mechanisms that can be used to measure and control these attributes are described.

However, prior to that, we describe the compaction process in detail based on one of the earliest models for roller compaction, the rolling sheet model, developed by Johanson [14]. Several researchers have used Johanson's model to predict the compaction pressure based on the properties of the blend and operating conditions. The model is based (1) on the Jenike–Shield [15] yield criteria for the blend in the slip regime and (2) on the compressibility of the blend in the nip region. In the slip region, the gradient of the stress is given as

$$\left(\frac{d\sigma}{dx}\right)_{slip} = \frac{4\sigma_\theta \left(\frac{\pi}{2} - \theta - v\right)\tan\delta_E}{\frac{D}{2}\left(1 + \frac{S}{D} - \cos\theta\right)\cot(A - \mu) - \cot(A + \mu)} \tag{5.1}$$

where

σ_θ = the mean normal stress at position θ, [MPa].

θ = the angular position with respect to the center of the roller, [°].

δ_E = is the effective angle of internal friction, [°], which can be measured for the blend using a shear cell or the FT4 system (Freeman Technology, Malvern, UK).

D = the roll diameter, [mm].

S = minimum gap thickness, [mm].

$$A = \frac{\theta + v + \frac{\pi}{2}}{2}$$

$$v = \frac{1}{2}\left(\pi - \sin^{-1}\frac{\sin\phi_w}{\sin\delta_E} - \phi_w\right)$$

$$\mu = \frac{\pi}{4} - \frac{\delta_E}{2}$$

The pressure gradient in the nip region is based on the compressibility of the blend which can be approximated based on the following equation.

$$\frac{\log p_1}{p_2} = K \cdot \frac{\log \rho_1}{\rho_2}, \tag{5.2}$$

where K is the compressibility factor of the blend. K can be directly measured from a uniaxial compaction experiment [10]; p is the pressure and ρ is the bulk density of the powder (i.e., ρ_1 and ρ_2 are the bulk densities when the applied pressures are p_1 and p_2, respectively).

The stress gradient in the nip region is given by

$$\left(\frac{d\sigma}{dx}\right)_{\text{Nip}} = \frac{K\sigma_\theta\left(2\cos\theta - 1 - \dfrac{S}{D}\right)\tan\theta}{\dfrac{D}{2}\left(1 + \dfrac{S}{D} - \cos\theta\right)\cos\theta}. \tag{5.3}$$

Johanson's model is based on the assumption that the stress gradient in the slip region is equal to the stress gradient in the nip region when $\theta = \alpha$ (the nip angle):

$$\left(\frac{d\sigma}{dx(\theta = \alpha)}\right)_{\text{slip}} = \left(\frac{d\sigma}{dx(\theta = \alpha)}\right)_{\text{nip}} \tag{5.4}$$

Johanson's model assumes a one-dimensional powder flow, where in reality, the flow of powder between the rollers is nonuniform. Due to the assumption of one-dimensional flow, the Johanson's model may overpredict the relative density of the ribbon. To account for this error, Liu and Wassgren [8] modified Johanson's model by including a mass correction factor along the rollers in the nip region and found a better prediction of the ribbon density. Johanson's model also does not account for the heterogeneity of ribbon density. Akseli and Iyer [3], using ultrasound measurement techniques, showed that the density of a ribbon is not constant along the width of the ribbon. They found that the ribbon has a higher density in the center region compared to the sides of the ribbon. This is due to the presence of friction between the powder and the side walls in the slip regime. The side wall friction inhibits the downward movement of the powder at the edges. They also found that the tablets made from the center part of the ribbon have lower mechanical strength. In addition, Johanson's model does not account for the speed of the rollers. Experiments by Al-Asady and Dhenge [11] show that the nip angle decreases as the speed of rotation is increased. They also found that the prediction of the nip angle based on Johanson's model is a good prediction for low rotational speeds. Based on these findings, it is clear that Johanson's model alone cannot be used to predict the compaction process accurately. Additional analysis is required based on experiments, finite element method (FEM) or discrete element method (DEM) simulations, and other analytic techniques [8,16].

3. Milling

Milling is an intrinsically continuous unit operation common to the direct compaction, wet granulation, and dry granulation routes for continuous pharmaceutical manufacturing. In the direct compaction route, mills are used for material delumping and for active pharmaceutical ingredient (API) silication, whereas in the granulation routes, mills are used for granule size reduction. In the wet granulation route, mills reduce the size of granules that are larger than the desired size, after the granulation and drying unit operations. In the case of dry granulation route, mills are used to break the compacted ribbons.

In this section, we mainly focus on mills as a particle size reduction tool. As mentioned in the previous section, the granulation process is primarily required in drug product manufacturing in order to alleviate issues related to powder handling such as poor flowability. However, granulation may also produce particles with undesirable size distributions. This may downplay the positive benefits of granulation and also affect further processing during tablet compaction. The presence of large particles might lead to formation of tablet with pitted surfaces and poor strength, whereas an excess amount of fines leads to poor granule flow and weight variability [17,18]. In addition, the PSD also impacts drug bioavailability [19]. Milling can also improve the dissolution of poorly soluble drugs, thereby improving their bioavailability [20,21]. In this case, API is milled into ultrafine (micronized) particles to increase the surface area, leading to improvement in dissolution kinetics. Hence, an understanding of the effects of mill design and operation is critical.

3.1 Types of mill

Milling equipment is generally classified on the basis of types of forces applied to break the particles—impact, attrition, and shear-compression [22]. The choice of equipment depends on the properties of the feed material (hardness, elasticity, etc.) and the finished product specifications such as particle size, particle shape, etc. [23]. Table 5.1 shows the most commonly used mills and their size reduction capacities. Only the impact mill and the shear-compression mill are discussed in detail, as these mill types are commonly integrated with roller compactors.

3.1.1 Impact mill

The main mode of breakage in an impact mill is via mechanically induced high force collisions. Examples of impact mills are hammer mill and pin mill. Hammer mills are capable of significant size reduction and can reduce the size of particle down to about 10 μm. The force imparted by the hammers, the feed rate, and the screen opening size are the critical parameters that control the degree of particle size reduction. Generally, the PSD of granules produced by impact mills are relatively narrow, with fewer fines, because of self-classification of particles in the screen. Pin mills operate similar to hammer mills, but typically with faster tip speeds and lower mechanical tolerances between rotating and stationary pins [25].

TABLE 5.1 Mills classified on the basis of milled particle size [24].

Mechanism		Impact		Attrition	Impact and attrition	Shear-compression	
Particle description	Size (μm)	Pin mills	Hammer mills	Jet mills	Ball mills	Conical screen mill	Oscillating granulator
Medium fine	500–1000	Yes	Yes	No	No	Yes	Yes
Fine	150–500	Yes	Yes	No	No	Yes	Yes
Very fine	50–150	Yes	Yes	No	No	Yes	Yes
Super fine	10–50	Yes	Yes	Yes	Yes	No	No
Ultra fine	<10	No	No	Yes	Yes	No	No
Colloidal	<1	No	No	No	No	No	No

3.1.2 Shear-compression mill

The conical screen mill (comil) is a popular type of shear-compression mill as it can be used for delumping as well as granule breakage. It has been found to be suitable for milling a wide range of products [26,27]. In a comil, the intense shear applied on the material in the gap between the impeller and the screen leads to creation of interparticle and particle-wall frictional contacts and subsequent reduction in size. The force imparted by the impeller also affects the breakage of particles. The resultanat smaller granules escape through the screen. The size and shape of the screen holes, the type of screen, the shape of the impeller, the fill level in the mill, and the speed of the impeller are the important mill parameters that control the quality attributes of milled granules [75,76]. Another type of shear-compression mill is an oscillating granulator [28], which is generally used for roller compacted ribbons, which are passed through an assembly of wire mesh screen and oscillating rotors. The particle size of the milled granules is controlled by the screen size, speed of the rotor, and rotational angle of rotors [29]. They generally produce coarser granules than the comil. There are many roller compactors available commercially with an oscillating granulator incorporated after the rolls. These machines provide an advantage as they allow continuous processing from powder blend to granules. For example, the Gerteis roller compactors employ one oscillating granulator under the compaction zone after the rolls, whereas the Alexanderwerk roller compactors employ two granulators for better control over the particle size. Often, a comil is integrated after an oscillating granulator for secondary milling.

4. Roller compaction characterization and micromechanical modeling

Several methods have been used to characterize the ribbons produced by roller compactors, including near-infrared (NIR) spectroscopy, ultrasound, and X-ray tomography [3,30]. These characterization techniques have been used to measure heterogeneity of material composition and porosity. In addition, several types of models have been developed to predict the properties of the ribbons. These models include theoretical models such as the Johanson, Dehont, and Heckel models [12,31,32], continuum computational models using the FEM, and discrete particle modeling methods. In this section we discuss details of NIR spectroscopy and discrete particles based analysis of ribbon properties.

4.1 Near-infrared spectroscopy—information on chemical composition and physical properties

NIR spectroscopy has been extensively reviewed in the literature and its use as a PAT (process analytical technology) method is discussed elsewhere in this book (internal cross-reference). In this chapter, we focus on NIR only in the context of roller compaction. NIR spectroscopy is capable of providing information on both the chemical composition and the physical properties of roller compacts and the milled material. The NIR spectrum is the result of interaction of electromagnetic radiation in the NIR region of the spectrum, with the vibrations of the chemical bonds in the particles that form the ribbon. These complex interactions contribute to the NIR spectrum, which depend on the physical properties and the chemical composition of materials. A series of studies have provided guidance on how to differentiate between the effects of physical properties and the contributions of chemical composition on the NIR spectrum [29,30,33]. However, NIR spectroscopy is not without its limitations, as physical and chemical information is not easily distinguished [34]. Scientists working with NIR spectroscopy require significant training to avoid confusing these two important contributions to the NIR spectrum.

4.1.1 Near-infrared spectroscopy in monitoring roller compaction

Physical and chemical effects are intimately linked for roller compacts. As shown in Fig. 5.2, the intensity of NIR bands will increase as the compaction pressure increases. As the pressure increases, particles come closer together and the radiation is better transmitted through the particles as the air–particle interface is reduced [33]. The radiation that is transmitted is not measured by the diffuse reflectance detector and is therefore considered to be absorbed. The baseline also increases as more pressure is used to form ribbons or tablets, and less radiation returns to the diffuse reflectance detector [30,35]. Tablets or

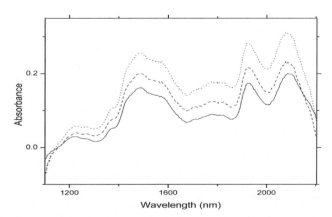

FIGURE 5.2 Near-infrared spectra without spectral pretreatment (1100–2205 nm) obtained from ribbons of MCC 200 produced at 15 (straight), 25 (dash), and 45 bar (dot). *From Acevedo D, et al. Evaluation of three approaches for real-time monitoring of roller compaction with near-infrared spectroscopy. AAPS Pharm Sci Tech 2012;13(3): 1005–1012.*

roller compacts at lower pressure have a greater surface area that increases the diffuse reflectance. Thus, at lower pressures, more diffusely reflected light comes back to the detector and the absorbance (log 1/R) decreases.

The slope and baseline of the NIR spectra increases with the compaction pressure, as shown in Fig. 5.2 [35,36]. A line drawn through the baseline of the different spectra would show a positive slope as the wavelength is increased, as shown in Fig. 5.2. The slope change is the result of physical changes as confirmed in one study where the ribbons were made with only one component, microcrystalline cellulose [37]. The slope of NIR spectra was also used as a measure of tablet relaxation [30].

The slope of NIR spectra baseline has been related to relative density, tensile strength, and the solid fraction of the compact [38,39]. At least two studies have estimated the bulk density of the ribbons through the measurement of the spectral slope [35,38]. Table 5.2 shows some of the reference methods that have been related to NIR spectral changes through the use of multivariate calibration models [40]. However, it is also possible to use the spectral slope to monitor roller compaction without relating the slope to an offline reference method such as tensile strength or solid fraction. The slope may also be used directly to monitor roller compaction. The periodicity of the slope variations was investigated by fast Fourier transform analysis and the dominant frequency was found to have a strong correlation with the roll speed [41]. The NIR spectrum may also be related to process parameters such as roll pressure without reference to an offline method. In-line NIR spectra were collected for rough surface ribbons for a process in which the roll pressure varied between 30 and 35 bar. This variation was considered to be within the expected process variation, and the spectra collected within this range were used as a training set for qualitative principal component analysis (PCA) calibration model [35]. The PCA model easily detected events where the roll pressure moved outside of this 30—35 bar range.

NIR spectroscopy has some limitations, as all analytical techniques. The NIR spectra do not represent the entire thickness of the roller compact. The

TABLE 5.2 Summary of different density methods that have been related to changes in near-infrared spectra of ribbons.

Density measurements	Study
True density or solid fraction with helium pycnometer	[41,44]
Envelope density	[40,44]
Digital caliper for thickness of ribbons	[40]
Laser sensor for in-line measurement of ribbon thickness	[40]

radiation that returns to the NIR spectrometer is mostly measuring the surface and some of the top ~ 2 mm of subsurface material. The NIR radiation does not penetrate equally at all wavelengths [42] and thus the mass analyzed can be estimated but is not exactly known. The effect of this limitation can be reduced by obtaining spectra from multiple areas of the ribbon to obtain a composite sample that will be more representative of the ribbon [43]. In general, the acquisition of only one spectrum of a ribbon or a pharmaceutical blend should not be considered as adequate sampling [44]. In spite of this limitation, NIR spectroscopy is valuable in monitoring roller compaction operations due to its ability to monitor the process in real time and provide information on both the chemical composition and physical properties of the ribbons obtained.

4.2 Computational modeling of compaction

In general, continuum and discrete particle models have been used to study compaction of powders [2,45−51]. As the discrete particle models account for the deformation of individual particles and the bonding between particles, they have great advantage over continuum models. The inputs to the DEM are directly the particle properties, whereas in continuum models, the inputs are bulk powder properties which are not readily available.

In this chapter, we will only present a brief discussion of compaction models; a detailed description of compaction models is presented in Chapter 8, while Chapter 11 discusses models most commonly used in the development of flowsheets. In the discrete particle models, the position and deformation of individual particles is considered. In addition, the bonding between the particles is considered. The deformation and bonding of particles depends on their plastic, elastic, and bonding properties. Also, the deformation and bonding depend on the applied pressure. The mechanisms of deformation and bond formation in compaction of a blend in roller compaction to form a ribbon and in compaction of granules (or blends) to form tablets are very similar. Therefore, discrete particle models that are developed for modeling tablet compaction can be applied directly to roller compaction.

However, there are some significant differences between tablets and ribbons. One of the most important differences is the porosity of the final products: ribbons have much higher porosity than tablets. Fig. 5.3 shows the typical relative density ranges for free flowing powders, ribbons, and tablets. The ribbons have lower density because the compaction pressure used in roller compactors is lower than that used in tablet presses. As a result, particles in ribbons undergo less deformation than particles in tablets. Correspondingly, the bonding force between particles in ribbons is smaller in magnitude than that of tablets. Fig. 5.4 shows the distribution of the interparticle bonding force (B) at different relative densities [45]. The average bonding force increases as

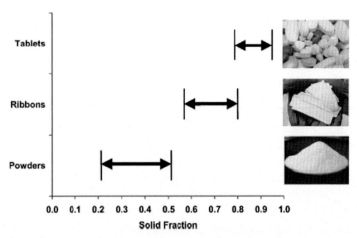

FIGURE 5.3   The range of relative densities (solid fractions) for loose powders, ribbons, and tablets [16].

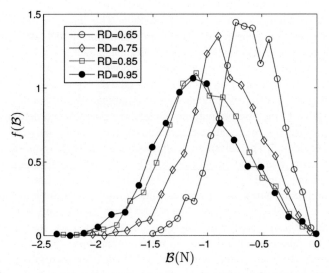

FIGURE 5.4   The distribution of bonding force between particles at different levels of relative density (RD) [50].

the relative density increases. However, it is important to note that the width of the distribution also increases as the relative density increases. The increase in the width of the distribution represents the heterogeneity of the microstructure in the powder compacts. The conclusion is that compacted tablets have more microstructure heterogeneity than ribbons.

In addition, tablets are compacted in a die, which provides confining pressure in the radial direction, while ribbons are compacted between rollers where there is minimum or no confinement in the lateral and axial directions.

5. Granule characterization after milling

As milled granules are further processed to manufacture tablets or capsules, it is important to characterize the milled granules. The milling process is generally assessed by its ability to achieve the required PSD. Sieve analysis and laser diffraction are the most common techniques used to measure particle size of granules [52], although vision-based methods are gaining growing acceptance.

5.1 Sieve analysis

Sieve analysis involves passing the granulated product by mechanical agitation through a series of sieves arranged in the order of decreasing sieve aperture size. The portion retained on each screen is then weighed and a mass-based PSD is obtained. Major advantages of sieve analysis are its cost-effectiveness and ease of use. However, it is time-consuming and requires large amount of sample [53]. The sample size used for sieve analysis depends upon the sieve diameter used, i.e., the diameter of the sieve pan. For a small sieve diameter, from a few grams to a few tens of grams can be used, while for a large sieve diameter, generally a sample mass of few hundred grams is required. Also, sieve analysis cannot be performed if the sample is cohesive, as this might lead to blockage of sieve apertures. In addition, friable materials can give unreliable results, as the granules can experience significant breakage as they pass through the sieves. Significant deviations from spherical particles may also result in poor measurement of the PSD and reproducibility issues in the measurement technique.

5.2 Laser diffraction

Unlike sieve analysis, laser diffraction requires very little sample, and the time required for analyzing the sample is short. In laser diffraction, a monochromatic laser light scattered by the particles in the sample is detected at various angles. A volume-based PSD is developed out of the scattered intensity data using a system of linear equations [54] subjected to assumptions of sphericity. The scattering pattern can be explained by Mie's theory or by Fraunhofer approximation based on the ratio of particle size to wavelength. Two widely used laser diffraction types of equipment are Malvern Mastersizer and Helos laser diffraction (Sympatec), which use Mie's theory and Fraunhofer approximation, respectively. This technique is able to measure particle sizes between 0.05 and 2000 µm [55], but as the particle size increases, there is a risk of choking the equipment, and photons increasingly experience multiple scattering, which might lead to inaccuracy in the measurement.

Sieve analysis and laser diffraction techniques measure the performance of the milling step in an intrusive manner, i.e., they require the user to collect samples and perform analysis on the collected samples. In contrast, the continuous mode of manufacturing requires adoption of innovative techniques that can monitor system performance and assess quality attributes in real time. To ensure efficient drug manufacturing, the USFDA launched a revision to the cGMPs (Current Good Manufacturing Practices) in the year 2002. This established more stringent regulatory quality controls, and developed expectations of a higher level of process understanding than previously achieved by many pharmaceutical companies [56]. QbD (Quality by Design) and PAT methods were endorsed as a part of this paradigm shift to promote superior understanding and to embrace novel techniques for better control on the process and timely quality checks.

Evolution of process analyzers such as NIR, Raman spectroscopy, and FBRM (focused beam reflectance measurement) facilitated the shift toward QbD [57].The primary goal of PAT is to help understand the effect of process/ equipment parameters by enabling timely checks of the critical material properties and quality attributes, thereby enhancing the understanding of the process. Over the last decade, many nonintrusive PAT techniques such as FBRM, Insitec, dynamic image analysis (DIA), and spatial filtering technique have been developed to measure PSD [22]. Nonintrusive methods have the advantage of allowing real-time analysis, as these methods do not interfere with the process and provide nearly instantaneous measurements. For these methods, a small material stream has to be separated from the process to prevent the main stream from flooding the detectors. Some of these methods are discussed in the next sections.

5.3 Laser diffraction (Insitec)

This technique overcomes the problem of multiple scattering at high particle concentration that limits the use of regular laser diffraction, which enhances its flexibility for online application [58]. In addition, it has the ability to measure particle concentration. Particles are generally assumed to be spherical in this method. The deviation from sphericity might be considerable in the case of micronized particles; in such instances, it is advised to incorporate particle shape into the optical model to obtain a more accurate estimate of the PSD [22].

5.4 Dynamic image analysis

Image analysis is a technique in which two-dimensional images of the sample are taken and converted into PSD. In a dynamic image analysis (DIA) system such as QICPIC, particles are accelerated to high speeds by a Venturi tube through the detection zone. In the detection zone, a xenon flashlight illuminates the particles and the camera captures the images of the fast moving particles. The particles are introduced into the system with the help of a vibratory chute. This method can measure particle sizes between 1 and 3000 µm. In addition to PSD, time-evolving size and shape analysis can also be applied to determine equilibrium reached during a continuous milling operation [22].

5.5 Focused beam reflectance measurement

In an FBRM system, reflected light from the particle is detected to determine the chord length of the particle. A rotating laser optics setup is used to shed light on the particles [59]. This method measures the chord length of the particles several times within a few seconds to provide chord length distribution [54]. A self-cleaning spinning disc can be employed in this setup in order to reduce the adherence of small particles to the probe window. The major advantage of this technique is that it does not assume the shape of the particles being measured [54], although the relationship between cord length distribution and PSD for nonspherical particles can be difficult to unravel.

During manufacturing, flowability of granules can affect the content uniformity of the final tablets. The bulk density and tap density of granules characterize their flowability. Friability and porosity of granules characterize the compactibility and hardness of the granules and thereby determine the hardness of the tablet. It is thus important to measure these properties.

5.6 Bulk density

The ratio of mass of loosely packed (untapped) granules to its volume is known as the bulk density. The bulk density of a sample is determined by measuring the volume of a known weight of the sample in a graduated cylinder [60]. This volume includes the volume of voids present between the particles. Hence, the bulk density depends on both the density of granules and the spatial arrangement of granules in the particle bed. Powder handling techniques affect the bulk properties of granules as well. Thus, the bulk density of granules is often difficult to measure with good reproducibility, in particular for cohesive powders that can contain large voids within a poured bed. If a large amount of sample (400g) is available, the gravitational displacement rheometer method provides a convenient and accurate alternative for this measurement [61].

5.7 Tap density

The tapped density is an increased bulk density attained after mechanically tapping a container containing the sample, which forces the particles to pack densely. After observing the initial sample volume and weight, the measuring cylinder is mechanically tapped a predetermined number of times, and volume readings are taken after each step. The tapping is stopped when the difference between two successive volume readings is lower than a certain value. The mechanical tapping is achieved by raising the cylinder or vessel and allowing it to drop under its own weight to a specified distance. To minimize any possible separation of particles during tapping, an equipment design that rotates the cylinder or vessel is preferred [60].

5.8 Compressibility index and Hausner ratio

The compressibility index is a measure of the propensity of the granules to be compressed and Hausner ratio represents the flow of granules. These are given by the following equations, where V_0 is the unsettled apparent volume and V_F is the final tapped volume. A value of Hausner ratio greater that 1.25 is an indication of poor flowability [62].

$$\text{Compressibility index} = 100 * \left(\frac{V_0 - V_F}{V_0} \right)$$

$$\text{Hausner ratio} = \left(\frac{V_F}{V_0} \right)$$

5.9 Friability

Granule friability characterizes granule strength. If the granules are very weak, they might crumble before they are tableted, which can further lead to segregation of the sample. On the other hand, if the granules are too hard, their compactibility is affected. Granule friability can be determined by the following methods.

Air jet sieving: Granules in air jet milling are subjected to mechanical stress due to the collisions of the particles against each other and against the wall and lid of the mill. These collisions occur due to the induced circular motion of the particles. Prior to determination of particle size, fines in the sample are removed and approximately 10 g of this modified sample is taken for experimentation. The friability is defined as the loss in weight of the sample after sieving it at a negative pressure of 2000 Pa for 10 min [63].

Friabilator: In this method, approximately 10 g (I_{wt}) of sample is conditioned by removing the fines prior to experimentation. Fines are defined as particles that have particle size lower than a predetermined screen size. The conditioned sample is then subjected to mechanical stress using a friabilator at 25 rpm for 10 min which contains 200 glass beads (mean diameter 4 mm). After the stress step, the glass beads are removed, and fines are also removed again. The weight retained on the predetermined screen is determined (F_{wt}) [64]. The friability of sample is calculated as

$$\text{Friability} = \left(\frac{I_{wt} - F_{wt}}{I_{wt}} \right) * 100$$

5.10 Porosity

Porosity gives insight into granule structure and strength, which affects the tablet compaction. Very low porosity values indicate dense granules, which could be difficult to compact. On the other hand, high porosity indicates brittle and weak granules which could break due to the stresses from tableting or packaging. Porosity is generally measured using pycnometers, described as following.

Helium and mercury pycnometry: A helium pycnometer is used to measure helium density or true density (ρ_{He}), which is the property of the material. The same granule sample is then used to determine the apparent density (ρ_{Hg}) using a mercury pycnometer at 400 kPa. In this method, only granules in the size range of 1120–1600 μm can be used, as there is a risk of smaller granules being sucked up into the dilatometer capillary and choking the capillary during the vacuum phase [63]. Porosity is calculated using the following formula:

$$\text{Porosity} = \left(1 - \frac{\rho_{Hg}}{\rho_{He}} \right) * 100\%$$

True density and envelope density: True density (ρ_t) is measured using helium pycnometer in this method as well. For envelope density measurement using GeoPyc 1360 (Micromeritics), a dry solid medium (Dryflow) comprising small and rigid spheres with high flowability displaces the void space and closely envelopes the particle surface, thereby giving the envelope density (ρ_e) [62]. Porosity is calculated using the formula:

$$\text{Porosity} = \left(1 - \frac{\rho_e}{\rho_t} \right) * 100\%$$

In addition to the above measurements, milled granules can be analyzed for tabletability by examining the tablets produced for weight variability, tabletability profile, and compactibility profile. In addition to mill type and PSD, the formulation composition has a significant effect on the compaction properties of the granules [65,66].

6. Models for milling

Models to simulate the milling process typically aim to predict the resulting milled product PSD. Limited models are published to predict other properties such as bulk density, friability, etc. Metta and Verstraeten [67] applied a population balance model (PBM) to predict the resulting PSD. A partial least squares modeling approach is then used to predict other CQAs such as bulk density, tapped density, etc. The PLS model takes the mill operating variables and milled granule PSD as input, thus establishing a comprehensive PBM-PLS modeling approach. In this section, the discussion is mainly focused on mechanistic and PB models for a milling process. Generally, the roller compactor unit is followed by a milling unit that includes a screen. The screen provides classification of granules and thus aids in achieving the required size distribution. Mechanistic and PB models for mills with screens such as comils will be discussed in particular and, where applicable, models relevant to general milling processes will be mentioned.

6.1 Population balance models

PBMs are used to track change in mass or number of particles of various sizes over time as shown in Eq. (5.5):

$$\frac{d\,M(w,t)}{dt} = R_{\text{form}}(w,t) - R_{\text{dep}}(w,t) + \dot{M}_{\text{in}}(w,t) - \dot{M}_{\text{out}}(w,t) \qquad (5.5)$$

where $M(w,t)$ represents mass of particles of volume w at time t, R_{form} and R_{dep} represent rates of formation and depletion of particles, respectively. $\dot{M}_{\text{in}}$ and $\dot{M}_{\text{out}}$ are the mass flow rates of particles entering and exiting the mill, respectively. Formation and depletion of particles occur in a mill due to the breakage process. The rate of formation R_{form} and the rate of depletion R_{dep} are formulated in the model as given in Eqs. (5.6) and (5.7), respectively, using a breakage kernel and a breakage distribution function.

$$R_{\text{dep}}(w,t) = K(w)M(w,t) \qquad (5.6)$$

$$R_{\text{form}}(u,t) = \int_{u}^{\infty} K(w)M(w,t)b(w,u)dw \qquad (5.7)$$

The breakage distribution function $b(w,u)$ represents the distribution of the daughter particles formed when a particle of volume w undergoes breakage.

Breakage distribution functions may take up several forms. For example, Barrasso and Oka [18] used a log-normal distribution function as given in Eq. (5.8):

$$b(w, u) = \frac{C(w)}{u\sigma} \exp \left[\frac{-\left(\log u - \log \left(\frac{w}{n} \right) \right)^2}{2\sigma^2} \right] \tag{5.8}$$

where $C(w)$ is introduced to ensure that mass conservation holds. Metta and Verstraeten [67] used a Hill-Ng distribution as given in Eq. (5.9):

$$b(w, u) = \frac{p \dfrac{u^{q-1}}{w} \left(1 - \dfrac{u}{w} \right)^{r-1}}{w \, B(q, r)} \tag{5.9}$$

where the parameter p is the number of daughter fragments formed, the parameter q represents the fragment size dependence, and $B(q, r)$ refers to beta function with q and r as the arguments $(r = q(p - 1))$ A large value of q indicates fragmentation, whereas a small value of q indicates chipping or erosion mechanism. Reynolds [68] used a generalized Hill-Ng distribution function where the fine phase is modeled using the feed powder PSD and the coarse phase is modeled using a Hill-Ng distribution. The breakage kernel $K(w)$ defines the probability that a particle of volume w undergoes breakage. The kernel usually takes semi-empirical forms such as a shear rate or impeller speed–based kernel as given in Eq. (5.10):

$$K(w) = P_1 G_{\text{shear}} w^{P_2} \tag{5.10}$$

where P_1 and P_2 are parameters estimated from experimental data. Reynolds [68] used a classification kernel as given in Eq. (5.11):

$$K(w) = \begin{cases} K & \text{for } w \geq \text{ critical size} \\ 0 & \text{else} \end{cases} \tag{5.11}$$

This kernel formulation aligns with theory published by Vogel and Peukert [69], i.e., a particle breaks only if the energy lost by it from collisions is greater than a certain "threshold energy," which is a material-specific property dependent on particle size. In this kernel, particles below a certain size are not considered to break as the energy imparted mechanically is considered insufficient. Another plausible theory is that small size particles exit the mill (with a screen) and hence do not undergo breakage. The size limit below which breakage kernel is assumed to be zero is obtained using a heuristic approach during model calibration.

The mass flow rate out of the mill through a screen, $\dot{M}_{out}(w,t)$, is formulated using a screen model as given in Eq. (5.12):

$$\dot{M}_{out}(w,t) = \left(R_{form}(w,t) - R_{dep}(w,t) + \gamma d_{in}(w,t)\right)(1 - f_d(w)) \qquad (5.12)$$

where feed PSD entering the mill is denoted by d_{in}. A parameter $\Delta = d_{screen}*\delta$ is used, where δ is referred to as critical screen size ratio and d_{screen} is the screen size. The critical screen size ratio reflects the size limit below which the particle exits the mill instantaneously. If the size of the particle is greater than the screen size, the particle does not exit the mill. Linear models similar to the one given in Eq. (5.13) are used to define the flow rate of particles of various sizes exiting the mill.

$$\text{where } f_d(x) = \begin{cases} 0 & \text{for } x \leq \Delta \\ \dfrac{x - \Delta}{d_{screen} - \Delta} & \text{for } \Delta \leq x \leq d_{screen} \\ 1 & \text{for } x > d_{screen} \end{cases} \qquad (5.13)$$

In addition, for a comil, a relationship between the critical screen size ratio δ and the impeller speed υ_{imp} is given in Eq. (5.14):

$$\delta = \varepsilon \left(\frac{\upsilon_{imp,min}}{\upsilon_{imp}}\right)^{\alpha} \qquad (5.14)$$

The relationship is reflective of the reduced apparent screen size available for the particle to exit the mill as the impeller speed increases.

In the PBM framework, various parameters introduced such as γ, ε, and α are to be estimated from experimental data. It is evident that this parameter set is different when the composition of the powder blend changes. The need to include material properties in the modeling framework has further propelled research in development of mechanistic models for milling processes, as the effect of material properties and operating conditions can be effectively captured through fundamental models such as DEM models.

6.2 Mechanistic models

Use of mechanistic models to simulate comminution processes is common in the mining industry. Weerasekara and Powell [70] provided a detailed review of the use of the DEM models used in particle breakage processes. In DEM simulations, equations to conserve energy and momentum of particles are evaluated over short timescales. The position and condition of every particle in the system is simulated accurately considering applied stress from the population of surrounding particles and the system geometry. This provides a means to effectively include the effect of material properties and processing conditions into the modeling framework. To this effect, a large amount of work has been published on mechanistic modeling of pharmaceutical unit operations [71]. However, as particle-scale phenomena are captured, DEM simulations are expensive to evaluate. The issue of higher computational expense is exacerbated for milling as the particle size and number changes and the computation power needed increases with the number of particles in the system. This limits its use in flowsheet models where the model needs to be evaluated quickly for dynamic prediction and control purposes. PBMs are more suitable for use in flowsheet modeling as they are relatively faster to evaluate compared to DEM. A hybrid model, i.e., a PBM that is mechanistically informed by DEM simulations, can thus capture advantages from both modeling approaches.

To establish a PBM-DEM framework, an energy-based breakage kernel developed by Capece and Bilgili [72] is utilized in Ref. [73] to model comilling. For the ith bin of diameter x_i, the breakage kernel K_i is given by Eq. (5.15):

$$K_i(t) = f_{mat} x_i f_{coll,i} \left(E_i - E_{i,min} \right) \tag{5.15}$$

where $f_{coll,i}$ is the frequency of collisions of particles in ith bin defined as number of collisions per particle per second as obtained from DEM simulations. Here, $E_{i,min} = \frac{E_{const}}{x_i}$, where x_i represents the size of particles in ith bin. E_i is the mass specific energy of particles with energy greater than the threshold energy. Only particles with "contributing energies," i.e., energies of particles greater than threshold energy, are considered in the kernel as these lead to breakage of particles.

With the formulation of the energy-based kernel as explained above, empirical kernels can be replaced by a kernel that captures particle-scale information from the DEM model. Metta and Ierapetritou [73] proposed an iterative algorithm to estimate the material-specific parameters in the kernel f_{mat} and E_{const}. Determination of these parameters from experimental data alone needs data from milling of mono-sized feeds. To obtain a mono-sized feed for lower size fractions, a large amount of material is required as the mass percentage of low size fractions in granulated product is relatively low. Circumvention of experiments with mono-sized feeds is provided by the iterative algorithm proposed. This is especially an advantage for pharmaceutical granules as the preparation of such feed is laborious as well as leads to wastage of large amounts of expensive pharmaceutical ingredients.

Through the use of the energy-based kernel, a DEM-PBM framework is established for a milling process. However, the hybrid DEM-PBM framework suffers from the computational expense that is incurred due to the inherent inclusion of high fidelity DEM in the combined framework. Reduced order-discrete element method (RO-DEM) methodology where reduced order modeling techniques, such as kriging, radial basis functions, etc., are used to represent the high fidelity DEM models can bridge this gap. Metta and

Ramachandran [77] used kriging and artificial neural networks to represent the mass specific energy and collision frequency data obtained from DEM simulations. These surrogate models not only efficiently represent the mechanistic data but also eliminate the need to run the DEM simulation at a new processing condition. The improvement in computational expense is drastic as a DEM simulation takes days to run whereas the developed surrogate model takes few seconds. Recent work published [74] also used mechanistic models and related the parameters of the generalized Hill-Ng breakage distribution function to particle surface energy, thus paving way toward a fundamental modeling of breakage processes.

7. Conclusions

This chapter provided a detailed review of the roller compaction and milling processes used in the continuous manufacturing of tablets through the dry granulation route. An attempt has been made to discuss various aspects of process monitoring, characterization, modeling, and quality control. Several online and offline characterization methods and detailed and relatively fast modeling approaches were discussed for predicting properties of the ribbons as well as the milled granules.

References

[1] Farber L, et al. Unified compaction curve model for tensile strength of tablets made by roller compaction and direct compression. Intl J Pharm 2008;346(1−2):17−24.

[2] Peter S, Lammens RF, Steffens KJ. Roller compaction/Dry granulation: use of the thin layer model for predicting densities and forces during roller compaction. Powder Technol 2010;199(2):165−75.

[3] Akseli I, et al. A quantitative correlation of the effect of density distributions in roller-compacted ribbons on the mechanical properties of tablets using ultrasonics and X-ray tomography. Aaps Pharmscitech 2011;12(3):834−53.

[4] von Eggelkraut-Gottanka SG, et al. Roller compaction and tabletting of St. John's wort plant dry extract using a gap width and force controlled roller compactor. I. Granulation and tabletting of eight different extract batches. Pharm Dev Technol 2002;7(4):433−45.

[5] Sajjia M, Albadarin AB, Walker G. Statistical analysis of industrial-scale roller compactor 'Freund TF-MINI model. Int J Pharm 2016;513(1−2):453−63.

[6] Miller RW. Roller compaction technology. In: Parikh DM, editor. Handbook of pharmaceutical granulation Technology. CRC Press; 2016.

[7] Hsu SH, Reklaitis GV, Venkatasubramanian V. Modeling and control of roller compaction for pharmaceutical manufacturing. Part I: process dynamics and control framework. J Pharm Innov 2010;5(1−2):14−23.

[8] Liu Y, Wassgren C. Modifications to Johanson's roll compaction model for improved relative density predictions. Powder Technol 2016;297:294−302.

[9] Perez-Gandarillas L, et al. Effect of roll-compaction and milling conditions on granules and tablet properties. Euro J Pharm Biopharm 2016;106:38−49.

[10] Yu S, et al. A comparative study of roll compaction of free-flowing and cohesive pharmaceutical powders. Int J Pharm 2012;428(1−2):39−47.

[11] Al-Asady RB, et al. Roller compactor: determining the nip angle and powder compaction progress by indentation of the pre-compacted body. Powder Technol 2016;300:107−19.

[12] Dehont FR, et al. Briquetting and granulation BY compaction new granulator compactor for the pharmaceutical-industry. Drug Dev Ind Pharm 1989;15(14−16):2245−63.

[13] Herting MG, Kleinebudde P. Studies on the reduction of tensile strength of tablets after roll compaction/dry granulation. Eur J Pharm Biopharm 2008;70(1):372−9.

[14] Johanson JR. A rolling theory for granular solids. J Appl Mech 1965;32(4):842−8.

[15] Jenike AW. On the plastic flow of coulomb solids beyond original failure. Appl Mech 1959;81:599−602.

[16] Zinchuk AV, Mullarney MP, Hancock BC. Simulation of roller compaction using a laboratory scale compaction simulator. Int J Pharm 2004;269(2):403−15.

[17] Samanta A, Ng K, Heng P. Cone milling of compacted flakes: process parameter selection by adopting the minimal fines approach. Int J Pharm 2012;422:17−23.

[18] Dana Barrasso SO, Muliadi A, Litster JD, Wassgren Carl, Ramachandran Rohit. Population balance model validation and prediction of CQAs for continuous milling processes: toward QbD in pharmaceutical drug product manufacturing. J Pharm Innov 2013;8(3):147−62.

[19] Yin SX, et al. Bioavailability enhancement of a COX-2 inhibitor, BMS-347070, from a nanocrystalline dispersion prepared by spray-drying. J Pharm Sci 2005;94(7):1598−607.

[20] Loh ZH, Samanta AK, Heng PWS. Overview of milling techniques for improving the solubility of poorly water-soluble drugs. Asian J Pharm Sci 2015;10:255−74.

[21] Ashford M. Assessment of biopharmaceutical properties. In: Aulton M, editor. Pharmaceutics: the science of dosage form design. London: Churchill Livingstone; 2002. p. 253−73.

[22] Naik S, Chauduri B. Quantifying dry milling in pharmaceutical processing: a review on experimental and modeling approaches. J Pharm Sci 2015;104:2401−13.

[23] S C, Purutyan H. Narrowing down equipments for particle size reduction of drug. Chem Eng Prog 2002;98:50.

[24] H H. Some notes on grinding research. J Imp Col Chem Eng Soc 1952;(6):1−12.

[25] Abdel-Magid AF, Caron S. Fundamentals of early clinical drug development: from synthesis design. 1 ed. New Jersey: Wiley; 2006.

[26] A V, et al. Effects of mill design and process parameters in milling of ceramic (alumina-magnesia) extrudates. 2012.

[27] B M. In: A L, H S, editors. Milling in Pharmaceutical dosage forms: unit operations and mechanical properties; 2008. p. 175−93.

[28] Yu S, Gururajan B, et al. Experimental investigation of milling of roller compacted ribbons. The British Library; 2011.

[29] EL P. Milling of pharmaceutical solids. J Pharm Sci 1974;63(6):813−29.

[30] Vanarase A, et al. Effects of mill design and process parameters in milling dry extrudates. Powder Technol 2015;278:84−93.

[31] Murugesu B. Milling. In: Augusburger LLaH, Hoag SW, editors. Pharmaceutical dosage forms: unit operations and mechanical properties. Informa Health Care; 2008.

[32] Yu S, et al. Experimental investigation of milling of roll compacted ribbons. In: Particulate materials: synthesis, characterisation, processing and modelling. The Royal Society of Chemistry; 2012. p. 158−66.

[33] Heywood H. Some notes on grinding research. 1950.

[34] Ropero J, et al. Near-infrared chemical imaging slope as a new method to study tablet compaction and tablet relaxation. Appl Spectros 2011;65(4):459—65.

[35] Miller CE. Chemical principles of near-infrared technology. Near-infrared Technol Agricult Food Indus 2001;2.

[36] Mayo DW, Miller FA, Hannah RW. Course notes on the interpretation of infrared and Raman spectra. John Wiley & Sons; 2004.

[37] Kleinebudde P. Roll compaction/dry granulation: pharmaceutical applications. Eur J Pharm Biopharm 2004;58(2):317—26.

[38] Olinger JM, Griffiths PR. Effects of sample dilution and particle size/morphology on diffuse reflection spectra of carbohydrate systems in the near- and mid-infrared. Part I: single analytes. Appl Spectros 1993;47(6):687—94.

[39] Siesler HW. Basic principles of near-infrared spectroscopy. In: Handbook of near-infrared analysis. 3rd ed. CRC press; 2007. p. 25—38.

[40] Acevedo D, et al. Evaluation of three approaches for real-time monitoring of roller compaction with near-infrared spectroscopy. AAPS Pharm Sci Tech 2012;13(3):1005—12.

[41] Gupta A, et al. Real-time near-infrared monitoring of content uniformity, moisture content, compact density, tensile strength, and young's modulus of roller compacted powder blends. J Pharm Sci 2005;94(7):1589—97.

[42] Kirsch JD, Drennen JK. Nondestructive tablet hardness testing by near-infrared spectroscopy: a new and robust spectral best-fit algorithm. J Pharm Biomed Anal 1999;19(3):351—62.

[43] Gupta A, et al. Nondestructive measurements of the compact strength and the particle-size distribution after milling of roller compacted powders by near-infrared spectroscopy. J Pharm Sci 2004;93(4):1047—53.

[44] Talwar S, et al. Understanding the impact of chemical variability and calibration algorithms on prediction of solid fraction of roller compacted ribbons using near-infrared (NIR) spectroscopy. Appl Spectros 2016;71(6):1209—21.

[45] Romañach RJ, Román-Ospino AD, Alcalà M. A procedure for developing quantitative near infrared (NIR) methods for pharmaceutical products. In: Process simulation and data modeling in solid oral drug development and manufacture. Springer; 2016. p. 133—58.

[46] Feng T, et al. Investigation of the variability of NIR in-line monitoring of roller compaction process by using Fast Fourier Transform (FFT) analysis. Aaps Pharmscitech 2008;9(2):419—24.

[47] Iyer M, Morris HR, Drennen JK. Solid dosage form analysis by near infrared spectroscopy: comparison of reflectance and transmittance measurements including the determination of effective sample mass. J Near Infrared Spectrosc 2002;10(4):233—45.

[48] Romañach RJ, Esbensen KH. Theory of sampling (TOS) for development of spectroscopic calibration models. Am Pharm Rev 2016;19(6).

[49] Esbensen KH, et al. Adequacy and verifiability of pharmaceutical mixtures and dose units by variographic analysis (Theory of Sampling) - a call for a regulatory paradigm shift. Int J Pharm 2016;499(1—2):156—74.

[50] Yohannes B, et al. Evolution of the microstructure during the process of consolidation and bonding in soft granular solids. Int J Pharm 2016;503(1—2):68—77.

[51] Yohannes B, et al. Discrete particle modeling and micromechanical characterization of bilayer tablet compaction. Int J Pharm 2017;529(1—2):597—607.

[52] Shekunov BY, et al. Particle size analysis in pharmaceutics: principles, methods and applications. Pharm Res 2007;24(2):203—27.

[53] HG M. Particle size measurements: fundamentals, practice, quality. 1 ed. New York: Springer; 2009.

[54] FT SA, et al. Particle sizing measurements in pharmaceutical applications: comparison of in-process methods versus off-line methods. Eur J Pharm Biopharm 2013;(85):1006−18.

[55] Chan LW, Tan LH, Heng PWS. Process analytical technology: application to particle sizing in spray drying. AAPS Pharm Sci Tech 2008;9(1):259−66.

[56] Sangshetti JN, et al. Quality by design approach: regulatory need. Arab J Chem 2014;10:3412−25. https://doi.org/10.1016/j.arabjc.2014.01.025.

[57] De Beer T, et al. Near infrared and Raman spectroscopy for the in-process monitoring of pharmaceutical production processes. Int J Pharm 2011;(417):32−47.

[58] Dan Hirleman E. Modeling of multiple scattering effects in Fraunhofer diffraction particle size analysis. Part Part Syst Charact 1988;(5):57−65.

[59] David G, et al. Measuring the particle size of a known distribution using the focused beam reflectance measurement technique. Chem Eng Sci 2008;(63):5410−9.

[60] Convention, U.S.P. United States pharmacopeia. In: Bulk density and tapped density of powders. United States Pharmacopeial Convention; 2015.

[61] Vasilenko A, Glasser BJ, Muzzio FJ. Shear and flow behavior of pharmaceutical blends — method comparison study. Powder Technol 2011;208(3):628−36.

[62] Meng W, et al. Statistical analysis and comparison of a continuous high shear granulator with a twin screw granulator: effect of process parameters on critical granule attributes and granulation mechanisms. Int J Pharm 2016;513:357−75.

[63] Djuric D, Kleinebudde P. Impact of screw elements on continuous granulation with a twin-screw extruder. J Pharm Sci 2008;97:4934−42.

[64] Vercruysse J, et al. Continuous twin screw granulation: influence of process variables on granule and tablet quality. Eur J Pharm Biopharm 2012;82:205−11.

[65] Vendola TA, Hancock BC. The effect of mill type on two dry-granulated placebo formulations. Pharm Technol 2008;32(11):72−86.

[66] Hancock BC, Vendola TA, Hancock BC, Vendola Thomas A. Pharm Technol 2008;32(11).

[67] Metta N, et al. Model development and prediction of particle size distribution, density and friability of a comilling operation in a continuous pharmaceutical manufacturing process. Int J Pharm 2018;549(1):271−82.

[68] Reynolds GK. Modelling of pharmaceutical granule size reduction in a conical screen mill. Chem Eng J 2010;164(2−3):383−92.

[69] Vogel L, Peukert W. Breakage behaviour of different materials—construction of a mastercurve for the breakage probability. Powder Technol 2003;129(1−3):101−10.

[70] Weerasekara NS, et al. The contribution of DEM to the science of comminution. Powder Technol 2013;248:3−24.

[71] Ketterhagen WR, am Ende MT, Hancock BC. Process modeling in the pharmaceutical industry using the discrete element method. J Pharm Sci 2009;98(2):442−70.

[72] Capece M, Bilgili E, Dave RN. Formulation of a physically motivated specific breakage rate parameter for ball milling via the discrete element method. Aiche J 2014;60(7):2404−15.

[73] Metta N, Ierapetritou M, Ramachandran R. A multiscale DEM-PBM approach for a continuous comilling process using a mechanistically developed breakage kernel. Chem Eng Sci 2018;178:211−21.

[74] Loreti S, et al. DEM-PBM modeling of impact dominated ribbon milling. Aiche J 2017;63(9):3692−705.

[75] Kotamarthy L, Metta N, Ramachandran R. Understanding the effect of granulation and milling process parameters on the quality attributes of milled granules. Processes 2020;8(6). https://doi.org/10.3390/pr8060683.

[76] Schenck LR, Plank RV. Impact milling of pharmaceutical agglomerates in the wet and dry states. Int J Pharm 2007;348(1). https://doi.org/10.1016/j.ijpharm.2007.07.029.

[77] Metta N, Ramachandran R, Ierapetritou M. A computationally efficient surrogate-based reduction of a multiscale comill process model. J Pharm Innov 2019:1—'21. https://doi.org/10.1007/s12247-019-09388-2.

Chapter 6

A modeling, control, sensing, and experimental overview of continuous wet granulation

Shashank Venkat Muddu, Rohit Ramachandran

Department of Chemical and Biochemical Engineering, Rutgers, The State University of New Jersey, Piscataway, NJ, United States

1. Introduction

Granulation is a size enlargement unit operation whereby the constituent chemical components of a desired product are mixed thoroughly along with a "binder" that holds the different components together in the small scale and prevents segregation/localization of any particular species [1]. Wet granulation is a granulation process wherein a liquid solvent is added to a preblended mixture of powder ingredients. The solvent is usually water or a polymer solution [2–5]. The binder is added either in a dry state mixed along with the formulation powders in the granulator or dissolved in the binder liquid solvent [6,7]. The addition of the binder in the wet granulation process facilitates the formation of liquid capillary bridges that hold the primary particles together via granulation mechanisms such as nucleation, aggregation, consolidation, and layering [8]. In addition, breakage mechanisms play a role in determining the resulting granule size distribution.

As wet granulation involves mixing of different components with different properties, and the process comprises several intertwined mechanisms operating at different rates and in different physical scales, it is nontrivial to control the output results or rectify them in case the granules are off the specified standards or requirements. Numerous material properties and process parameters can affect the outcome. This calls for a Quality by Design (QbD) approach, rather than a Quality by Testing approach, leading to a science-based understanding of the design and operation of the process [9].

For many decades, the batch mode of manufacturing has been the norm for the wet granulation operation in the pharmaceutical industry [9]. However, as described elsewhere in this book, in recent years there has been an accelerating

How to Design and Implement Powder-to-Tablet Continuous Manufacturing Systems
https://doi.org/10.1016/B978-0-12-813479-5.00015-X

119

shift toward continuous manufacturing, partly driven by the growing need for industries to reduce the time to market for products, while maintaining strict quality requirements. Furthermore, the pharmaceutical industry has found staunch support from the regulatory authorities, and most clearly from the US Food and Drug Administration, to shift toward continuous processes given the higher degree of quality assurance associated with continuous processes. This has resulted in extensive investigation of continuous wet granulation processes and the associated processing equipment. Continuous high shear mixers and twin screw extruders are the two major classes of equipment that facilitate continuous wet granulation, and discussions in this chapter will be limited to these two pieces of processing equipment, although we mention in passing that basket granulators and fluid bed processors have also been used and hold promise for future applications. Wet granulation is carried out continuously in all of these instruments by employing gravimetric feeders in order to maintain steady inlet flow into the system, and the liquid is pumped at the desired flow rates into the mixers in a controlled flow ratio with respect to the granulating material.

A better process understanding and characterization of the continuous granulation unit operation necessitate mechanistic analysis of the physical processes occurring within the unit operation. The typical critical quality attributes of a wet granulation process are the exit particle size distributions (PSDs), densities, and the distribution of the formulation material components within the granules of different sizes exiting at different times. To this effect, the aim of this chapter is to provide a holistic overview of the state of the art with respect to process modeling, experimental investigation, sensing, and control of continuous wet granulation. This chapter, in alignment with the objective of the book, also aims to provide a set of guidelines for design and characterization of continuous wet granulation processes.

2. Experimental design

Continuous wet granulation is performed with one or more powder feeders, a pump to inject liquid, and the continuous granulator. The combination of solid components may or may not be blended together prior to being fed to the granulator using loss-in-weight feeders to meter the powder. Along with the powder, distilled or deionized water (with or without a polymer binder dissolved in it) is added to the system in accordance with the desired liquid-to-solid (L/S) ratios. The liquid is discharged through nozzles to achieve uniform and proper wetting of the powder particles. A typical wet granulation setup, comprising of a continuous twin screw granulator (TSG), is illustrated in Fig. 6.1.

One can achieve granules with different attributes by varying the powder feed rate, L/S ratio, rotation speed of the granulator shaft (RPM), and configuration of the granulator blades/screws. Often, the powder feed rate is

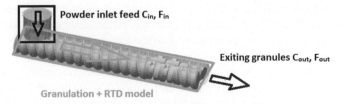

FIGURE 6.1 Typical schematic of a continuous granulation process where C indicates concentration and F indicates flow rate. Residence time distribution (RTD) represents a RTD model. The subscripts in and out represent input and output, respectively.

dictated by demand or constrained by capacity of other unit operations. The powder feed rate is thus rarely varied or varied within a small range. The screw configuration in a twin screw granulation system (or the blade arrangement in a high shear granulator) significantly affects the properties of the resulting granules. These effects have been quantitatively studied by Djuric et al. [10]. It is thus recommended that process characterization begins with understanding the role of unit geometry on granule properties. Once the optimal geometry has been identified, the role of impeller speed, L/S ratio, and powder flow rate (generally over a smaller range) can be investigated. The cubic centric design (CCD) is often used to determine the experimental runs of the granulation experiments for the aforementioned process variables. In a design space with low, high, and moderate settings for the three variables, the CCD generates 17 points, which includes 14 face center and corner points, akin to the positions of a face centered cubic unit cell and 3 additional runs of the body center. The center point is repeated thrice in order to test the replicability of the experimental setup. The schematic representation of the DOE is shown in Fig. 6.2.

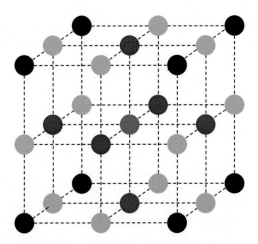

FIGURE 6.2 The points chosen in the cubic centric design experimentation: corner points, face center points, and three body center replicates of a cubic lattice.

Resulting granules are dried and characterized to examine the effect of process variables. Granules sampled from the experiments are first dried and sieved to obtain their size distribution. Particle size image analysis may also be utilized in-line or off-line.

Apart from PSDs, the granules are also characterized by other quality attributes such as porosity and content uniformity. The method for these characterizations have been well documented in Granberg et al. [11], Kaspar et al. [12], and Oka et al. [13].

The central composite design enables one to develop response surface methodologies for experimental metrics such as mean residence time (MRT), variance/standard deviation, and skewness of the residence time distribution (RTD) profile. Moreover, the CCD in three factors and three levels enables one to obtain the second-order behavior response to variables and interactions (if any) in the aforementioned metrics.

Kumar et al. [14] showed that increasing the L/S ratio increased the mean size of the granules exiting a twin screw granulation process, in agreement with literature in batch granulation and other continuous granulation processes. Moreover, increasing the energy imparted to the process by increasing shaft rotation speed reduced the particle size range and led to a narrower distribution. The authors suggested that while a process map was developed for a particular formulation and equipment, similar studies, if extended to other formulations and equipment, would lead to the development of generic scale—independent regime maps that can be applied to any continuous wet granulation process.

Meng et al. [15] performed granulation experiments on a continuous high shear CM-5 Lödige granulator and Thermo Scientific Pharma 11 TSG. The experiments showed that both processing types generated granules with controllable attributes and a very short residence time. The Lödige granulator generated granules that had smaller size variance and finer granule structure. The sizes of the granules created by the Lödige granulation process showed less variance. Moreover, as it was seen that the Lödige was responsive to the process parameters, the equipment was able to be operated in a wide design space. It was reported that the minimum boundary for L/S ratio could be plausibly extended well below the chosen DOE lower bound of 0.35, with shaft rotation speed ranging roughly between 1500 and 2500 RPM. On the other hand, the TSG produced granules with multimodal distributions containing ungranulated materials. This suggested suboptimal liquid dispersion and mixing due to a limited residence time in the granulator.

Meng et al. [16] performed another set of experiments on a Glatt GCG-70 high shear granulator and observed that process parameters again influenced the granule properties. However, as opposed to the results in the Lödige experiments, the granules had broad PSDs and a significant amount of ungranulated fines were observed. The mode of binder delivery method was identified as the potential opportunity for improvement, which could lead to narrower PSD and more efficient nucleation.

2.1 Residence time distribution in continuous wet granulation

Experimental RTD studies are required to develop and validate an RTD model, which is useful for process design, material traceability, control, and optimization. The fundamentals of the RTD and their role in continuous manufacturing have been discussed in detail in Chapter 6 and will not be repeated here.

A popular method to measure the RTD of continuous wet granulation systems is by the use of a colored dye, which tracks the bulk material within the system. The dye in the solid powder form (usually less than 1 g, depending on the intensity of outlet granules) is added at the entrance of the granulator by means of funnel. The colored dye is blended and distributed inside the granulator according to the conveying and mixing profile of the system. The RTD profiles are dependent on the effective internal volume of the granulator, flow rates of the solid powder streams, the L/S ratios, the rotation speed of the shaft, the configuration of the shaft elements (conveying/kneading/distributive elements), and the material properties such as the compressibility of the components of the solid powder blend. The colored granules exiting the system are collected at different time intervals from the start of the addition of the tracer into the granulator. The granules are dried and dissolved in distilled water, and the concentrations of the tracer in the various samples are measured using ultraviolet-visible spectrophotometry.

The second method to run RTD pulse experiments is to use one of the components (usually the key component of interest such as the API—active pharmaceutical ingredient) as the tracer. In-line sensing equipment such as a near-infrared (NIR) spectrometer can be installed at the end of the granulator in order to measure the concentration of the tracer in the granules online. As the NIR spectrometer reads the spectrum of the key component, a suitable calibration model needs to be built.

It is often desired to understand the effects of one unit operation on the behavior of the subsequent unit operations in any simulation of a continuous manufacturing line. Wet granulation is an intermediate step in a continuous pharmaceutical manufacturing process and is generally preceded directly by feeding/blending and followed by drying, milling, tablet compaction, and dissolution. The effects of disturbances caused in any unit need to be mitigated to achieve final tablets of desired CQAs within tolerance limits. Therefore, process control models are integrated with continuous manufacturing models to predict the overall behavior of equipment in pilot plant/manufacturing scale settings.

RTD models are useful in predicting the flow behavior of the materials within the system. The internal distribution is estimated by measuring the exit age distribution functions, also known as RTD functions, of the granulating material. The RTD models find immense application in real-time estimation of the exiting material concentration in a continuous operation scheme. The main application of any RTD model is to observe the effect of fluctuations and disturbances in the inlet/upstream unit operations. The development of an accurate model of the RTD of a system would enable the experimenter to build a proper network of controllers to reduce input noise and drive the operations toward an overall QbD approach, as opposed to the traditional Quality by Testing approach. Using a Laplace transform, the output pulse response of the RTD function $E(t)$ can be converted into a transfer function in the "s" domain of the process control variable and the effect can be studied after linking it with the controller models. The RTD models can further play a role in the optimization of the process by optimizing the parameters of the model to reduce the relative error between the model CQAs (granules PSDs, RTD profiles, etc.) of the process and the target/experimental values. The most common RTD models that can be used to characterize a wet granulation system have been described next.

RTD models have been extensively developed for continuous flow reactors such as continuous stirred tank reactors (CSTRs) and plug flow reactors (PFRs). These models were fitted to nonreacting systems using experimental data in different industries such as food, petrochemicals, pharmaceuticals, and even in environmental engineering applications [17–19]. However, a model developed for a specific system needs to be re-validated (and often its parameters need to be recalculated) for a different piece of equipment and/or a different formulation. In general, processes involving wet powder flow behavior require close examination. Therefore, it is the overall aim of the researchers working in this field to develop mechanistic models accounting for the chosen vessel's geometry (length of screws, pitch of screws, screw element configuration), its operating parameters (screw speeds and rotation direction of screws), and process parameters (material feed rates and the rate of liquid addition).

The tanks-in-series (TIS) model consists of a single stream of CSTRs (all of equal volume) in series with each other. The pictorial representation of the same has been shown in Fig. 6.3A.

The RTD function, $E(t)$, of a pulse input for a TIS model is given as follows:

$$E(t) = \frac{n^n \theta^{n-1} e^{-\theta}}{t_m(n-1)!}$$

$$\theta = \frac{t}{t_m}$$

where t is time, t_m is the MRT of the system, n is the number of CSTRs, and θ is the dimensionless time with respect to the MRT of the entire system.

The parallel stream model for one CSTR in each stream [20] is depicted subsequently in Fig. 6.3B.

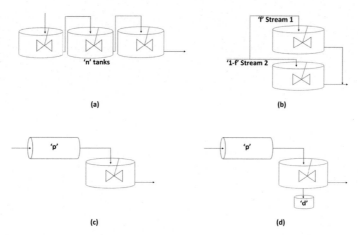

FIGURE 6.3 Pictorial depictions of various reactor schemes used for modeling the residence times of granulation streams. (A) Schematic of "*n*" finite equal volume stirred tanks connected in series. (B) Schematic of two equal volume stirred tanks connected in parallel with the fraction of material going to each tank "*f*" and "(1 − *f*)," respectively. (C) Schematic of a plug flow vessel of volume fraction "*p*" of the total system volume in series with a stirred tank. (D) Addition of a dead volume with fraction "*d*" to the PFR-CSTR schematic.

The RTD function of a pulse input for the same is given below:

$$E(t) = \frac{f\beta}{t_m}e^{-f\beta\theta} + \frac{(1-f)\beta}{t_m\alpha}e^{-\frac{(1-f)\beta}{\alpha}\theta}$$

$$\beta = f + (1-f)\alpha$$

where f is the fraction of material going into the first stream and α is the ratio of MRT in Stream 2 to MRT in Stream 1.

The diagram for a PFR in series with one CSTR is shown in Fig. 6.3C. The pulse RTD for the system is described as follows:

$$E(t) = \frac{(\theta - p)e^{-\frac{(\theta-p)}{(1-p)}}}{t_m(1-p)}$$

where p is the fraction of the PFR volume in the reactor scheme.

The schematic for a PFR in series with one CSTR along with dead volume fraction is given as follows in Fig. 6.3D.

The RTD pulse response function for the same is given accordingly:

$$E(t) = \frac{(\theta - p)e^{-\frac{(\theta-p)}{(1-p)(1-d)}}}{t_m(1-p)(1-d)}$$

where d is the fraction of the dead volume in the CSTR.

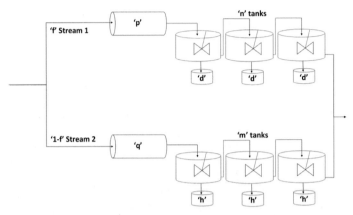

FIGURE 6.4 Schematic of model depicting a combined reactor scheme for residence time distribution (RTD) prediction in a twin-screw granulator.

The above models could be combined to model two streams of reactors, each one containing a PFR fraction and dead volume fraction for the CSTRs. The overall final scheme is given in Fig. 6.4. The RTD function of such a model is given as follows.

$$E(t) = \frac{f\beta}{t_m(n-1)!} \frac{(\theta - p)^n}{(1-p)(1-d)} e^{-f\beta \frac{(\theta-p)}{(1-p)(1-d)}}$$
$$+ \frac{(1-f)\beta}{t_m\alpha(m-1)!} \frac{(\theta - q)}{(1-q)(1-h)} e^{-\frac{(1-f)\beta}{\alpha} \frac{(\theta-q)}{(1-q)(1-h)}}$$

where n, m, p, q, d, h, α, and f hold same interpretations as described in the previous models. The tuning parameters variables could be varied for each data run, and a unique $E(t)$ function would be obtained using the model equation, respectively.

Kumar et al. [21] developed an RTD model for a twin screw granulation process using steady state expressions to describe the exit age distribution profile of the tracer, as opposed to a differential equation. The authors observed that the steady state results were fairly good at predicting the experimental trends. The authors inferred from the experimental and fitting results that a moderate amount of throughput force and conveying rate are necessary for good axial mixing of the powder blend and reduction in the dead zones of the granulator.

3. Process modeling

Performing comprehensive experimentation is time-consuming and resource-intensive. During early development, the active ingredient is scarce, which prevents intensive experimental investigation. Well-developed models, which are representatives of the system, can decrease the need for experimentation and provide a practitioner with experiment-like data. In granulation, like with other pharmaceutical unit operations, modeling and simulation tools have been successfully employed to predict and validate granule properties. As it is also the case for other systems where PSDs change as a result of processing conditions, population balance modeling (PBM) has been the most popular approach for modeling granulation processes, both batch and continuous. In a PBM, particles are grouped according to their key attributes such as size, liquid content, porosity, and spatial positions. Population balance equations are formulated as first-order partial integro-differential equations where the rate of growth/ death of each particle class is a function of the number of particles currently available of the said class and other classes such as those forming the particles. The generic expression of a PBM is as follows:

$$\frac{\partial N(x,s,l,g,t)}{\partial t} + \frac{\partial N(x,s,l,g,t)}{\partial x}\frac{\partial x}{\partial t} + \frac{\partial N(x,s,l,g,t)}{\partial s}\frac{\partial s}{\partial t} + \frac{\partial N(x,s,l,g,t)}{\partial l}\frac{\partial l}{\partial t}$$
$$+ \frac{\partial N(x,s,l,g,t)}{\partial g}\frac{\partial g}{\partial t} = R_{\text{Aggregation}} + R_{\text{Breakage}} + R_{\text{Nucleation}}$$

For a continuous granulation system, the terms on the left-hand side of the equation represents the rate of change with respect to time, the axial position in the granulator, solid content, moisture content and gas porosity in the granules. The terms on the right-hand side of the equation indicate the rate processes such as aggregation, breakage, and nucleation. There are several empirical parameters in the rate expressions. Different theories and model variations express differently these rates processes, with ultimate objective of achieving fully mechanistic representation of the rate processes.

PBMs have been extensively developed and are still being researched and updated in order to completely and accurately characterize continuous wet granulation processes [22,23].

Ramachandran and Chaudhury [24] developed a model to design and control a continuous drum granulation process. The study presented a compartmentalized PBM for a pilot plant scale simulation. The simulation results showed that the average diameter, moisture content, and bulk density of the outlet granules could be controlled by manipulating the nozzle spray rates of the liquid binder and the feed rates of the inflow solid powder blends. A model was developed to control the PSDs of the granules, with binder distribution on the powder particles in the granulator as a new manipulated variable.

Barrasso et al. [25] presented a continuous PBM that simulated the difference in PSDs and composition for a two component system (API and excipient), liquid binder content and the porosity of granules. The results showed good agreement with experimental trends. In a subsequent work by Barrasso et al. [26], a calibrated and validated PBM was developed where the empirical rate constants and parameters were determined using experimental data.

Kumar et al. [27] presented a one-dimensional PBM that included aggregation and breakage processes for a twin screw granulation process. The model parameters and their respective 95% confidence range were estimated using experimentally measured PSDs. The model was accordingly used for predicting granulation outcomes within the design space of the experiments. Moreover, operating conditions were identified, where the different granulation mechanism regimes could be separated in distinct compartments in the granulator, which would enable one to control each mechanism accordingly and influence the granule size distribution as needed.

Another modeling tool, discrete element modeling (DEM), is often employed when the rate constants in the models are empirical and there are limited experimental data to develop statistical or PBMs. The method is based on solving Newton's equations of motion for a system of particles, along with interactions between the particles and the vessel geometry. The generic DEM equation can be written as:

$$m\frac{d^2x}{dt^2} = \sum F_{ext}$$

where the product of mass and net acceleration of any particle is given by the net external forces acting on the particle. In DEM simulations, several expressions are formalized to approximate the forces acting on the particles based on the particle properties and processing conditions.

Firstly, for building a DEM for unit operations, one constructs the equipment geometry in CAD software and then imports it into the DEM software. Some common DEM platforms include EDEM, LIGGHTS, and STAR-CCM+. In a DEM simulation, the properties of particles are inputted, and their individual interactions are calculated to obtain the collision frequencies and particle trajectories. From the collision frequencies and aggregation/breakage rates, empirical parameters of the rate constants in the PBMs can be approximated. Coupling DEM simulations with PBM has been more successful for granulation processes [28] than the use of either platform on its own. It is to be noted that in DEM, each particle is given individual treatment and not approximated to behave similarly as others in same size, position, and attribute class. To model the system at scale would mean simulating billions of particles and thus it is computationally expensive. Therefore, the number of particles is first scaled down by scaling up the size of the particles in the similar geometry dimensions. The particles are given initial position, velocity coordinates as per processing conditions and the subsequent positions, and velocities are calculated by observing their interactions.

One must note that both the processes, PBM and DEM, have vastly different timescales of simulations. In cases where coupled studies are required, the two modules are often run intermittently one after the other, after transferring necessary information back and forth between them, or the DEM simulation is run for different conditions for small time spans approximating the data to pseudo—steady state conditions. The DEM data are then transferred to the PBM module for further simulations. The dynamics of the interaction between the solid and liquid phases is better understood by employing computational fluid dynamics.

Barrasso et al. [29] developed a multiscale model for a twin screw granulation process that combined PBM and DEM techniques to predict the effects of equipment design, material properties, and process parameters on the CQAs of the granulated product. The model results were consistent with experimental trends and found that the screw element configuration strongly influenced the final product attributes. It showed that the mixing elements rather than the conveying elements resulted in more aggregation, breakage, and consolidation, thereby leading to larger and denser particles.

Barrasso and Ramachandran [30] further validated this fact and added that the offset angle of the kneading elements in the granulator influenced the residence time of the material with forward angles imparting more conveying characteristics to the moving granules.

Kulju et al. [31], through further DEM-PBM studies, showed that shaft speed was a critical process parameter in a high shear granulator. The simulations showed that lower rotations lead to an increased residence time of the granulation material in the spray zone. Increased RTD caused increased wetting, nucleation, and growth—all three of which are crucial rate processes for granule formation, thereby leading to formation of larger granules.

In the work of Boukouvala et al. [32], an integrated continuous wet granulation configuration was simulated consisting of powder feeders, blender, wet granulator and dryer, a milling unit, tablet compaction, and a dissolution testing unit (Fig. 6.5). The work was novel in demonstrating a continuous process model consisting of constant powder input and tablets out. It captured the effects of disturbances in each unit operation such as feed rate of materials and step changes in critical performance parameters such as binder addition and milling speed. However, the authors acknowledged that there were challenges in validating the entire line model with experimental data.

Singh et al. [33] further extended the control model for a wet granulation flowsheet line by adding single-loop and cascade feedback PID controllers. The work simulated the entire flowsheet model along with the control model in gPROMS software integrated with a DeltaV process control software (Fig. 6.6).

4. Case studies

4.1 Twin screw granulator

The Thermo Pharma 11 TSG consists of two rotating shafts with various screw conveying, mixing, and kneading elements mounted on the shafts. The arrangement of the elements on the shafts is modular and can be arranged in

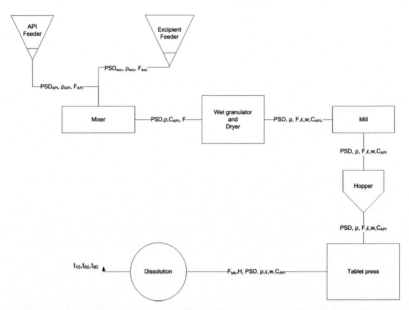

FIGURE 6.5   A schematic process of an integrated continuous wet granulation process. *Adapted from Boukouvala F, Chaudhury A, Sen M, Zhou R, Mioduszewski L, Ierapetritou MG, Ramachandran R. Computer-aided flowsheet simulation of a pharmaceutical tablet manufacturing process incorporating wet granulation. J Pharm Innov 2013;8(1):11–27.*

any manner as desired. The entire setup is contained within a temperature-controlled shell. Pulse tracer technique experiments were carried out in-house by the authors on the Thermo Pharma 11 TSG equipment. In the presented work, the API used was anhydrous caffeine; excipients were microcrystalline cellulose (MCC) PH 101 (Avicel PH101, FMC Corporation, USA) and lactose monohydrate (Foremost Corporation, USA); and the binder used was polyvinylpyrrolidone (PVP) (Acros Organics, USA). The concentration of the API was measured at the exit of the granulator using in-line NIR technique. In order to measure and predict the concentration of API online and in real time, an NIR calibration was previously built. Calibration blends were prepared in a LabRam Acoustic Mixer. The details of the main blend for running the RTD pulse experiments and the ones used for the calibration of NIR are shown as follows in Table 6.1:

The effects of powder feed rate, L/S ratio, and screws' rotation speeds on the RTD were studied through pulse input experiments. The design variables were chosen as follows: the feed rates of the powders in the granulator were 0.4, 0.8, and 1.2 kg/hr; the L/S ratios (liquid water pumped to the solid powder feed) of the system were 0.35, 0.45, and 0.55; and the granulator shaft rotation speeds were 350, 500, and 650 RPM. The experiments were run in a CC design within the range of the parameters.

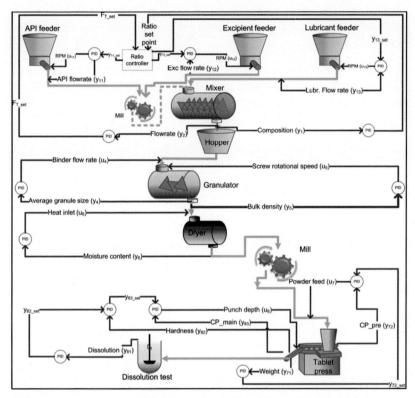

FIGURE 6.6   Control system for the continuous tablet manufacturing process as implemented in a flowsheet model. *Adapted from Singh R, Dana B, Chaudhury A, Sen M, Ierapetritou M, Ramachandran R. Closed-loop feedback control of a continuous pharmaceutical tablet manufacturing process via wet granulation. J Pharm Innov 2014;9(1):16–37.*

TABLE 6.1 Composition of the blends used to build the NIR calibration model.

Blend No.	Caffeine (%)	Microcrystalline cellulose (%)	Lactose (%)	Binder (%)	Total (%)
1	11.60	43.91	40.20	4.29	100
2	10.40	37.11	49.49	3.00	100
3	9.20	48.78	39.02	3.00	100
4	8.00	44.00	44.00	4.00	100
5	6.80	38.66	51.54	3.00	100
6	5.60	39.17	52.23	3.00	100
7	4.40	50.33	40.27	5.00	100

TABLE 6.2 The CC DOE for the residence time distribution experiments on a Pharma 11 TSG.

Run No.	Throughput kg/hr	Liquid to Solid ratio	RPM	Liquid rate kg/hr	Total kg/hr
1	0.4	0.35	350	0.14	0.54
2	1.2	0.35	350	0.42	1.62
3	0.8	0.45	350	0.36	1.16
4	0.4	0.55	350	0.22	0.62
5	1.2	0.55	350	0.66	1.86
6	0.8	0.35	500	0.28	1.08
7	0.4	0.45	500	0.18	0.58
8,9,10	0.8	0.45	500	0.36	1.16
11	1.2	0.45	500	0.54	1.74
12	0.8	0.55	500	0.44	1.24
13	0.4	0.35	650	0.14	0.54
14	1.2	0.35	650	0.42	1.62
15	0.8	0.45	650	0.36	1.16
16	0.4	0.55	650	0.22	0.62
17	1.2	0.55	650	0.66	1.86

Table 6.2 illustrates the experimental design.

Fig. 6.7 illustrates the effects of powder feed rate on the MRT of the system. The x-axis shows the value of the powder feed rate in kg/hr. The y-axis shows the value of the MRT for that run in seconds.

One observes that for moderate and high powder feed rate, the MRT remains more or less the same, averaging around 75 s. However, the observed MRT is higher at the low powder feed rate of 0.4 kg/hr. The plausible reason for this phenomenon is that for powder feed rate beyond a cutoff value, the size and amount of granules forming and exiting the equipment is more significant than the amount of ungranulated powder particles and fine granules. The larger granules more likely fill up the volume in the screws and thus come out easily. The smaller granules and ungranulated fines on the other hand would fill up the screw volume much less and thus lead to greater MRTs due to them traversing greater distances within the screws.

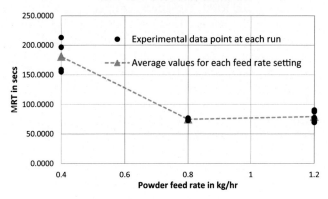

FIGURE 6.7   Effect of powder feed rate on the mean residence time (MRT) of the process stream.

It was also observed that the span of the MRT values was much higher (~ 60 s) at low powder feed rate of 0.4 kg/hr than (~ 20 s) at high powder feed rate of 1.2 kg/hr. The MRTs were in the higher end of the span when both the L/S ratio and RPM were in the lowest settings (0.35, 350) or highest settings (0.55, 650) at the same time. The MRT was high when both the L/S ratio and the shaft RPM were low because the powder/fine granule particles took more time to exit the equipment as they traversed the screws. At the higher shaft speed (650 RPM) and higher L/S ratio (0.55), the L/S ratio was more dominant on the flow behavior of the material, resulting in more holdup. The result is consistent with the one previously presented by Dhenge et al. [34], where the effect of L/S ratio on the MRT during wet granulation was observed on a TSG equipment. It was theorized that as the liquid flow rate was increased, the material became more paste-like, thus leading to longer residence time.

4.2 High shear granulator

The CM-5 Lödige High Shear Granulator (HSG) consists of a single rotating shaft with two blades at the ends and mixing elements mounted on it. The entire setup is contained within a cylindrical shell. Pulse tracer technique experiments were carried out on the CM-5 Lödige equipment. The API used was semi-fine acetaminophen (APAP) (Mallinckrodt Inc. USA); excipients were Avicel PH101 (FMC Corporation, USA) and lactose monohydrate (Foremost Corporation, USA); and the binder used was PVP (Acros Organics, USA). The tracer used, Nigrosin (Sigma–Aldrich, USA), was water soluble. Prior to running the experiment on the CM-5 Lödige, a preblend of the granulation ingredients was prepared consisting of 2.5% PVP, 8% APAP, 44.75% MCC, and 44.75% lactose. The preblend was prepared in a Glatt 40 L

Twin Axle Blender. In each run, 6.7 kg of blend was prepared, and the blender was run for 30 min at 100 RPM. The effects of three parameters were studied on the RTD, namely, the powder feed rate, L/S ratio, and shaft revolution speed. The values of the design variables were as follows: the feed rates of the powders in the granulator were 10, 15, and 20 kg/hr; the L/S ratios (liquid water pumped to the solid powder feed) of the system were 0.35, 0.45, and 0.55; and the granulator shaft rotation speeds were 1000, 2000, and 3000 RPM. The HSG experiments were run in a CC design, as in the TSG case, though without a repeat of the center points.

The full design is shown in Table 6.3:

From Fig. 6.8, one sees that the MRT decreases with increasing RPM. Moreover, the trend fits perfectly to a linear relation. The reasoning for the observation is that the material conveys through the granulator more quickly as the shaft speed increases, thereby reducing stagnation in the equipment.

From Fig. 6.9, one sees that the MRT increases with increasing L/S ratio. The plausible reasoning for the observation is that the holdup volume increases in the granulator with increasing liquid flow rate, thereby leading to greater residence time of the material inside.

TABLE 6.3 The design of the runs for the residence time distribution pulse experiments on the Lödige HSG equipment.

Run name	Throughput kg/hr	Liquid to Solid ratio	RPM
A	10	0.35	1000
B	20	0.35	1000
C	15	0.45	1000
D	10	0.55	1000
E	20	0.55	1000
F	15	0.35	2000
G	10	0.45	2000
H	15	0.45	2000
I	20	0.45	2000
J	15	0.55	2000
K	10	0.35	3000
L	20	0.35	3000
M	15	0.45	3000
N	10	0.55	3000
O	20	0.55	3000

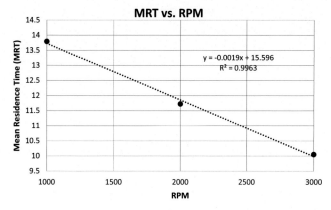

FIGURE 6.8 Effect of shaft rotation speed on the mean residence time (MRT) of the process stream.

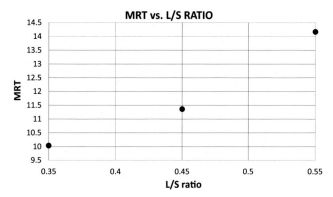

FIGURE 6.9 Effect of liquid to solid (L/S) ratio on the mean residence time (MRT) of the process stream.

From Fig. 6.10, the MRT decreases with increasing powder throughput rate. Like the case in TSG, the MRT is very high in low feed rate of the respective DOE, and the difference is less between the intermediate flow rate and high flow rate.

From the results in Fig. 6.11, it can be seen that there were some runs (C, F, and L) where the RTD showed distinct multiple peaks. These results validate the incorporation of parallel streams in the developed RTD model (Section 3, Fig. 6.4).

Therefore, it can be concluded that the parallel streams model with dead volumes and plug flow fractions shows promise in predicting the RTD of a continuous granulation system. It is intended to extend this work to other

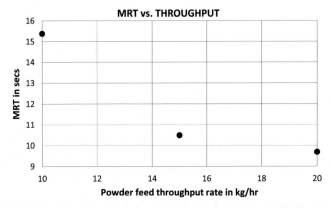

FIGURE 6.10   Effect of powder feed rate on the mean residence time (MRT) of the process stream.

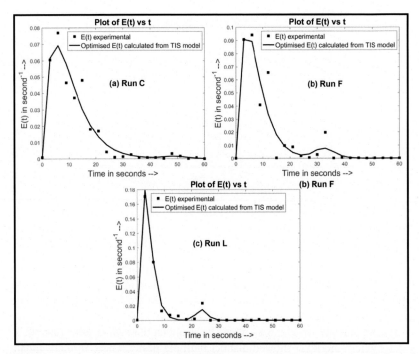

FIGURE 6.11   Residence time distribution (RTD) experimental pulse results and predictions of the same using the combined model for (A) Run C, (B) Run F, and (C) Run L.

systems and develop wholly mechanistic RTDs that can be directly incorporated in other continuous granulation models as predictive tools for estimating the variation in concentrations due to changes in either process parameters or operational settings.

5. Conclusions

This book chapter began with the explanation of the granulation phenomenon and the various mechanisms that lead to the formation of granules. The reasons for push toward continuous mode of pharmaceutical manufacturing by regulatory authorities are briefly discussed. An overview of the various experimentation and modeling techniques is reviewed, with focus on RTD characterization of continuous wet granulation unit operation. A special emphasis is given on the RTD experimental work and modeling that has been done in the field of continuous wet granulation owing to the potential of employing RTD models in studying content uniformity of granules in multicomponent granulations. The employment of PBMs, DEMs, and their coupled approaches are briefly described in the context of characterizing the growth trends of the particles dynamically along the profile of the granulation vessels. Furthermore, experimental findings and simulation results of aforementioned PBMs and DEMs in previously published literature are paraphrased. Special emphasis has been given to the studies on high shear mixer granulators and TSGs, as these are by far the most common devices employed in continuous mode of granulation mechanisms. In these subsections, some of the in-house results by the authors have been presented and key trends briefly touched upon. Therefore, it is hoped and recommended that this review book chapter be studied both by novice students looking to pursue and commence research in this area and by veteran researchers in this field of both academia and industry looking for a concise and quick reference guide on continuous wet granulation.

References

[1] Iveson SM, Litster JD, Hapgood K, Ennis BJ. Nucleation, growth and breakage phenomena in agitated wet granulation processes: a review. Powder Technol 2001;117(1):3−39.

[2] Kinoshita R, Ohta T, Koji S, Higashi K, Moribe K. Effects of wet-granulation process parameters on the dissolution and physical stability of a solid dispersion. Int J Pharm 2017;524(1):304−11.

[3] Morkhade DM. Comparative impact of different binder addition methods, binders and diluents on resulting granule and tablet attributes via high shear wet granulation. Powder Technol 2017;320:114−24.

[4] Nie H, Xu W, Lynne S, Taylor PJM, Byrn SR. Crystalline solid dispersion- a strategy to slowdown salt disproportionation in solid-state formulations during storage and wet granulation. Int J Pharm 2017;517(1):203−15.

[5] Oka S, Smrčka D, Kataria A, Emady H, Muzzio F, Štěpánek F, et al. Analysis of the origins of content non-uniformity in high-shear wet granulation. Int J Pharm 2017;528(1−2):578−85.

[6] Chaturbedi A, Bandi CK, Reddy D, Pandey P, Narang A, Bindra D, et al. Compartment based population balance model development of high shear wet granulation process via dry and wet binder addition. Chem Eng Res Des 2017;123:187−200.

[7] Jia D, Staufenbiel S, Hao S, Wang B, Dashevskiy A, Bodmeier R. Development of a discriminative biphasic in vitro dissolution test and correlation with in vivo pharmacokinetic studies for differently formulated racecadotril granules. J Control Release 2017;255:202−9.

[8] Chaudhury A, Tamrakar A, Schongut M, Smrcka D, Stepanek F, Ramachandran R. Multidimensional population balance model development and validation of a reactive detergent granulation process. Ind Eng Chem Res 2015;54(3):842−57.

[9] Sangshetti JN, Deshpande M, Zaheer Z, Shinde DB, Arote R. Quality by design approach: regulatory need. Arab J Chem 2017;10:S3412−25.

[10] Djuric D, Kleinebudde P. Impact of screw elements on continuous granulation with a twin-screw extruder. J Pharm Sci November 1, 2008;97(11):4934−42.

[11] Granberg RA, Rasmuson ÅC. Solubility of paracetamol in pure solvents. J Chem Eng Data November 11, 1999;44(6):1391−5.

[12] Kašpar O, Tokárová V, Oka S, Sowrirajan K, Ramachandran R, Štěpánek F. Combined UV/vis and micro-tomography investigation of acetaminophen dissolution from granules. Int J Pharm December 31, 2013;458(2):272−81.

[13] Oka S, Kašpar O, Tokárová V, Sowrirajan K, Wu H, Khan M, Muzzio F, Štěpánek F, Ramachandran R. A quantitative study of the effect of process parameters on key granule characteristics in a high shear wet granulation process involving a two component pharmaceutical blend. Adv Powder Technol January 1, 2015;26(1):315−22.

[14] Kumar A, Dhondt J, Vercruysse J, De Leersnyder F, Vanhoorne V, Vervaet C, Paul Remon J, Gernaey KV, De Beer T, Nopens I. Development of a process map: a step towards a regime map for steady-state high shear wet twin screw granulation. Powder Technol 2016;300:73−82.

[15] Meng W, Kotamarthy L, Panikar S, Sen M, Pradhan S, Marc M, et al. Statistical analysis and comparison of a continuous high shear granulator with a twin screw granulator: effect of process parameters on critical granule attributes and granulation mechanisms. Int J Pharm 2016;513(1−2):357−75.

[16] Meng W, Oka S, Liu X, Omer T, Ramachandran R, Fernando J. Muzzio. Effects of process and design parameters on granule size distribution in a continuous high shear granulation process. J Pharm Innov 2017;12(4):283−95.

[17] Effects of wetland depth and flow rate on residence time distribution characteristics. Ecol Eng 2004;23(3):189−203.

[18] Kumar A, Ganjyal GM, Jones DD, Hanna MA. Modelling residence time distribution in a twin-screw extruder as a series of ideal steady-state flow reactors. J Food Eng 2008;84(3):441−8.

[19] Yeh A-I, Jaw Y-M. Predicting residence time distributions in a single-screw extruder from operating conditions. J Food Eng 1999;39(1):81−9.

[20] Himmelblau DM, Bischoff KB. Process analysis and simulation. deterministic systems; 1968.

[21] Kumar A, Vercruysse J, Vanhoorne V, Toiviainen M, Panouillot P-E, Juuti M, et al. Conceptual framework for model based analysis of residence time distribution in twin-screw granulation. Eur J Pharm Sci 2015;71:25−34.

[22] Chaudhury A, Kapadia A, Prakash AV, Dana B, Ramachandran R. An extended cell-average technique for a multi-dimensional population balance of granulation describing aggregation and breakage. Adv Powder Technol 2013;24(6):962−71.

[23] Chaudhury A, Ramachandran R. Integrated population balance model development and validation of a granulation process. Part Sci Technol 2013;31(4):407−18.

[24] Ramachandran R, Chaudhury A. Model-based design and control of a continuous drum granulation process. Chem Eng Res Des 2012;90(8):1063–73.

[25] Dana B, Walia S, Ramachandran R. Multi-component population balance modelling of continuous granulation processes: a parametric study and comparison with experimental trends. Powder Technol 2013;241:85–97.

[26] Barrasso D, Hagrasy A El, Litster JD, Ramachandran R. Multi-dimensional population balance model development and validation for a twin screw granulation process. Powder Technol 2015;270 B:612–21.

[27] Kumar A, Vercruysse J, Séverine T, Mortier FC, Vervaet C, Paul Remon J, Gernaey KV, et al. Model-based analysis of a twin-screw wet granulation system for continuous solid dosage manufacturing. Comput Chem Eng 2016;89(9):62–70.

[28] Gantt JA, Cameron IT, Litster JD, Gatzke EP. Determination of coalescence kernels for high-shear granulation using DEM simulations. Powder Technol 2006;170(2):53–63.

[29] Dana B, Eppinger T, Pereira FE, Aglave R, Debuse K, Bermingham SK, Ramachandran R. A multi-scale, mechanistic model of a wet granulation process using a novel bi-directional PBM–DEM coupling algorithm. Chem Eng Sci 2015;123:500–13.

[30] Dana B, Ramachandran R. Multi-scale modeling of granulation processes: Bi-directional coupling of PBM with DEM via collision frequencies. Chem Eng Res Des 2015;93:304–17.

[31] Kulju T, Paavola M, Spittka H, Keiski RL, Juuso E, Leiviskä K, Muurinen E. Modeling continuous high-shear wet granulation with DEM-PB. Chem Eng Sci 2016;142:19–200.

[32] Boukouvala F, Chaudhury A, Sen M, Zhou R, Mioduszewski L, Ierapetritou MG, Ramachandran R. Computer-aided flowsheet simulation of a pharmaceutical tablet manufacturing process incorporating wet granulation. J Pharm Innov 2013;8(1):11–27.

[33] Singh R, Dana B, Chaudhury A, Sen M, Ierapetritou M, Ramachandran R. Closed-loop feedback control of a continuous pharmaceutical tablet manufacturing process via wet granulation. J Pharm Innov 2014;9(1):16–37.

[34] Dhenge RM, Cartwright JJ, Hounslow MJ, Salman AD. Twin screw wet granulation: effects of properties of granulation liquid. Powder Technol October 1, 2012;229:126–36.

Chapter 7

Continuous fluid bed processing

Stephen Sirabian

Equipment & Engineering Division, Glatt Air Techniques, Inc., Ramsey, NJ, United States

1. Introduction

Continuous processing has been called the "future of pharmaceutical manufacturing," and as of this writing, this principle has achieved widespread acceptance, having been well established. In order to understand the future, it is said one must first understand the past. It follows that by examining the birth of pharmaceutical oral sold dosage processing, we can see how continuous processing evolved, drawing upon existing technologies from other industries along the way.

2. Basics of fluidized beds

A fluidized bed comes into existence when a gas, such as air, passes through a sitting mass of solid material. Depending on the velocity of the gas, different states of fluidization can be achieved, which then drives the processing performance of the system. The behavior of a fluidized bed is shown by means of a simple example illustrated in Fig. 7.1.

We will assume a bulk material with a very narrow particle size distribution is filled into a vessel. The bottom of the vessel consists of a porous plate, known as the bottom screen. A gas stream enters this vessel moving upward through the bottom screen. At low flow velocities, the material rests on the bottom screen without any particle movement. This is the fixed bed state.

As the gas flow rate is gradually increased, a gas velocity is reached, which creates the minimum fluidization state. This is characterized by the beginning of particle movement above the bottom screen, and the bed of bulk material is no longer behaving as a single solid mass. This level of movement represents the beginning of a fluidized bed (see Fig. 7.1A).

With a continued increase in the gas flow rate, the intensity of the particle motion is also increased and the velocity of particle movement changes dramatically. In comparison to the gas consumption during the minimum

How to Design and Implement Powder-to-Tablet Continuous Manufacturing Systems
https://doi.org/10.1016/B978-0-12-813479-5.00010-0

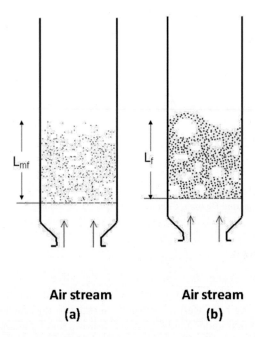

Air stream   **Air stream**
(a)   **(b)**

FIGURE 7.1   (A) Incipient fluidization. (B) The formation of bubbles contributes to mixing of particles within the fluidized bed.

fluidization state, the extra gas volume flows through the fluidized bed in the form of bubbles. These bubbles are the reason that the product (particles) inside the bubbling fluidized beds is well mixed (see Fig. 7.1B). It is this mixing performance which forms the foundation for consistent final product quality.

The maximum gas flow rate is determined by the terminal velocity of the material being fluidized. At this velocity the particles are pneumatically transported out through the top of the vessel.

In a typical manufacturing operation, the starting bulk material in the fluidized bed layer is characterized by a wide particle size distribution (powder or particle). However, the adjustable intensity of turbulence in fluidized beds provides effective particle mixing, leading to excellent mass and heat transfer. This makes the fluid bed process a highly efficient dryer and, when a binder liquid is introduced via a spray nozzle, an equally efficient granulator with a tight, predictable particle size distribution.

Typically, a certain amount of dust from both the raw material and attrition from the particle motion will be present within the processing chamber. This dust along with some coarse particles will be transported upward by the gas flow. Therefore, dedusting systems are always necessary for fluidized bed operations. In most cases, the filter system is integrated into the fluid bed processor directly above the expansion zone or granulation zone; there may not be filter elements above the downstream cooling zone of the processor as no fine material can be agglomerated there.

3. Drying background and theory

For our purposes, drying can be defined as the removal of moisture from a dense bed of particles. This removal process consists of two mechanisms: mass transfer and heat transfer. In a fluidized bed process, this is accomplished with a heated air stream flowing upward into the particle bed. This air stream suspends the particles and exposes all surfaces of the particle. The air can then efficiently heat the moisture to separate it from the solid particles and provide a physical means to remove the freed moisture from the drying chamber.

Moisture, solvent or water, is generally considered either bound or free. Free moisture resides on the surface of the particle and readily evaporates when exposed to a given temperature and humidity. The moisture retained in the capillaries is also considered free moisture, but the rate of its removal is governed by diffusion through the particle, as opposed to the drying capacity of the air stream. Bound moisture occurs on the molecular level, where a water molecule is bound to a molecule in the particle. This water molecule can only be removed if there is enough energy available to overcome the bond. This energy level is measured in terms of temperature. For example, air at 85°C cannot break the water bonded to lactose, but air at 105°C can. Some caution must be exercised when using higher temperatures to overcome the bond; the actual temperature used may be limited by the material properties subject to degradation at elevated temperatures.

A brief summary of the internal drying mechanisms follows:

a. Capillary flow: Moisture is held in the interstitial spaces of the solid resides on the surface or within cell cavities. Liquid flow is driven by moisture vapor pressure above the equilibrium moisture content at atmospheric pressure.
b. Vapor diffusion: Moisture movement through a solid driven by the temperature gradient from the heated air, creating a vapor pressure gradient.
c. Liquid diffusion: This is most relevant to the later stages of drying where liquid movement is restricted by the equilibrium moisture content being below the point of atmospheric saturation.

Pragmatically speaking, for pharmaceutical drying in a fluidized bed environment, heated air is employed as the purge gas acting as both the heat and mass transfer medium. As the rate of liquid vaporization depends upon the vapor concentration within the drying chamber, both inlet air temperature and humidity become key variables along with the mass flow of air.

Drying curves: Drying curves plot moisture content over time and have been used to characterize batch drying phenomenon, but they apply to continuous processing as well. The difference is that for continuous processes, the various drying phases occur over the length of the drying chamber independently of time, once steady state has been achieved. Batch operations are constant throughout the drying chamber, but the phases vary over time. The difference in inlet temperature and bed temperature is an indication of drying rate, and, therefore, by knowing the bed temperature at several points along the continuous fluid bed, the drying rate can be determined at each point.

The drying curve in a fluid bed processor has four distinct zones. An example of a drying curve is illustrated in Fig. 7.2. The warm up period is the initial stage when the heat transfer from the inlet airflow raises the temperature of the substrate while the drying rate increases. This is followed by the constant rate period where, at the surface of the particle, the mass transfer and heat transfer are balanced as long as the particle surface remains at a saturated condition. Once sufficient moisture has been removed and the surface is no longer saturated, the rate of drying is reduced. This marks the beginning of the falling rate period characterized by unsaturated surface drying. As there is less moisture to vaporize at the particle surface, the remaining energy from the air stream will heat the particle, increasing the bed temperature. Following this, the second phase of the falling rate period sees internal moisture movement.

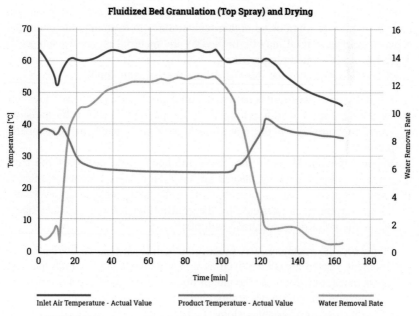

FIGURE 7.2   An example of a drying curve with four distinct drying zones.

The length of this last phase will determine the final moisture content. Increasing the inlet air temperature at this point will not increase the drying rate.

4. Granulation drying background and theory

Agglomeration is the bringing together of two or more materials to form an aggregate. In the pharmaceutical industry, the term granulation has been universally adopted to describe this unit operation which entails an API (active pharmaceutical ingredient), excipient(s), fillers, and other materials as needed such as dispersing agents or even a second API. Key target properties of the granules are size, porosity/density, and strength.

Granulation processes are used for size enlargement, dedusting, and enhanced dissolution and dispersibility of powdered solids. During a fluid bed granulation process, liquid or a liquid binder solution is applied to the raw material particles, which are in a fluidized state. These liquids wet the particle surface or apply the binder for binding the particles together. For distributing the liquid into the bed mass, a binary nozzle design is employed in which atomization is created with compressed gas. This liquid injection can be done from the top of the processing chamber (top spray) spraying downward onto the bed; from the bottom of the processing chamber (bottom spray) spraying upward directly into the fluidized bed; or tangentially from the side or bottom (tangential spray). Batch processors historically have used the top spray method for granualtion while continuous processors can use either top or bottom spray.

The intensive mixing of the solid material (powder and already formed agglomerates) ensures that a constant exchange of particles occurs in the area of the fluid bed where the spray liquid is injected. This in turn ensures the spray is uniformly applied onto the surface of the solid particles, which initiates the particle enlargement process.

The fluidizing gas is preconditioned to a specified temperature and humidity and the gas flow rate is also defined, which allows precise control of material and heat transfer between the fluidizing gas, the solid particles, and the atomized binder liquid. The atomized liquid droplets are dried onto the particles as the water evaporates and is removed by the fluidizing gas, leaving a newly formed granule behind. This process repeats itself multiple times depending upon the residence time (the amount of time the particles stay in the fluidized bed chamber), after which the granules are discharged from the process chamber, having attained the target particle size and density.

The joining of particles together is accomplished by four mechanisms, which are illustrated in Fig. 7.3. They all start with the fluidized bed processor, the fundamental workings of which have previously been explained. A nozzle is introduced to spray a binder solution onto the individual particles as they collide and attach to each other as aggregates or granules. The initiation of the

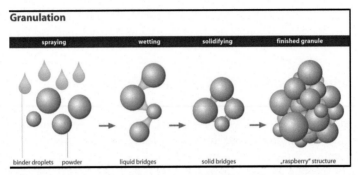

FIGURE 7.3 Mechanism of granule formation during fluidized bed processing. *Figure courtesy of Glatt Group.*

liquid spray into the fluidized bed is known as the wetting phase, where the binder is introduced, and the overall bed moisture is increased. As wetted primary particles, known as the nuclei, collide, they coalesce in the region where a droplet binds two primary particles together. The collision of nuclei to form granules is known as the growth phase. Eventually, the increased density and fluidization pattern movement create compaction forces on the granules, driving the final phase of consolidation. One of the key benefits of fluid bed granulation is that controlling the inlet temperature, inlet air humidity, fluidization air velocity, liquid spray rate, and droplet size/pattern allows precise control of the final granule properties.

5. Commercial application

Modern-day batch processing for pharmaceutical solid dosage forms began in the 1950s along with many of the post-war manufacturing industries. Circa 1954, the need to remove moisture from batches of pharmaceutical granulations was targeted as a critical unit operation for pharmaceutical manufacturing. Drying ovens had been the default technology at the time, but they were notoriously slow and inefficient, with inconsistent final moisture profiles and occasional case hardening. To address these issues, rather than pouring the material onto solid trays, a cylindrical container with a perforated/screened bottom plate was fabricated and a fan was used to pull warm air through the drying enclosure. This first convection oven was known as a box dryer and significantly improved drying performance.

A further refinement was achieved when, rather than placing the product container within a boxed enclosure, a tall cylindrical chamber was added above the product container with a fan above that, thus pulling 100% of the heated air though the moist material. By directing all the air through the wet material, efficiency was enhanced further, but there was still room for improvement. It was found that once the air passing through the material lying on the bottom screen reached a certain velocity, the individual particles would be lifted until gravity overcame their upward velocity and then the particles would fall back to the screen where they would be entrained in the upward airflow once again. This recirculating motion of the powder resembled a liquid, giving rise to the term "fluidized bed drying." Now with each particle surrounded on all sides with warm air, drying was faster and more efficient with a highly uniform final moisture profile.

Still driven by the need for manufacturing efficiency, the concept of combining the granulation step and drying step into a single unit operation was borne. Rather than granulate in a separate step, both steps were performed inside the fluid bed dryer, now renamed as a fluid bed granulator. The granulation process took advantage of the suspended particles inside the fluid bed drying chamber and used a high-pressure air to atomize the granulation binder liquid into fine droplets. As these droplets contacted the suspended particles, they would form bridges between multiple articles as they collided while being fluidized. Besides the inherent efficiency of two simultaneous unit operations, users saw benefits in the final product quality in terms of homogeneity, precision, and density.

6. Why batch

Batch processing began more by default than design. The earliest records of the practice of the apothecary as a profession date back to 2600 BCE in ancient Babylon. Formulations were mixed on a benchtop with the mortar and pestle serving as the technology of the day. That basic philosophy of producing a batch from a combination of ingredients prevailed through the 20th century. The advantages of batch processing were many, but simplicity and controllability are perhaps the most often cited. It is these two issues that are the center of the push into continuous solid dosage processing that began in earnest in the United States in the first decade of the 21st century.

Batch compounding allowed pharmacists to mix up any formulation they desired, adjusting the recipe as they went along. Even with today's automated recipe-driven manufacturing, product properties can be readily sampled throughout the batch as a means of quality assurance as well as allowing adjustments to ensure the final batch parameters meet quality specifications. Certainly, the ability to adjust the recipe and process parameters is limited to the R&D phase, but then those processes and recipes are transferred to manufacturing as a batch process. Taking drying as the simplest example, sampling can be conducted throughout the process, and if the final moisture specification has not been achieved, there are several adjustments that can easily be made to get the batch into specification, such as increased air temperature, longer batch drying time, or increased airflow. Thus, there was usually a way to "save a batch" rather than waste costly ingredients.

Despite all this, the potential efficiency of continuous manufacturing drove interest in the first decade of the 21st century, which led to some hybrid solutions. One of the most fundamental forms was simply linking a series of individual batch processors into a single unit. This so-called batch—continuous approach maintained the actual processing chambers of a batch machine, but by joining them together with a constant feed stream of material, they produced a constant output. Another approach was rapid cycling of batch processes, with fast automated material transfer in and out of the processing unit, so that over the course of a day, the machine was in process mode a high percentage of the time. Still, the need for true continuous operation asserted itself and the technology began to take shape.

It became clear that although continuous processing was initially viewed with some well-known concerns, notably batch definition and quality control, these were successfully addressed by both the FDA and the industry. As of 2019, there were five drugs on the market produced using this technology and the technical concerns have been addressed. No one would argue that there is universal agreement about the move into continuous manufacturing, but it is quickly becoming a preferred manufacturing mode for many pharmaceutical companies.

7. Continuous processes in other industries

While new to pharmaceuticals, continuous processing has long been a staple for the chemical and food industries, particularly for detergents, enzymes fertilizers, and food ingredients. Here the high production volumes go well beyond standard batch sizes, making the number of batch units necessary impractical both financially and operationally. A large batch granulator, for example, might be up to about 2000 liters in volume, while continuous lines routinely handle flow rates of thousands of liters per hour. Moreover, rotary calciners used in the construction, mining, and petrochemical industries can handle hundreds of tons per hour. These continuous machines for installations outside the pharmaceutical industry have been around for decades, and much of today's expertise is built upon this experience.

For pharmaceutical applications, the small volume requirements coupled with the highly oriented R&D focus led to a different execution of the same basic principles successfully deployed elsewhere. Not only was a far lower throughput needed, but also the ability to isolate and separately control various unit operations was seen as a critical design parameter in order to control and verify quality. Advances in process analytical technology (PAT) and acceptance of Quality by Design principles have addressed the quality issue. That, along with the efficiencies of real-time release and the FDA's explicit support of continuous manufacturing have overcome historical reticence toward this technology.

8. Traditional continuous fluid bed design

Fig. 7.4 shows a flow diagram with raw material feed and product discharge on a continuous fluidized bed granulation line. In the figure, a dosing screw is used to feed and dispense a solid raw material (for instance, an API powder) into the granulator. A fixed speed rotary valve ensures that the pressure inside the fluid bed processing chamber is isolated from atmosphere. The continuous product discharge is controlled by a variable speed rotary valve mounted on a discharge chute. This rotary valve can be used to adjust the volumetric discharge rate depending on the feed rates of solid and liquid raw materials to satisfy the mass balance.

To ensure a constant height of the bed inside the granulator, simple overflow designs in different executions have been developed. Fig. 7.5 depicts two examples. In both cases, fixed speed rotary valves are used for atmospheric isolation or separation from downstream equipment. The left sketch shows a typical setup for horizontal fluidized beds with a rectangular cross section of the process chamber. On the outlet side of the apparatus, an end weir is mounted which divides the fluidized bed section from the product discharge chute.

When the fluidized bed height is increased, such as when there is an increase in raw material feed rate, solids flow over the weir and into the discharge chute at an increased rate. One must consider that a fluidized bed also expands (meaning the bed height increases) when the fluidization gas velocity increases. This means that there is a relationship between the holdup mass and the bubbling behavior of the fluidized bed.

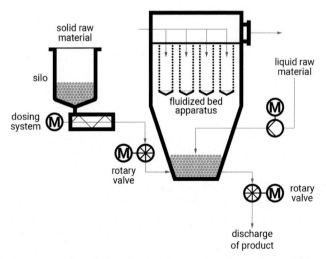

FIGURE 7.4   A process flow diagram showing the operation of a continuous fluidized bed with raw material feed and product discharge.

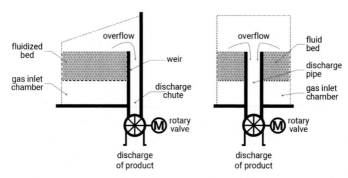

FIGURE 7.5 Overflow configurations to ensure fixed bed height in fluidized bed processing.

The discharge principles shown in Fig. 7.5 all discharge the entire particle size range of the fluidized bed mass. Consequently, any dust or undersized grains created by the granulation process leave the apparatus. These fractions can be recycled to the granulator/dryer to maintain extremely high yields, a common practice in many non-pharmaceutical applications. To control this aspect of the granulation line, size classification devices are utilized at the product outlets. Outside of the pharmaceutical industry, the undersized particles along with milled oversized particles from the classifier are recycled into the raw material feed to the fluid bed granulator. However, in the pharmaceutical arena, quality and material traceability issues have not allowed this recycling, which makes the unrecycled product yield more critical for process efficiency and economic viability.

Traditionally, horizontal fluidized bed units characterized by a rectangular design of the process chamber have been the norm. As described below, this has been adapted into a round configuration for pharmaceutical R&D and pilot scale applications that offer some distinct advantages.

In Fig. 7.6, a typical setup of a horizontal fluidized bed apparatus is depicted. In this example, a simple inlet chute is used to charge the fluidized bed with solids in a continuous processing operation. A rotary valve is used as the physical disconnection to the upstream processing steps. The discharge of product is accomplished with a discharge chute equipped with a variable speed rotary valve on the outlet. Both top and bottom spray configurations can be used for the granulation/binder liquid. Internal filter systems are typically used to keep any dust and fines in the apparatus, which minimizes the need for peripheral equipment for outlet gas handling, although such external devices can be used for specific applications.

While a single air supply can be used for fluidization in all sections of the granulator, multiple independent air streams can also be used to feed different inlet gas flows. Each of these flows can have individually defined parameters (e.g., temperature, flow rate, and humidity), thereby allowing different

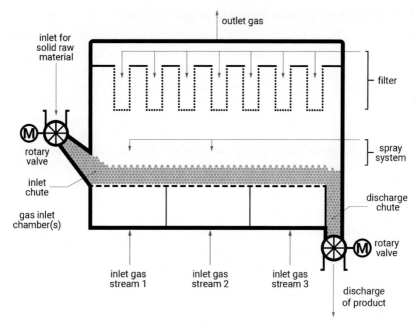

FIGURE 7.6   Setup of a typical horizontal continuous fluidized bed.

processing steps to be carried out in one single unit. For instance, spray granulation can take place upstream in the processor followed by a drying or cooling zone. Moreover, processing conditions of single-step processes can be optimized in the fluidized bed by adjusting, for instance, inlet gas flows, inlet gas temperatures, and binder liquid spray rates.

The dynamic behavior of particles in a circular fluid bed is different compared to rectangular geometries. Rectangular fluid beds (see Fig. 7.6) allow predetermination of residence time distribution through selection of the bottom screen aspect ratio (length and width), process throughput, and bed mass. A horizontal fluidized bed type shows a material movement closer to plug flow behavior, instead of the well-mixed "continuous stirred tank" behavior that is typical of circular fluidized bed apparatuses. Chapter 6 provides details on residence time behavior of materials inside continuous processing units.

Typically, external filters, cyclones, or wet scrubbers are used downstream to clean the outlet gas, and the fines and dust collected in filters or cyclones can be recycled to the process as seeds for further particle growth for certain nonpharmaceutical processes.

9. Adaptation to pharmaceutical processing

For pharmaceutical applications, the throughputs are obviously much lower than in other industries, so the equipment design involves different considerations. As a new technology, much of the initial interest has been focused on the R&D arena, where space is a consideration as well as the need for processing small quantities, which are often limited and expensive for new APIs. Consequently, the long rectangular cross section used for the bottom screen in the preceding section has been adapted into a circular format, as shown in Fig. 7.7. This saves space and is adequate for equipment with throughputs from under 1 kg/h up to 50 kg/h or so, which accommodates both R&D and pilot scale work. As shown in Fig. 7.7, the inlet and discharge from Fig. 7.6 in the preceding section are also implemented in this circular design along with the plug flow particle motion.

An additional advantage to such designs is that the exterior dimensions of the circular configuration mimic that of batch processors with the same capacity (i.e., R&D and pilot) so that a single fluid processor can be multitasked for both batch and continuous operations. This allows retrofitting a continuous processing chamber to a preexisting batch processor's skeleton, saving space and capital resources. Generally speaking, the transition to the rectangular design occurs somewhere between 50 and 100 kg/h.

The processing chamber can have either static or dynamic inner baffles. The static baffles rely upon material flow around the axis as material moves under the baffles, which create a uniform residence time. Dynamic baffles

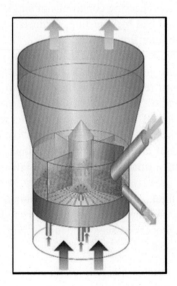

FIGURE 7.7 A circular processing chamber adapted for continuous processing.

FIGURE 7.8 A rotating drying chamber commonly used in continuous fluidized bed processing systems.

(Fig. 7.8) rotate and seal against the floor of the process chamber as an alternative means of enforcing the desired residence time distribution (closest to plug flow movement).

Other design details have impact on the integration of the full system. Much of this is covered elsewhere in the book, so this chapter identifies them in the context of fluid bed processing to provide an overall system understanding. Fig. 7.9 shows a photograph of a system capable of achieving a throughput of 50 kg/h of dry granulation. Starting with the feeders in the upper left hand corner of the photograph, materials drop through a continuous blender. In this case, the blended material is granulated in a twin screw extruder for high shear granulation, followed by fluid bed drying and then milling prior to open feed into a tablet press with an integrated deduster. This process train was selected to highlight the number of different unit operations that must be interfaced in continuous operations. While batch systems operate each unit operation independently, in a continuous process a master control system must be used to make sure the various operations are coordinated, as they must operate simultaneously and at a constant flow rate. Adding to this, at least one of the several PAT devices will likely be employed, such as the unit shown at the dryer outlet to measure particle size and distribution. Meaningful PAT integration is essential for efficient closed loop control and to enable real-time release.

FIGURE 7.9 A pilot scale continuous granulation line with open transfer.

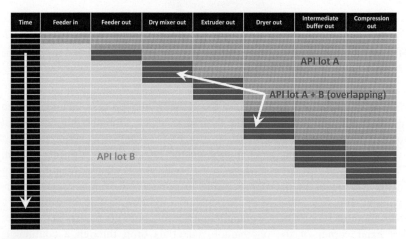

FIGURE 7.10 Tracking performed by a control system enables the identification of the lots of material that constitute the final product.

10. Traceability

In addition to the normal control system requirements within a regulated in-dustry, the continuous control system must track materials over time. Where batch systems have an inherent definition of what constitutes a batch, continuous systems must define its own version of a batch, commonly referred to as a "lot." It should be noted that the FDA's definition of a batch only refers to a specific quantity of a drug, which, in combination with other materials, is intended to have uniform character and quality within specified limits. Details of batch definitions and considerations around each definition can be found in Chapter 16. Defining a batch for a continuous manufacturing process can utilize several approaches, including run time, volume produced, API lot, etc. As different ingredients are received in different quantities, a finished lot will likely pull from multiple source lots for the various ingredients being processed. This traceability of materials becomes a major consideration for continuous control systems. An example of a system tracking two source lots (lots A and B) of the same API against various unit operations is shown in Fig. 7.10.

The key for traceability is the use of the RTD (residence time distribution) to track product flow through the system. In other words, the system must know when a given raw material lot is being depleted in a given unit operation and when the changeover to the next raw material lot is occurring. When tracking raw material and finished lots, it is important to be able to go in both directions. That is, if a given raw material lot is known, you should be able to determine which finished product lot(s) it ended up in, but if you know the finished product lot, you should be also able to determine which raw material lots went into that finished product lot.

Another important function delegated to the control system is detecting and managing OOS (out-of-specification) material. This requires defining CPP (critical process parameters) for various unit operations and monitoring the CPP and PAT information accordingly. Once operating under a state of control, the goal is to keep the system running in that state and avoid shutting down for an OOS event. In such cases, the OOS portion of the lot is tagged as waste and diverted. This approach to tracking different raw material lots of two excipients (A and B) along with a single API lot along with the finished product lot and waste portion of the lots is depicted in Fig. 7.11.

PAT plays a critical role in any and all of the forgoing. The various PAT options and technologies are covered in depth in Chapter 12, but any discussion of granulation must acknowledge that PAT serves as the bedrock of the control strategy for any continuous granulation process.

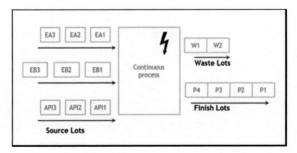

FIGURE 7.11   Raw material and finished product tracking.

11. Other continuous granulation methods

This chapter has focused upon fluidized bed granulation thus far, but there are other methods that have been used successfully in pharmaceutical applications that should be identified for completeness. Most common is the use of a twin screw extruder to wet granulate the materials. This technology and roller compaction are inherently continuous unit operations. They were used by pharmaceutical manufacturers as discrete unit operations for many years before their incorporation into continuous process trains. Development and characterization of twin screw granulation (see Fig. 7.12) [see Chaper 6] and roller compaction processes [see Chapter 5] have been discussed elsewhere in the book.

Extrusion is inherently a high shear wet granulation process which typically produces a denser granule than a fluidized bed (but less dense than a typical high shear batch granulator). The extrudate produced by a twin screw extruder is discharged into a fluid bed processor to dry it to the desired moisture content before being conveyed to downstream process steps, such as milling, blending, and tablet compression. The dried granulate can also be filled into capsules.

Intermeshing twin screws are the most prevalent extruders for these applications. The extruder (see Fig. 7.13) uses two co-rotating shafts onto which a series of conveying, mixing, and kneading elements can be mounted, allowing the user to configure the shear profile needed to achieve the target final granule properties. In addition, one can select various L/Ds (the ratio of screw shaft length to its diameter), which has a direct impact on the residence time and also can impact the shear profile and thus ultimately the processing characteristics of the system.

Material flow from the weigh/dispense equipment (e.g., feeders, blenders, mills, etc.) is fed into the mouth of the extruder and then binder liquid is

FIGURE 7.12 An R&D twin screw continuous granulator integrated with a single-celled fluidized bed dryer.

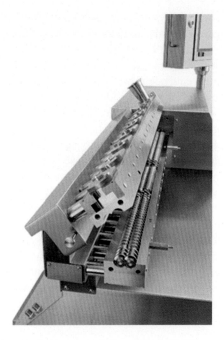

FIGURE 7.13 Interior of twin screw extruder/granulator with different granulation elements.

introduced partway along the length of the processing chamber along with (if needed) additional solid ingredients. As material moves along the length of the processing chamber, the binder and any added ingredients are subject to various mechanical forces by the screw configuration which creates the desired granule.

Roller compaction is another higher shear granulation process in which the compaction force is created by passing a flow of powder between two counter-rotating rolls. The operation is "dry," meaning it does not involve the use of a binder fluid suitable for active ingredients that are sensitive to moisture. A full description of this technology is provided in Chapter 5.

12. Summary and conclusion

The science of forming granules does not change when moving into the continuous arena. The binder—particle interaction and growth phenomena take place regardless of the process running in batch or continuous mode. The difference is in the ancillary processes of introducing materials into each unit operation and transferring them through the entire process stream with a focus upon residence time. In addition, continuous granulation must constantly operate within specification which makes PAT such a critical component, unlike batch process development where the granulation process is more prone to manual manipulation in order to achieve the target final properties.

Typically, there is much less material in each process chamber at a given time for continuous processes, so the equipment size and footprint can be much smaller than the equivalent batch processing equipment, even with all the support operations taken into consideration. In addition, there is generally more flexibility in varying the product volume or throughput with continuous systems to the point that pilot and production quantities can be run in the same equipment. Lastly, because PAT is a necessity and therefore always included in a continuous manufacturing line and because continuous systems can be manipulated dynamically to keep the process at the desired control set point, there is an enhanced level of process control and quality inherent in these systems with an obvious benefit when considering the regulatory aspect of pharmaceutical manufacturing.

There can be no doubt that continuous processes for pharmaceutical applications have come into technological maturity. Not only are there several commercial products being manufactured in this manner (with a much larger number of products anticipated in the next few years), but also there are a number of equipment and technology manufacturers offering viable systems.

Chapter 8

Continuous tableting

Sonia M. Razavi[1], Bereket Yohannes[2], Ravendra Singh[1],
Marcial Gonzalez[3], Hwahsiung P. Lee[2], Fernando J. Muzzio[1] and
Alberto M. Cuitiño[2]

[1]*Engineering Research Center for Structured Organic Particulate Systems (C-SOPS), Department of Chemical and Biochemical Engineering, Rutgers, The State University of New Jersey, Piscataway, NJ, United States;* [2]*Department of Mechanical and Aerospace Engineering, Rutgers, The State University of New Jersey, Piscataway, NJ, United States;* [3]*School of Mechanical Engineering, Purdue University, West Lafayette, IN, United States*

1. Fundamentals of tableting

Tableting is the stage in a pharmaceutical tablet manufacturing prior to coating. If coating is not required, compaction is the last stage in making a tablet. High-speed tablet production can be best achieved using rotary tablet presses, which consist of a rotating turret with variable number of punches and dies. The properties of the final product (i.e., tablet) are of great importance and are linked directly to the inherent powder properties and the effects of processing prior to and during this stage.

Tablet compaction starts with the die filling stage, where intricate powder flow phenomenon is involved [1]. Gravity-fed and force-fed options are the two most commonly used die filling methods. In the most common continuous force-fed option, the powder mixture is guided through a pipe to a hopper connected to the feed frame. A feed frame then uses rotating paddles to force the powder into dies and can be regarded as a separate unit operation, as feed frames have their own dynamics. They exert a separate effect on the powder characteristics and exhibit different designs of the paddles (e.g., number, size, and shape of paddled wheels) and the chamber containing them (e.g., shape and volume) [2]. The same applies to the hopper design, where it has been shown that hopper shape and size are important design parameters in impacting the powder flow and discharge rate [3]. However, hoppers and feed frames generally are considered as part of the tablet press unit operation because most of the commercially available tablet presses provide feed frames of fixed design. The operator can optimize the feed frame speed and keep the fill level in the hopper fixed to avoid disparities.

How to Design and Implement Powder-to-Tablet Continuous Manufacturing Systems
https://doi.org/10.1016/B978-0-12-813479-5.00009-4
159

Variability in the die filling stage, which is a major source of quality issues affecting content uniformity, tablet hardness, and dissolution, may be linked to powder properties such as cohesion, bulk density, particle size, and shape and/or process parameters such as turret speed, feed frame speed, and die size [4−6]. The residence time of the powder in the feed frame, which is a function of the powder density, the feed frame size and speed, and the turret speed [7], imparts shear to the powder, affecting the hardness and dissolution rate of tablets [2].

After the filling stage, the weight of the powder is adjusted by changing the die volume. Excess powder is pushed out of the die due to the action of the fill cam in a brief stage called the metering stage. This is followed by the compression stage, which usually consists of two steps: precompression and main compression. In the precompression step, a low compression force is applied to the powder to release air trapped between the powder particles. In the main compression stage, the punches get closer and the thickness of the powder bed decreases until it reaches its minimum, where the maximum compaction pressure is applied. When the force is released, the unloading stage starts and some of the mechanical energy is recovered as the tablet expands, mainly axially.

There mainly exists two kinds of rotary tablet presses: displacement-controlled and force-controlled. In displacement-controlled presses, the gap between the punches can be controlled and powders are compressed to a constant target thickness. In force-controlled presses, the force exerted on the powder can be controlled using hydraulic pressure applied by air compensators. Depending on which type of tablet press is in use, the controlling process parameters change. Focusing on a displacement control tablet press, the adjustable process parameters are turret speed, feed frame speed, fill depth, displacement at precompression, displacement at main compression stage, and tooling geometry, which they need to be installed before operation. In some presses, for cohesive blends, the height of the powder column in the feed chute has also been observed to play a role.

Critical parameter identification, reduction of the experimental design, and optimization of the compaction stage are all areas worth exploring in continuous tableting. Compaction modeling plays an important role in achieving these objectives, which help us understand the phenomena of powder compaction and the physics behind it and its effects on the process.

2. Phenomenological modeling of compaction

The transformation of a powder bed, confined inside a rigid die, into a compressed solid compact by the sole application of a compaction force engages multiple physical mechanisms. Typically, the initial stage of this process is characterized by particle rearrangement that leads to the formation of a closely packed system. In the subsequent stage, as higher pressure is applied, the porosity cannot be further reduced by particle rearrangement and therefore particles undergo brittle fracture, plastic deformation, or both [8,9]. These dissipative and irreversible processes, in which the volume of the powder bed is reduced, give rise to the formation of a solid tablet. Specifically, fracture and permanent deformation generate particle-to-particle contact surface and thus the opportunity for bond formation and, under large deformations, for solid bridge formation—which is driven by processes such as sintering, melting, crystallization of amorphous solids, or chemical reactions [10]. Particle size, shape, and roughness affect the initial stage of compaction, but it is fragmentation and plastic deformation that dominate the formation of highly dense compacts of many pharmaceutical blends [11].

A phenomenological description of compaction at the particle scale then requires the development of loading—unloading contact laws for elastoplastic particles with formation of solid bridges. Loading contact laws for elastoplastic spheres have been developed by Storakers and coworkers [12,13]. These contact laws for elastoplastic spheres are successful in simulating the deformation of pharmaceutical excipients, such as microcrystalline cellulose and lactose monohydrate [14,15]. Unloading contact laws for elastoplastic spheres with bonding strength, or adhesion, have been developed by Mesarovic and Johnson [16]. This formulation exhibits a discontinuity at the onset of unloading when particles form solid bridges during plastic deformation, a limitation that has been overcome by introducing a regularization term [17]. Specifically, this generalized contact law considers two elastoplastic spherical particles of radius R, Young's modulus E, Poisson's ratios ν, plastic stiffness κ, and plastic law exponent m, which deform plastically under loading and relax elastically under unloading. The contact radius a of particles located at relative position γ is given by

$$a^2 = \begin{cases} c^2 R \, \gamma = a_P^2 & \text{plastic loading} \\[2ex] \left[a_P^2 - \left(\dfrac{2E \left(\gamma_p - \gamma \right)}{3(1 - \nu^2) \, n_P a_P^{1/m}} \right)^2 \right]_+ & \text{elastic (un)loading} \end{cases}$$

where $n_P = \pi k R^{-1/m} \kappa$ and $a_P^2 = c^2 R \gamma_P$, with $k = 3 \times 6^{-1/m}$, $c^2 = 1.43 \, e^{-0.97/m}$, and $[\cdot]_+ = \max\{ \cdot, 0 \}$. The elastoplastic spherical particles are capable of forming a solid bridge characterized by its fracture toughness $K_{Ic} = \sqrt{\omega E/(1 - \nu^2)}$, with ω being the interfacial fracture energy. Therefore, the plastic and elastic (un)loading force is defined by

$$P = \begin{cases} n_p \, a^{2+1/m} & \text{plastic loading} \\[2ex] \dfrac{2 n_P}{\pi} a_P^{2+1/m} \left[\arcsin\left(\dfrac{a}{a_P} \right) - \dfrac{a}{a_P} \sqrt{1 - \left(\dfrac{a}{a_P} \right)^2} \right] \\[2ex] -2 \, K_{Ic} \pi^{1/2} a^{3/2} \dfrac{(1 + \xi_B)^2 [a_B - a]_+}{(1 + \xi_B) a_B - a} & \text{elastic (un)loading} \end{cases}$$

where ξ_B is a regularization parameter and a_B is the radius of the bonded area, or solid bridge, which is equal to a_P if the solid bridge is formed and is equal to zero if the solid bridge is broken. The regularization term [17] endows the contact law with the property of being continuous at the onset of unloading after formation of a solid bridge and at the onset of plastic loading after breakage of a solid bridge or bonded surface (see Fig. 8.1).

The phenomenological modeling of tablet also requires the development of a computational modeling approach capable of describing each individual particle in the powder bed and the collective rearrangement and deformation of the particles that result in a compacted specimen. To this end, a particle mechanics approach for granular systems under high confinement [18,19] was developed and it has been used to effectively predict the microstructure evolution during die compaction of elastic spherical particles up to relative densities close to one [14,15,20,21]. Fig. 8.1 shows three-dimensional particle mechanics static calculations of die compaction, unloading, and ejection of elastoplastic powders with bonding strength.

3. Characterization of compaction operations

Typically, characterization of the compaction operation requires a significant effort in finding the relationship between the applied compaction pressure and speed, the resulting tablet density, and the relationship between compaction pressure and tensile strength of tablets. Thus, it is desirable to develop a standard procedure that can minimize this effort. Using an instrumented tablet press along with mechanistic modeling and a material database could significantly reduce the compaction characterization processes. Compaction simulators (e.g., MCC's Presster, MCC, East Hanover, NJ) and a recently introduced single punch benchtop tablet press (Gamlen GTP-1, Gamlen Tableting, United Kingdom) are good candidates for small-scale production and research purposes. In addition, mechanistic models discussed earlier in Section 2 can be used to develop robust procedures for characterization of powder compaction.

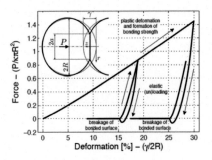 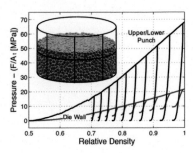

FIGURE 8.1 Left: Generalized loading–unloading contact laws for elastoplastic spheres with bonding strength. Right: Particle mechanics results of punch and die-wall pressures as a function of relative density during the tableting process. A regularization parameter of $\xi_B = 0.01$ has been adopted in the simulations [17].

For powder compaction application, the level of compaction is often measured based on the relative density. Thus, the relationship between the compaction pressure and relative density (i.e., compressibility) is desired. More importantly, the stress path (the stress–strain relationship) during the application of the compaction force is unique for a given material.

Therefore, by accurately reading punch force and displacement, the compression process can be characterized more efficiently. This is possible by placing strain gauges on the compression roll pins to measure force and a linear displacement transducer connected to each punch to measure the displacement relative to the stage. Based on the measured forces and the displacement readings, it is possible to plot the force-powder bed thickness profile during the compaction process, which includes the loading and unloading stages. A large amount of information is gained by studying the powder compaction profile. For example, the work done to compress the powder can be directly computed by calculating the area under the force–displacement curve.

As the loading path for a given powder bed is the same, a single compaction curve is sufficient to establish the relationship between the relative density and compaction pressure. It is preferred if this one compaction profile, that is going to be used for the characterization of compaction, is for a compaction process where the relative density approaches 1 (i.e., zero porosity). Fig. 8.2 shows the compaction pressure versus the relative density for different levels of compaction for lactose powder (spray-dried mixture of crystalline and amorphous lactose, Foremost Farms, Baraboo, WI). The compaction curve at higher level of compaction includes the entire compaction curve of the lower levels of compaction during the loading stage. In this instance, the compaction curve at the higher level of compaction is the only curve that can be evaluated to determine the material properties.

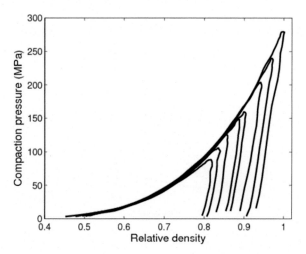

FIGURE 8.2 Compaction pressure versus relative density for lactose powder.

Once the compaction curve is extracted from an experiment, the mechanical parameters can be determined based on the compaction equations, such as the Kawakita model [22], or the mechanistic model discussed in Section 2. The Kawakita model requires only fitting the compaction curve to determine the model parameters. Although extracting the Kawakita parameters is easy and much faster than the mechanistic model, the Kawakita model works only for the loading stage. Previously, Yohannes et al. [14] have shown that the loading stage is mostly influenced by the plastic properties of the particles, while the unloading stage is influenced by the elastic and surface bonding properties of the particles. Therefore, the parameters for the Kawakita model are mostly related to the plastic behavior of the powders. Hence, the mechanistic model [14,15] is recommended for characterizing the compaction processes as it provides more information about the particle properties.

Determining the mechanical properties relevant to powder compaction using the mechanistic model involves two major steps [14]. The first step is determining the plastic properties based on one experimental compaction curve. As the plastic properties are not known a priori, initially the plastic properties are guessed. From the simulation, the compaction versus relative density plot is compared to that of the experiment. If the curves are not similar, which is mostly the case for the first trial, the plastic properties are adjusted and a second simulation is run. The simulations are repeated until the compaction curve from the simulation is close enough to that of the experiment. Fig. 8.3 shows several simulations attempted to get the compaction curve for the lactose powder (spray-dried mixture of crystalline and

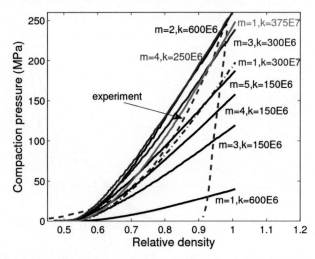

FIGURE 8.3 Compaction pressure versus relative density curve. Several simulations were run to determine the plastic parameter of lactose powder. Least mean square method is used to determine the best fitting plastic parameters [14].

amorphous lactose, Foremost Farms, Baraboo, WI). Performing several simulations to determine the plastic parameters for one powder may seem tedious, especially if one intends to characterize the properties of several powders. However, the results (compaction pressure versus relative density data) of all simulations can be stored and used in the future to quickly estimate the properties of other materials. With sufficient data from earlier simulations, only a couple of additional simulations will be needed to determine the plastic properties of the powders.

The second step in the characterization technique is determining the elastic and bonding properties of the powder. As in the first step, this requires several simulations. The plastic properties determined during the loading stage are used for the unloading simulations. The best fitting elastic and bonding properties are determined using a similar procedure as in the first step. Fig. 8.4 shows the comparison of the unloading stage of the simulations and experiment. The simulation results can be saved to the database for the purposes of determining elastic and bonding properties of other powders.

As shown in Fig. 8.4, the plastic, elastic, and bonding properties determined based on simulations at high compaction pressure can be used to predict the loading and unloading at lower compaction pressures. The benefit of this procedure is further demonstrated in its applicability to predict the tensile strength of tablets [14]. Therefore, if several simulations are available in the database, the compaction behavior and the tensile strength of the tablets can be quickly and easily determined. Overall, the compactability behavior of a powder, which is described with four parameters ($m, k, E,$ and ω), can be added to the material characterization database as discussed in Chapter 3. In addition, the database can be expanded by adding other material properties such as friction and viscosity to account for lubrication and rate-dependent properties, respectively.

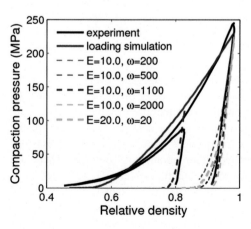

FIGURE 8.4 Compaction curve of the experiment and simulations, which are run to determine the elastic and bonding properties of the powder. The best fitting elastic and bonding properties are determined using least mean square method. Once the plastic, elastic, and bonding parameters are determined for high compaction pressure, the simulations can predict the loading and unloading at lower compaction pressure [14].

4. Characterization of tablets in continuous manufacturing

Before a tablet is released into the market, it needs to meet certain predefined specifications emphasized by the Food and Drug Administration (FDA) called the critical quality attributes (CQAs) [23]. The CQAs for a tablet are typically defined as its weight, hardness, drug potency and potency variability (respectively, "assay" and "content uniformity"), and dissolution profile. Thus, tablets are usually sampled from a batch and are subjected to a set of tests to measure each of these CQAs. Traditionally, and as is still often the case, a small number of tablets are sampled and tested offline for their CQAs in an analytical lab containing benchtop devices. In recent years, post-tableting devices that house some of these tests can be found at-line. Such a device can be integrated fully into the in-process control system. Devices such as the TANDEM from Bruker, AT-4 and HT-100 from SOTAX, and Checkmaster from Fette have some or all of the following tests performed on tablets: weight, hardness, thickness, diameter or length, and chemical composition using near-infrared (NIR) spectroscopy. The at-line testing of tablets has proved to be enabler of continuous manufacturing and real-time release testing.

Over a decade ago, nondestructive techniques such as NIR spectroscopy, NIR chemical imaging, Raman spectroscopy, ultrasound (US) and photo-acoustic measurements, and terahertz pulsed imaging were explored and studied to predict CQAs. As there is an inherent probability of tablet-to-tablet variability (even for tablets ejected consecutively), using nondestructive methods permits gathering information about all the CQAs on a single tablet. This helps examination of the relationship between different measurements and improves the existing models, which can then be used as soft sensors for some of the same CQAs. In all of the aforementioned devices, the hardness test involves calculating the breaking force, which inevitably results in the tablet getting destroyed, rendering it unusable for any additional tests, if required. This latter aspect also causes it to be unavailable for the final attribute measurement—tablet dissolution, which is also destructive. Thus, alternative methods for hardness and dissolution that avoid destruction of the tablet are highly desirable.

Moreover, in real-time release (RTR) testing, and even more so for continuous process control, it is crucial to measure or predict the tablet CQAs quickly. Faster data collection allows a larger number of tablets to be tested, enhancing quality assurance. The time spent for a tablet to undergo all the designated tests in the existing at-line devices is in the order of several minutes, which is not suitable for an online monitoring process. There are different means to gather information about the tablet CQAs: direct measurement, indirect measurement, or manipulation of tablet press response parameters. There are some simple measuring techniques that can be used to directly measure the property of interest. For example, laser triangulation is a fast, noncontact, easy-to-use method to directly measure tablet thickness and diameter. For some CQAs, such as tablet weight, there may be no need to use any measuring tool, given that there is a strong relationship between the precompression force, or precompression thickness, which are readings of the tablet press, and the tablet weight [24]. All of the other tests that involve measuring secondary properties need to be converted to give the primary property of interest. Thus, predictive modeling is an important area in quality control strategy. In the next sections, models to predict chemical composition, hardness, and dissolution profiles are discussed briefly.

4.1 Models for composition

NIR spectroscopy can be utilized to determine the composition of a tablet. Multivariate data analysis tools are utilized to convert the raw spectral data into quantifiable value. This transformation will include building statistical models to get amounts of some or all ingredients in the formulation. The models should be constructed offline and evaluated using well-known metrics before being applied as an at-line or an online measuring system. Often, an NIR sensor is placed on a pipe attached to the tablet press, measuring the powder blend before it enters the feed frame of the tablet press. Therefore, if there are inconsistencies in the composition of tablets, the cause should be examined in the tablet press; for example, whether particle size segregation is occurring in the feed frame [25]. Also, in some cases, the NIR measurements in the feed chute can be inaccurate, either if the composition near the pipe walls is different than in the bulk of the blend (far away from the sensor) or if the sensor is being coated by some blend ingredients. To circumvent these problems, several researchers have also placed the NIR sensors within the feed frame, which enables the measurement of powder composition immediately before it enters the tablet die [26,27]. The measurement has proven to be an excellent predictor of content uniformity of the final tablets.

4.2 Models for hardness prediction

4.2.1 Ultrasound testing

US testing is a rapid and nondestructive technique to gain information about the microstructure and mechanical state of tablets [28−30]. US testing can be used as an online compaction monitoring tool [31−33] or as an at-line tool to characterize the mechanical properties of tablets, specifically hardness of tablets. For the out-of-die measurements, the tablet is placed in direct contact between two piezoelectric transducers. A square electrical pulse from the pulser/receiver unit is launched into the transmitting transducer. The electrical signal is converted into US and propagates through the tablet as mechanical waves and is captured by the receiving transducer at the other end and digitized as a waveform by an oscilloscope.

Among US testing vast applications, measuring the time of flight (TOF) of the sound wave propagating in a tablet can be used to predict its hardness. The Young's modulus, E, can be extracted by measuring the longitudinal speed of sound (c) of this transmitted US signal for a known tablet thickness (D) according to Eq. (8.1).

$$c = \sqrt{\frac{E}{\rho}} = \frac{D}{ToF} \tag{8.1}$$

TOF, c, and E values of each tablet are stored. A model is needed to predict hardness based on the US data. This model is typically prepared and evaluated for a certain formulation before being used as an at-line measuring tool.

A limited number of tablets (~ 20 tablets) are sufficient to construct the hardness model. First, US testing is carried out to evaluate Young's modulus and then, on the same tablets, destructive hardness measurements are conducted. If the shape of the tablets is relatively simple (e.g., cylindrical tablets), the breaking force (hardness) is translated to tensile strength, which is the characteristic of the material. For some odd shaped tablets, there does not exist an analytical solution to calculate tensile strength, and breaking force values are used instead. However, understanding the diametrical stress state in the tablet can be explored using numerical analysis.

There is a one-to-one relationship between these two aforementioned measurements. The Young's modulus and the tensile strength are both functions of tablet porosity, and this dependency is different for each. The experimental data are fitted to Eqs. (8.2) [34] and 8.3 [35]. There are other existing models describing E and σ_t as a function of their zero-porosity values.

$$E = E_0 \left(1 - \left(\frac{1}{\phi_{c,E}} \right) \phi \right) \tag{8.2}$$

$$\sigma_t = \sigma_0 \left[1 - \left(\frac{\phi}{\phi_{c,\sigma_t}} \right) e^{(\phi_c - \phi)} \right] \tag{8.3}$$

E_0, σ_0, $\phi_{c,E}$, and ϕ_{c,σ_t} are the coefficient parameters. E_0 and σ_0 are Young's modulus and tensile strength at zero porosity, respectively. $\phi_{c,E}$ and ϕ_{c,σ_t} are the critical porosities at which E and σ_t vanish. The model can be improved by studying the effects of different processing parameters. For example, the lubricant mixing time or the lubricant concentration impacts the mechanical properties of the tablets [30,36]. Different E_0 and σ_0 are found for each condition. In order to relate these two properties, a model between E_0 and σ_0 needs to be elucidated.

4.2.2 Infrared thermography

Compaction of tablet from powder is always accompanied by the irreversible conversion of mechanical work of compaction into heat. Heat is generated by friction between powder particles, particles and the die wall, plastic deformation of particles, bonding, and other irreversible processes. The resulting temperature increase could significantly affect tablet performance (e.g., mechanical properties, disintegration times, and drug release profiles).

The primary source of infrared (IR) radiation is heat. IR signatures are picked up by utilizing IR thermography as a nondestructive and noncontact tool that allows accurate tablet temperature fields acquisition in real time during tablet compaction from the tablet press in laboratory experiments. The technique can also be implemented as in-line PAT tool for quality control.

The IR technique can discriminate between similar compositions of tablets produced at similar compaction force but experienced different shear strain conditions (Fig. 8.5). From the IR measurement, a clear correlation between tablet's rate of cooling right after its ejection from tablet press, tablet tensile strength, and relative density is found, which can be applied toward predicting tensile strength of the tablet [37]. The procedure is similar to what was discussed in the previous section (Section 4.2.1). Compared to US testing, IR thermography takes longer time for measurement and data acquisition.

4.2.3 Models for dissolution prediction

Dissolution tests are prolonged and tedious, and thus in CM, it is impossible to use them for real-time control. Tablets need to be subjected to dissolution measurements off-line. However, recently, work has been performed to predict dissolution behavior of tablets using nondestructive, namely, NIR methods.

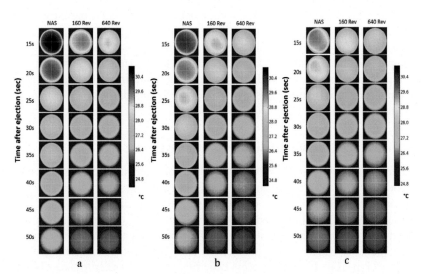

FIGURE 8.5 Infrared images of tablets with similar formulation but subjected to different shear strains (no additional shear [NAS], 160 Rev sheared, and 640 Rev sheared) compressed at (A) 24 ± 0.5 kN, (B) 20 ± 0.5 kN, and (C) 16 ± 0.5 kN.

The method involves building a predictive model offline, but once predictive methodologies are built, the dissolution performance of a tablet can be predicted using an at-line nondestructive measurement. Details of these methods have been discussed in Chapter 10.

5. Control

Most of the tablet presses commercially available are based on the real-time feedback control of the main compression force. It is normally assumed that, for a desired tablet weight and hardness, a specific main compression force is needed, and thus, the variation in the main compression force is a primary indication for the variation in the tablet weight and hardness. For many (but not all) formulations, the variation in the tablet hardness is one of the main causes of variation in the tablet dissolution and disintegration time, and therefore it affects the real-time release of the drug components and its effectiveness.

5.1 Inbuilt tablet press control strategy

Most modern, commercially available tablet presses include an inbuilt control system. The basic control loop involves the direct control of the main compression force through manipulation of fill depth. If a tablet weight and hardness measurement tool is available, two other control loops can be applied in addition to the previously described loop. The first one controls tablet weight and main compression force simultaneously through adjustments in tablet thickness and fill depth. The second one adjusts the tablet hardness while simultaneously controlling tablet weight and main compression force.

This control strategy usually does not take into account the material being compressed and does not allow the user to tune the controller, which can result in a poor performance. For this reason, it might be desirable to implement an external advanced control system. This implementation is described in the following section.

5.2 Advanced model predictive control system

Model predictive control (MPC) is a closed loop optimization-based control algorithm. It is an effective and proven strategy that has been widely used in process industries such as oil refining, bulk chemical production, and aerodynamics [38,39]. The advantages of using MPC over other controllers are that it can be easily adjusted to handle the complex process dynamics; it can efficiently handle strong interactions among the process variables; it can easily compensate for large process dead time; and it is easier to tune [38,40]. Several articles describing the development and implementation of this algorithm in a continuous pharmaceutical manufacturing setting [41,42] show the improved performance of the MPC. However, the increased efficiency comes at a price: MPC implementation is computationally more expensive and complex in comparison to a traditional PID (proportional, integral, derivative) controller.

5.3 Design of an advanced model predictive control system for a tablet press

Dynamic models play a major role in process dynamics and control. When designing and implementing a control strategy for any process, it is highly desirable to have a control-relevant model. These models can be useful to improve the understanding of the process, train operating personal through plant simulations, develop a control strategy for the process, and optimize operating conditions [43]. The models used in process control must be able to represent the dynamics of the process while requiring relatively low computational power, enabling fast execution of the models, when compared to traditional models used for powder dynamic simulations, such as population balance models and discrete element models. In operations such as tablet compaction, where general relations between variables can be easily grasped, the dynamic model of choice is usually semi-empirical in nature. This class of dynamic models is based on a mix of first principles and fitted parameters based on experimental data.

The first step in developing a control-relevant model is to determine the general interactions between the process parameters and CQAs. This relation can be determined from previous process knowledge and understanding, historical process data, or a global sensitivity analysis, where the changes are applied on the actuators and variations in critical quality attributes are observed. Once these interactions are understood, the overall control structure of the model is derived. Step change experiments are then conducted to obtain dynamic process data and the individual parameters of the model are then fitted. The fitted model is validated against experimental data, and if the validation is successful, the model is ready for use. The modeling process is an adaptive iterative procedure; models can always be expanded and improved as new process data become available.

It is important to consider that the compaction process has variables, such as the compression forces, that are characterized by nonlinear behavior. When possible, nonlinearities should be captured by the control relevant model, as they can heavily affect the performance of a controller.

Models can be a powerful tool in the development of control systems, as they allow for rapid prototyping, tuning, and evaluation of the designed control system. Multiple studies such as dynamic sensitivity analysis, comparison between different control algorithms and strategies, and system stability analysis can be conducted if a model is available. A complete example, from model development to application of the developed model for control strategy design and evaluation, has been presented by Barros et al. [42].

The design of a control strategy starts with the definition of which variables need to be controlled. Not all variables can be controlled simultaneously, so appropriately defining the controlled variables is essential for the controller performance. In the case of the tablet press, the variables that are available to be controlled are precompression and main compression forces (or displacements), tablet weight, and breaking force (hardness). Once the controlled variables have been defined, it is necessary to determine the actuators for each controller variable. Ideally, actuators should be selected to minimize the interaction between controlled variables and avoid nonlinearities. If this is not feasible, these interactions can be resolved by the controller itself. An example of possible control strategies, as well as the controlled variable/actuator pairs for a tablet compaction operation can be seen in the control superstructure presented in Fig. 8.6.

The next step in the design of a control system is to select the control algorithms to be used in the system (PID, ratio control, MPC). Regardless of the implemented control algorithm, the operational point (conditions at which the controller is tuned) also needs to be defined. This is specifically important for nonlinear systems. Once all the final steps have been fulfilled, the controllers can be tuned based on the dynamic process model. The control strategy is then implemented in the actual plant and fine-tuning of the controller can be conducted.

Set point tracking and disturbance rejection experiments are used to challenge the designed control algorithm. Standard controller performance metrics, such as integral squared error, integral absolute error, and integral time-weighted absolute error, can be used in the performance evaluation of the controllers.

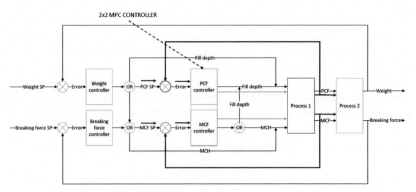

FIGURE 8.6 Tablet press control superstructure. *Adopted from Liu JF, et al. Real-time in-die compaction monitoring of dry-coated tablets. Int J Pharm 2011;414(1−2):171−178.*

5.4 Implementation of advanced model predictive control system into tablet press

The integration of the software and hardware is the first step in implementing the control system. In the case of direct compaction, the manufacturing process can be integrated to the control system using various communication protocols. In the works described here, the tablet press was integrated with the control platform using OPC (OLE process control). The important subsequent step is the tuning of the chosen control algorithm for the control of the process. In this case, MPC on the control platform was developed and tuned through step tests. These tests were manually generated to ensure that adequate data were obtained. The MPC model was autogenerated using the control platform.

The tuned MPC is then used for testing the control strategy. The performance evaluation can be conducted by checking whether the controller tracks the set point efficiently and if the disturbances are rejected by appropriate manipulations of the process parameters. One example has been highlighted in Fig. 8.6, and the performance of this control strategy has been illustrated in Fig. 8.7, where the precompression force and main compression force have been simultaneously controlled. The manipulating variables are the main compression height and the fill depth, respectively. This strategy relies on the control of the main compression force to indirectly control the hardness of a tablet and the control of precompression force to control the tablet weight. One of the advantages of the proposed control strategy is that it decouples the tablet hardness and weight control and therefore both CQAs can be (indirectly) controlled independently. As shown in Fig. 8.6, the set point of the main compression force has been changed while keeping the precompression force set point constant. The controller is able to control both variables around the set point. Subsequently, the precompression force set point has been changed while keeping the main compression force set

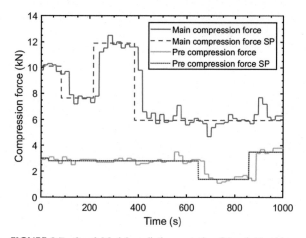

FIGURE 8.7   2 × 2 Model predictive control − Set point tracking.

point constant. The result shows that the implemented advanced MPC system performs well. Further details of advanced MPC system implementation into a tablet press and its performance evaluation are reported elsewhere [41].

5.5 Supervisory control system to integrate tablet press with CM line

Supervisory control systems are also needed in order to integrate the tablet press with the remaining unit operations in the continuous manufacturing process. The desired supervisory control system has been described in Chapter [cross-reference Integrated control chapter]. One of the supervisory control loops essential for continuous manufacturing is the chute (transfer pipe) powder level control loop. This control loop acts as a bridge between inlet and outlet flow rates of the line and is essential to avoid the scenarios where the tablet press would be overfilled with materials or would be running empty. An MPC system has been implemented into the direct compaction tablet manufacturing process for powder level control [44]. The implementation procedure has been described in Chapter [cross-reference Integrated control chapter].

6. Designing an experimental plan for continuous tableting

When dealing with a new formulation, we have to consider both the material properties and process parameters affecting the integrated process control and the final properties of tablets. Thus, establishing an effective experimental design is crucial in identifying the significant parameters and optimizing the process. For a thorough analysis, the experimental plan consists of two phases. Phase 1, which is called the screening design of experiment (DOE), is to identify the most significant factors affecting process and product quality. This DOE, which consists of many factors (e.g., blender speed, turret speed, feed frame speed, mill screen size, fill depth, etc.), is not designed to build predictive models, as each factor has only two levels. In phase 2, an optimization DOE is conducted with a smaller number of factors but additional levels to investigate the relationship between the response variables and input parameters. These models, when validated, can then be used as soft sensors to predict the CQA. Ideally, empirical models that are dependent on material properties and need calibration for every new formulation should, over time, be replaced with mechanistic and first-principle models.

7. Conclusions

This chapter focused on compaction integration, modeling and characterization using simulations, in-line/at-line tablet characterization, and control system used for the continuous tableting stage. Simulation of the tableting process is of great value, which helps to understand the behavior of processing parameters (i.e., tooling properties, lubrication, and compaction kinematics) on the performance of the product enabling product engineering.

References

[1]   Sinka IC, Schneider LCR, Cocks ACF. Measurement of the flow properties of powders with special reference to die fill. Int J Pharm 2004;280(1–2):27–38.

[2]   Mendez R, Muzzio F, Velazquez C. Study of the effects of feed frames on powder blend properties during the filling of tablet press dies. Powder Technol 2010;200(3):105–16.

[3]   Ketterhagen WR, Hancock BC. Optimizing the design of eccentric feed hoppers for tablet presses using DEM. Comput Chem Eng 2010;34(7):1072–81.

[4]   Mehrotra A, et al. A modeling approach for understanding effects of powder flow properties on tablet weight variability. Powder Technol 2009;188(3):295–300.

[5]   Sinka I, et al. The effect of processing parameters on pharmaceutical tablet properties. Powder Technol 2009;189(2):276–84.

[6]   Wu C-Y, Cocks A. Flow behaviour of powders during die filling. Powder Metall 2004;47(2):127–36.

[7]   Boukouvala FV, Niotis R, Ramachandran FJ, Muzzio, Ierapetritou M. An integrated approach for dynamic flowsheet modeling and sensitivity analysis of a continuous tablet manufacturing process. Comput Chem Eng 2012;42(0):30–47.

[8]   Çelik M. Pharmaceutical powder compaction technology. CRC Press; 2016.

[9]   Alderborn G, Nystrom C. Pharmaceutical powder compaction technology. Marcel Dekker, Inc; 1996.

[10]   Rumpf H. Basic principles and methods of granulation. III. Survey of technical granulation processes. Chem Ing Tech 1958;30:329–36.

[11]   Duberg M, Nyström C. Studies on direct compression of tablets XII. The consolidation and bonding properties of some pharmaceutical compounds and their mixtures with Avicel 105. Int J Pharm Tech Prod Manuf 1985;6(2):17–25.

[12]   Storåkers B. Local contact behaviour of viscoplastic particles. In: IUTAM symposium on mechanics of granular and porous materials. Springer; 1997.

[13]   Storåkers B, Biwa S, Larsson P-L. Similarity analysis of inelastic contact. Int J Solids Struct 1997;34(24):3061–83.

[14]   Yohannes B, et al. Evolution of the microstructure during the process of consolidation and bonding in soft granular solids. Int J Pharm 2016;503(1–2):68–77.

[15]   Yohannes B, et al. Discrete particle modeling and micromechanical characterization of bilayer tablet compaction. Int J Pharm 2017;529(1–2):597–607.

[16]   Mesarovic SD, Johnson K. Adhesive contact of elastic–plastic spheres. J Mech Phys Solids 2000;48(10):2009–33.

[17]   Gonzalez M. Generalized loading-unloading contact laws for elasto-plastic spheres with bonding strength. J Mech Phys Solids 2018;122:633–56.

[18]   Gonzalez M, Cuitino AM. A nonlocal contact formulation for confined granular systems. J Mech Phys Solids 2012;60(2):333–50.

[19]   Gonzalez M, Cuitino AM. Microstructure evolution of compressible granular systems under large deformations. J Mech Phys Solids 2016;93:44–56.

[20]   Yohannes B, Gonzalez M, Cuitiño AM. Discrete numerical simulations of the strength and microstructure evolution during compaction of layered granular solids. In: From microstructure investigations to multiscale modeling: bridging the gap; 2017. p. 123–41.

[21] Gonzalez M, et al. Statistical characterization of microstructure evolution during compaction of granular systems composed of spheres with hardening plastic behavior. Mech Res Commun 2018;92:131−6.

[22] Kawakita K, Lüdde K-H. Some considerations on powder compression equations. Powder Technol 1971;4(2):61−8.

[23] ICH. ICH harmonised tripartite guideline pharmaceutical development Q8(R2). 2008. www.ich.org.

[24] GEA. 2015. Available from: https://www.gea.com/en/binaries/pharma-tablet-compression-brochure-2015-05-EN_tcm11-25547.pdf.

[25] Mateo-Ortiz D, Muzzio FJ, Mendez R. Particle size segregation promoted by powder flow in confined space: the die filling process case. Powder Technol 2014;262:215−22.

[26] Mateo-Ortiz D, et al. Analysis of powder phenomena inside a Fette 3090 feed frame using in-line NIR spectroscopy. J Pharm Biomed Anal 2014;100:40−9.

[27] Sierra-Vega NO, et al. *In* line monitoring of the powder flow behavior and drug content in a Fette 3090 feed frame at different operating conditions using Near Infrared spectroscopy. J Pharm Biomed Anal 2018;154:384−96.

[28] Akseli I, Hancock BC, Cetinkaya C. Non-destructive determination of anisotropic mechanical properties of pharmaceutical solid dosage forms. Int J Pharm 2009;377(1−2):35−44.

[29] Simonaho SP, et al. Ultrasound transmission measurements for tensile strength evaluation of tablets. Int J Pharm 2011;409(1−2):104−10.

[30] Razavi SM, et al. Toward predicting tensile strength of pharmaceutical tablets by ultrasound measurement in continuous manufacturing. Int J Pharm 2016;507(1−2):83−9.

[31] Akseli I, Libordi C, Cetinkaya C. Real-time acoustic elastic property monitoring of compacts during compaction. J Pharm Innov 2008;3(2):134−40.

[32] Leskinen JTT, et al. In-line ultrasound measurement system for detecting tablet integrity. Int J Pharm 2010;400(1−2):104−13.

[33] Liu JF, et al. Real-time in-die compaction monitoring of dry-coated tablets. Int J Pharm 2011;414(1−2):171−8.

[34] Rossi R. Prediction of the elastic moduli of composites. J Am Ceram Soc 1968;51(8):433−40.

[35] Kuentz M, Leuenberger H. A new model for the hardness of a compacted particle system, applied to tablets of pharmaceutical polymers. Powder Technol 2000;111(1−2):145−53.

[36] Razavi SM, Gonzalez M, Cuitino AM. Quantification of lubrication and particle size distribution effects on tensile strength and stiffness of tablets. Powder Technol 2018;336:360−74.

[37] Lee HP, Gulak Y, Cuitino AM. Transient temperature monitoring of pharmaceutical tablets during compaction using infrared thermography. AAPS PharmSciTech 2018:1−8.

[38] Singh R, Ierapetritou M, Ramachandran R. System-wide hybrid MPC-PID control of a continuous pharmaceutical tablet manufacturing process via direct compaction. Eur J Pharm Biopharm 2013;85(3 Pt B):1164−82.

[39] PhRMA, Seborg D, Edgar T, Mellichamp D. Process Dynamics and Control. 2nd ed. Wiley, 2016; 2004 Retrieved from: http://phrma-docs.phrma.org/sites/default/files/pdf/biopharmaceutical-industry-profile.pdf.

[40] Garcia CE, Prett DM, Morari M. Model predictive control - theory and practice - a survey. Automatica 1989;25(3):335−48.

[41] Bhaskar A, Barros FN, Singh R. Development and implementation of an advanced model predictive control system into continuous pharmaceutical tablet compaction process. Int J Pharm 2017;534(1–2):159–78.

[42] Nunes de Barros F, Bhaskar A, Singh R. A validated model for design and evaluation of control architectures for a continuous tablet compaction process. Processes 2017;5(4):76.

[43] Seborg D, Edgar T, Mellichamp D. Process dynamics and control. John Wiley & Sons. Inc.; 2004. p. 312–3.

[44] Singh R. A novel continuous pharmaceutical manufacturing pilot-plant: advanced model predictive control. Pharma 2017;28:58–62.

Chapter 9

Continuous film coating within continuous oral solid dose manufacturing

Oliver Nohynek

Driam USA Inc., Coating Technology, Spartanburg, SC, United States

Film coating is usually the final step in the manufacture of a tablet in an end-to-end continuous manufacturing (CM) system. In the continuous film coating process, uncoated tablets enter a coating drum as coated tablets are discharged. Various continuous coating systems are available to meet the requirements of different applications and processes. This chapter presents details about various continuous coating applications, the types of continuous coating systems used, and the aspects that should be considered in selecting or designing a continuous coating process.

In addition to providing an overview of continuous coating technologies, this chapter focuses on the key considerations involved in ensuring the coating system will attain a drug product's critical quality attributes (CQAs). It also discusses the various critical process parameters (CPPs) that must be considered in the continuous coating process in order to attain these CQAs. CPPs can be controlled at various levels, depending on the complexity of the continuous coating system.

The chapter also discusses the pros and cons of continuous coating versus batch coating in order to evaluate the different possible approaches when implementing a coating process into a solid dose CM system. While continuous coating is more complex than batch operations, it offers major advantages. This chapter outlines these advantages and provides a glimpse into the future as coating technology continues to evolve.

1. Fundamentals of continuous coating within continuous manufacturing

Since the FDA began encouraging the pharma industry to adopt CM in 2004, the topic has generated substantial study and discussion. The goal of applying

How to Design and Implement Powder-to-Tablet Continuous Manufacturing Systems
https://doi.org/10.1016/B978-0-12-813479-5.00001-X

CM to pharmaceuticals—as expressed by Janet Woodcock, director of FDA's *CDER*—*is to improve the industry's agility and flexibility. "Continuous manufacturing will permit increased production volume without the current problems related to scale-up"* and *"may have a big payoff"* if it enables manufacturers to meet market needs—changes to product recipes or increases in capacity—more quickly [1]. A subcommittee for advanced manufacturing of the national science and technology council of the US presidential office stated this in 2016 [2]. Industry and academic professionals have cited additional benefits of adopting CM in the pharmaceutical industry over the last two decades, if not earlier.

Case studies and other presentations [2] have shown important and convincing benefits of CM, including

- Integrated processing with fewer steps
- No manual handling for increased safety
- Shorter processing times
- Better efficiency
- Smaller equipment and footprint
- Less loss of material and other resources
- More flexible operation
- Reduced inventory
- Lower capital costs because of less work in progress
- Less energy required

Furthermore, the development of new electronics, sensors, process analytical technology (PAT) systems, and IT, along with better process know-how and understanding have enabled pharmaceutical manufacturers to

- Perform online monitoring and control that increase product quality in real time
- Obtain more consistent product quality
- Succeed in applying a Quality by Design approach
- Add flexibility in recipes, campaign size, and delivery time.

Various continuous concepts, pilot plants, and production solutions have been used in the United States and elsewhere to make active pharmaceutical ingredients and solid dose products. Looking at continuous production of oral solid dose (OSD) forms today, it is possible to find systems where stand-alone equipment handles some process steps, process steps are combined, or manufacturing is completely continuous from powder blending to film coating. Fig. 9.1 illustrates an example of an integrated CM line from loss in weight to continuous coating.

2. Goals of continuous film coating

Film coating is usually the last process step in an OSD manufacturing line of film-coated tablets before they are packaged. Prior to coating, the powder

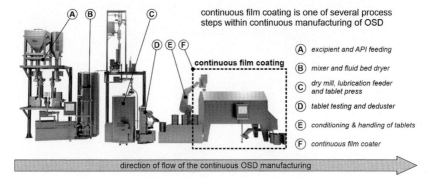

continuous film coating is one of several process steps within continuous manufacturing of OSD

continuous film coating

(A) excipient and API feeding

(B) mixer and fluid bed dryer

(C) dry mill, lubrication feeder and tablet press

(D) tablet testing and deduster

(E) conditioning & handling of tablets

(F) continuous film coater

direction of flow of the continuous OSD manufacturing

FIGURE 9.1   MODCOS continuous oral solid dose line with a continuous film coater. *Image courtesy Glatt Binzen.*

material must be blended, possibly granulated, compacted on a press, and dedusted. Most OSD products are coated to achieve certain properties. There are two main reasons to apply coatings to tablets: protection/esthetics and functionality (Fig. 9.2).

2.1 Cosmetic coatings

These are general purpose coatings that add value to the tablet by improving its robustness against moisture, light, heat, and/or mechanical impact. The coatings also facilitate printing/marking on the surface, which makes the tablet easier to identify. They are easier to pack, too, and they make the tablet easier to swallow. In addition, a tablet coated in blue, red, green, or any other color just looks better and can be identified. These cosmetic coatings are generally formulated to have minimum or no effects on drug release.

2.2 Functional coatings

These coatings affect how the drug product works, typically how the tablet releases its active ingredient(s). Enteric coatings, for example, apply a polymer barrier around the tablet that makes it resistant to specific gastric juices of the human or animal body, delaying release of drugs until the tablet is in the intestine. In this way, formulators can control when and where the tablet will dissolve and how much active ingredient is transferred through the gastrointestinal tract to other areas of the body. These are complex coatings with a specific pharmacokinetic effect.

In any coating that is applied, the process inputs, settings, and results must be tracked and recorded per 21 CFR 210.3, which states that a "lot means a batch, or a specific identified portion of a batch, having uniform character and quality within specified limits; or, in the case of a drug product produced by

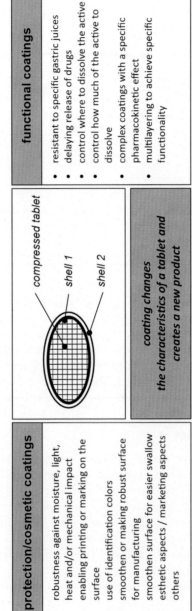

protection/cosmetic coatings

- robustness against moisture, light, heat and/or mechanical impact
- enabling printing or marking on the surface
- use of identification colors
- smoothen or making robust surface for manufacturing
- smoothen surface for easier swallow
- esthetic aspects / marketing aspects
- others

compressed tablet

shell 1

shell 2

coating changes the characteristics of a tablet and creates a new product

functional coatings

- resistant to specific gastric juices
- delaying release of drugs
- control where to dissolve the active
- control how much of the active to dissolve
- complex coatings with a specific pharmacokinetic effect
- multilayering to achieve specific functionality

FIGURE 9.2   Film coating as a protection/cosmetic coating versus functional coating.

FDA – 21 CFR 210.3 Definitions

(2) **Batch means** a specific quantity of a drug or other material that is intended to have uniform character and quality, within specified limits, and is produced according to a single manufacturing order during the **same cycle of manufacture**.

(10) **Lot means** a batch, or a specific identified portion of a batch, having uniform character and quality within specified limits; or, in the case of a drug product produced by continuous process, it is a specific identified amount **produced in a unit of time or quantity** in a manner that assures its having uniform

---

FIGURE 9.3 Definition by the FDA to describe the difference between a batch and a lot.

continuous process, it is a specific identified amount produced in a unit of time or quantity in a manner that assures its having uniform character and quality within specified limits" (Fig. 9.3) [1].

This indicates that as long as the product is within the specified time and quantity limits, it makes no difference whether it was applied to the lot in a batch or in a continuous coating machine. Nonetheless, continuous coating systems have advantages over batch systems depending on the manufacturing strategy and goals.

2.3 Basics of the film coating process

In the realm of film coating, the pharmaceutical industry is largely focused on cosmetic and enteric coatings. Some coatings offer a combination of the two, while others are used to seal or polish certain tablets.

There are various reasons to coat a tablet by depositing a thin, uniform functional film. Successful application of a uniform film coating that completely encloses the tablet in a pleasant-looking shell raises the perceived quality of the final product. A coating that includes a function is a more challenging task. To achieve such a shell, three components must be evaluated (Fig. 9.4): the unique process (recipe), the ideal coating solution/suspension (formulation), and the equipment platform (coater).

2.3.1 The process (recipe)

Film coating can be a complex process. Depending on the goal, the physical characteristics of the final film-coated tablet must remain within narrow limits. Doing so requires that the process steps be defined, worked out, and controlled during the entire process. Once properly set up to work within its defined limits, the process can be validated. The media—the coating solution and drying air—have a big influence on the process, as do the coating machine's mechanical (kinetic) aspects, such as the mixing action of the coating pan on the bed of tablets and the forces applied to the tablet. Processing time and the quality of the starting tablets are also important factors.

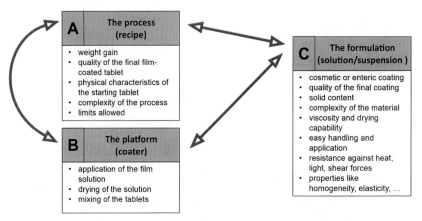

FIGURE 9.4 The process, the platform, and the formulation build are dependent on each other.

2.3.2 The platform (coater)

The coater provides the platform for the three basic steps (Fig. 9.5) of creating a shell around a tablet: (1) application of the film solution onto the tablet, (2) drying the solution on the tablet surface, and (3) the tablet mixing process.

2.3.2.1 Application of the film solution

The shell is formed by spraying a film coating material, which in its basic form comprises polymer, pigments, and plasticizers. The material is sprayed from an array of nozzles above the moving tablet bed. The nozzles, or guns, use

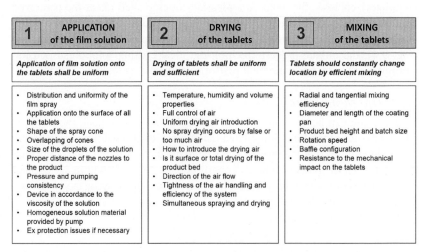

FIGURE 9.5 A proper defined application of the solution, the drying, and the mixing results in perfect coating.

atomizing air at the nozzle tips to disperse a solution delivered at low pressure, or they forego air atomization in favor of high-pressure delivery and dispersion. In both cases, little droplets of the solution impinge on the surface of the tablets as they pass through the spray zone.

The more uniform the spray, the better its distribution of the material. A low-viscosity solution allows fast distribution of the material over the surface of all the tablets. The rate of application depends on the characteristics of the spray formula and how readily the product bed can accept it. This, in turn, hinges on how much tablet surface area within the product bed passes through the cone-shaped spray zone as a function of time and how well the tablets pick up the droplets and the drying and mixing efficiency of the coater.

In addition to creating droplets of a specific size, the nozzles must be the proper distance from the product bed. A nozzle that is too close creates local overwetting. A nozzle too distant may allow the droplets to dry too much before coating the tablets, a result known as spray drying, or to cause the coating to display "orange peel" texture.

The goal is to distribute droplets of uniform size that are wet enough to reach the tablet surface according to the capacity of the product bed and to spread on the surface before drying excessively.

2.3.2.2 Drying the tablets

To create a coating shell, the film solution's solid material must remain on the surface of the tablet. The solvent/liquid carrier must be removed using conditioned process air. This important drying step requires defined and controlled air properties, including temperature, humidity, and volume. Warm and dry air is better able to evaporate and remove humidity from the product than air that is cold and/or humid. The amount of air also has an impact on drying efficiency. Too much air might lead to spray drying or tablet coatings with unequal or rough properties. Too little air allows overwetting and sticking of tablets to each other or to the equipment surfaces, which can also result in defects in the coating. The specific character of the coating material can also affect the process and may call for special air treatment. Moreover, all process air must be filtered before it is used in the process.

How the drying air is introduced to the tablet surface also affects efficiency. Some coaters direct the air only toward the upper surface of the tablet bed, while others can dry across the entire product bed, which is more effective. For the drying air to flow through the product bed, the pan must be the perforated type, which allows the air to move through the bed itself and then to the exhaust area below. Because they can dry tablets uniformly and sufficiently, most of today's coaters use perforated pans.

2.3.2.3 Mixing the tablets

Each tablet in the product bed will be exposed to various conditions during processing according to its location at any given moment. Moreover, the nozzles typically spray the coating solution onto the upper surface of the bed in a nonuniform manner. For these reasons, tablets should constantly change location by flow and efficient mixing. This is a fundamental factor that affects the final uniformity of the coating onto the tablets in the batch. The more uniform the treatment of all tablets, the better the result, typically quantified as a smaller relative standard deviation (RSD) value in tablet weight gain.

Mixing action, in turn, is closely related to the design of the coating pan. Its diameter and length determine the volume of tablets it can process (batch size), while its rotation speed and baffle configuration affect mixing.

Importantly, during scale-up, if a pan's diameter is increased while its length remains constant, the product bed will deepen. This creates a disadvantage because the tablets will circulate in closed-loop patterns within the product bed and experience slow mixing (Fig. 9.6). A deeper bed causes more recirculation, hindering mixing and decreasing uniformity of coating application, and the quality of the product may be adversely affected. This recirculating motion can only be interrupted by baffles, which help to mix the tablets radially and tangentially.

Thus, a shallower product bed is better at mixing and increases the frequency of an individual tablet to be exposed to the spray cone. The more equal the overall treatment of the tablets, the smaller the RSD in coating thickness (i.e., tablet weight gain).

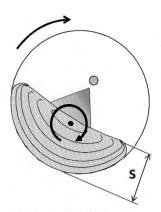

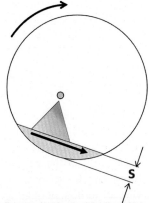

Due to the height of the product bed most batch coaters have a mixing motion of the product were the tablets rotate around a center point

A shallow product bed guides the tablet through the spray cone more frequently which results in better coating uniformity and high quality

FIGURE 9.6 Movement of the tablets in a batch with a deep or shallow product bed.

2.3.3 The formulation (solution/suspension)

2.3.3.1 Solutions for cosmetic film coating

The main goal of cosmetic coating is to create an attractive-looking product. All the tablets should look the same and therefore the covering of the product should be homogenous. Ideally, the coating solution will provide a shell that is stable even when exposed to heat, light, or moisture. A cosmetic coating should have no pharmacologic activity, should be easy to apply, and should be elastic enough that it does not crack. The shell should also be identifiable and printable.

2.3.3.2 Solutions and suspensions for enteric coating

As discussed above, enteric coatings protect the product against gastric fluids (and in some cases, protect the stomach from damage by some drugs) and enable formulators to achieve a defined release profile for the active component(s). While they perform differently from cosmetic coatings, they are applied similarly using standard pumps, mixers, spray bars, etc. Similar to cosmetic solutions, enteric coatings must be compatible with the drug substance, have a long shelf-life, and readily accept printing/marking.

Material suppliers have performed substantial development work on optimizing film coating materials. For instance, it has been demonstrated that the process efficiency remains high even when applying coatings that have a solids content as high as 35%, while still achieving excellent uniform tablet appearance. In addition, the coating industry has also developed elastic materials to avoid cracks caused by thermal relaxation of the coated tablets.

3. Expectations of continuous coaters

While CM is standard in other industries, it is a recent development in the pharmaceutical industry, which is only catching up in the last decade. Progress is uneven; while some brand-driven companies are moving quickly toward adopting continuous concepts, other companies lag behind. In the last few years, the leaders in the field have implemented complete continuous OSD lines with and without an integrated film coating process.

3.1 Change in manufacturing strategies

Going from batch to continuous operation is a substantial manufacturing strategy change and it is no different from the film coating process. Transforming production into a continuous process raises many questions about throughput, range of flexibility, and response time when changing to another product or changing the capacity of the coater. Cost of changing or implementing such a strategy is also an issue.

While companies may differ in their coating process strategy, the transition from batch to continuous coating has allowed companies to reap the benefits of being able to produce low- to high-volume products on the same equipment.

The pharmaceutical industry prefers continuous coaters that have the ability to produce various products with reliable consistency. The throughput of such coaters must adapt flexibly to the capacity requirements of upstream and downstream unit operations. All parameters of the continuous coating process must be controlled and monitored in all phases to achieve high-quality products.

3.2 Supporting factors

Various factors have driven the industry further toward continuous production. FDA is currently encouraging companies to adopt CM and supporting their efforts. But even more important than this regulatory support is the ability to use better electronics, sensors, and IT to control the production process constantly, which enables manufacturers to predict, influence, and improve the complete production line and to keep it on track and steady. More sophisticated equipment is helping, as is a much more detailed process understanding.

This trend has been evident for some years now, and therefore many upstream units have been transformed into continuous operation. Even the coating process, which is known to be complex, has received increased attention and at the present time it appears that continuous coating systems will be increasingly integrated into end-to-end continuous lines in the near future.

3.3 Partnering on the continuous manufacturing coating projects

Moving from batch to continuous processing successfully requires a joint effort from many participants. First, the pharmaceutical manufacturer must define and be prepared to implement a CM strategy. This affects the product flow as well as all the components of the production line. But most important is the full understanding of the coating process.

Second, the equipment supplier needs to supply a coater that can serve as the continuous process platform. The coater must be integrated to other process steps and must be a "never-stop" system operating at the same mass flow rate as the preceding process unit (typically the tablet press), unlike a batch coater, that produces a full pan load of tablets as a batch processing step. A continuous coater must generate coated product at a steady rate until the desired amount of product is manufactured. Given the synchronization of the continuous coater with other upstream units of operation, a well-thought-out monitoring and control system must exist to operate the integrated system in a mechanically reliable machine.

Last but not least is the coatings material supplier, whose materials must provide defined and homogeneous characteristics within a small window of allowable variability and must enable the requirement of constantly feeding accurately small amounts of the coating material to the coating process. Ideally, the material is adapted to the coating platform in order to achieve the highest efficiency and performance.

3.4 Special demands of the continuous coating process

Like batch coating, the first concern in continuous coating is the quality of the shell. Furthermore, the needs of the product and process will dominate the design and operation of the process. There is a big difference in applying a cosmetic coating at a weight gain of 1.5%−3.0% and applying coating with complex functional properties at a weight gain of 12%−20%. Both can be achieved in batch and continuous coaters. The big difference is the adjustment of the coating unit to the upstream and downstream components of the continuous OSD line. Essentially, higher weight gain typically requires a lengthier process, and if a single coating pan should accommodate both types of coating process, it must be designed to enable the resulting impact on residence time. Variation in input and output of material within a given processing interval is therefore critical and will affect the design and operation of such a film coater.

The goal is to overcome these challenges and to maintain efficiency and flexibility as capacity needs and recipes change from product to product. This has been achieved in different ways over the last few years. Overall, CM is growing rapidly and is increasingly becoming a recognized and accepted approach. Collaborative efforts from academia, industry, and regulatory organizations are certain to lead to more successful continuous film coating projects in the future.

4. Types of batch and continuous coaters used in continuous processes

As discussed above, the coating process—batch or continuous—is influenced by many factors. They include the coating solution and the method of its application, how the drying air is introduced, and how well the coater as the manufacturing platform handles and mixes the tablets.

The first and greatest benefit of a continuous coater is its ability to "go small and steady." In other words, the amount of coated tablets produced per minute is a small fraction of what a batch coater would produce in a single batch, which would require several hours. This small, continuously produced amount requires precise process control to ensure continuous and accurate dosing and steady tablet treatment within narrow limits.

The second benefit of continuous coating is its flexibility in scale-up. Once the coater starts, it will continue producing coated tablets as needed. The transfer of the coating process from R&D to production scale is seamless.

Last but not least, any error or mishap during continuous coating affects only a smaller amount of product instead of a full batch load. Any part of the continuous product stream that fails to meet quality specifications can be identified and diverted for further investigation.

The major challenges of continuous coating are the greater need for stability of the process, compared to batch production, and the need to synchronize the coating process with processing components upstream and downstream. So the rate of product flow—the amount of tablets coated per unit of time—is critical, as is the mass flow of the coating material and media introduced to the process. The material—the coating suspension—can be applied in smaller doses and therefore the mass additions need to be very precise and steady. While a batch process performs one process step after another on the entire batch, every phase of the continuous process happens simultaneously while the individual tablet is transferred from the start to the end of the process. As the FDA requires, all products in a lot are to be treated in the same way.

The quality of continuously produced tablets is directly linked to throughput. As mentioned, applying a complex coating at a large weight gain will require more residence time compared to a cosmetic coating, and throughput will thus be lower or holdup will be higher. In addition to weight gain, throughput is dependent on the type of film solution applied, RSD required, and desired surface appearance.

To assess which continuous film coating technology suits a given application, consider the following questions:

- What does the recipe require? Is it a cosmetic coating or a functional one, such as an enteric coating? What weight gain is required?
- If an enteric coating is applied, can different coating solutions be applied in a single coating pan and through flow?
- What is the biggest and the smallest throughput needed for a certain product? How much do the recipes differ?
- Is there an overall control strategy involved? Are the parameters for materials handling, process control, and process documentation managed by a central main control system?
- What level of PAT is needed?
- How are the individual components of the recipe applied? Can they be handled in small quantities over time and can they be sufficiently distributed?
- What should the final product look like and how will the quality be assessed? What is the acceptable coating RSD?

- What size are the campaigns or runs?
- Is preconditioning or relaxation of the tablet required? Are the relevant parameters known?

All these questions are easier to answer with batch coaters because they entail fewer interactions with other units of operation than a continuous OSD manufacturing line. In fact, some pharmaceutical manufactures have integreated batch coaters for their continuous production lines. Other options are fully continuous film coaters and hybrid film coaters. Each coater type has advantages and drawbacks depending on the application.

4.1 Traditional "batch" coaters in continuous manufacturing

Some CM lines incorporate traditional noncontinuous film coating units, typically at the end of the manufacturing line before printing or packaging. This reversion back to batch operation from the continuous mode creates a batch prior and after the coating process and eliminates some of the benefits of complete continuous production.

The commercial lines that use batch coating units have either one standard coater or a set of coaters that interact at the end of the line. In either case, when working with this setup, it is important to create the starting batch for the batch film coater first. With product coming continuously from an upstream tablet press or deduster, one or more buffer bins or hoppers are needed to collect and hold the specified amount of product, a "load," which is then taken from the buffers, weighed, registered, and loaded into the batch coater.

At that point, the heretofore continuous process becomes a classic batch coating operation, with all the disadvantages of a conventional noncontinuous operation, including the need for additional product handling and storage.

While one coater is discharged or loaded, the others would be at different process stages, thus smoothening the batch operation to allow the line's truly continuous upstream processes to generate a steady product flow (Fig. 9.7).

Any standard batch coater can perform this process step. The challenge is to design a buffering system that can deal with the different properties of the individual products. They may need time to cool or rest (relax) following compression, for example. The longer the storage period, the more uniform the properties of the starting tablets become. In reality, most tablets do not need a relaxation period to prevent cracks from forming in the tablet shell. While the expansion of the tablet due to the formulation may occur, there are highly elastic coating materials available for handling products that exhibit shape change during relaxation. The length of relaxation time is often unknown to manufacturers switching from batch to continuous operation because relaxation, if any, occurs naturally in the time between batch process steps.

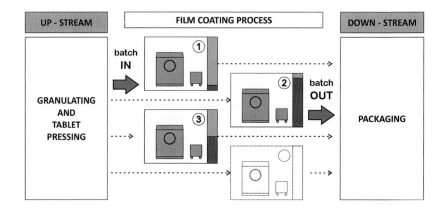

When working with batch coaters in commercial continuous lines the mode has to be reconverted from continuous into batch. A set of interacting coaters as well as product handling and storage capacity is required. Higher throughput likely increases the number of batch coaters handling effort.

FIGURE 9.7   A special situation is the interaction between more than one coater to even out the overall flow in the manufacturing.

If cracked shells are observed, such problems can often be addressed by optimizing the film coating materials. Suppliers of coating materials have developed a variety of formulations with increased flexibility to help minimize such defects.

4.2 The GEA ConsiGma coater

The ConsiGma coater is a unique batch coater, which processes tablets differently from traditional batch coaters and has been used in small and pilot size continuous OSD manufacturing lines. It works in small batches using a pan that processes them very quickly, which makes it feasible to integrate this coater into a continuous line. The coater is relatively small and uses an 18-inch-diameter pan that rotates very quickly at up to 100 rpm (as compared to 4—15 rpm in standard batch coaters) and can coat 1.5—3.0 kg of tablets per load.

In operation, uncoated tablets enter from the top. Once filled, the pan rotates quickly until all the tablets form a ring around outer periphery of the pan due to the centrifugal force of rotation. Once this is achieved, the pan slows its rotation slightly and air knives blow the tablet away from the pan wall to create a cascade or free-flight arc from one side of the pan to the other (Fig. 9.8). As they cascade, the tablets pass through a spray cone generated by nozzles located at the pan's center. At the same time, drying air is introduced through perforations in the pan, thereby simultaneously coating and drying the tablets.

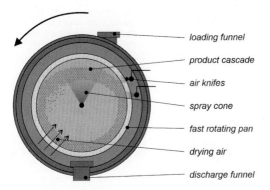

loading funnel

product cascade

air knifes

spray cone

fast rotating pan

drying air

discharge funnel

The GEA-Consigma coater using in small size batches creates a cascade or free-flight arc within the pan while the tablets pass through a spray cone.

FIGURE 9.8 The GEA ConsiGma coater used in small size batches creates a cascade or free-flight arc within the pan while the tablets pass through a spray cone.

The tablets pass through the spray cone very frequently and it is reported that this execution offers a fast batch process that results in coating uniformity.

As the load size of the coater is relatively small, processes designed for high throughputs would require using more than one coating unit. In cases where several coaters are required for commercial-scale manufacturing, they would work in tandem or several units, manufacturing all loads in identical manner [3].

4.3 Classic high-throughput continuous coaters

By definition, truly continuous coaters are fed continuously and at a nearly uniform rate with tablets and film solution. Typically, these coaters use long rotating drums, like tubes, that mix and transport the tablets from one end to the other as they are sprayed with a film coating. All operations occur simultaneously for as long as required to produce the required amount of coated tablets.

All the process steps of coating described earlier take place in consecutive zones along the cylinder axis and the tablets travel from one zone to the next until they are properly coated. The residence time of the tablets in the zones depends on the coater's length and its ability to hold the tablets before they move to the next zone (Fig. 9.9). The longer the pan, the longer the coating process can last.

The process achieves a certain quality, which is defined by the pan's design and its ability to handle the media (coating solution and drying air), as well as its ability to accept a constant feed of tablets. The unique behaviors of various tablet types—how they move within and along the product bed—as well as the

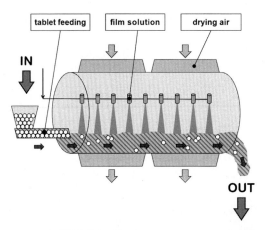

Classic continuous coaters
The tables are feed at one side and travel to the opposite side. While that happens the goal is to treat them in a uniform way to archive an uniform coating.

- As the traveling time and exposure time to the spray of each tablet varies the coating quality varies too
- The ability to hold the tablets for a most equal time in the zones where the spraying and drying happens defines the uniformity and quality
- Processes are used for coatings tend to be of cosmetic or a simple technical nature

FIGURE 9.9 Classic continuous coaters used for specific coating processes.

time they are exposed to the process define the quality and the achievable RSD. Obviously, as the traveling time and exposure time can vary, coating deviations in the tablets can occur and need to be quantified [4].

Due to this potential for deviation, the pan is usually designed to meet basic cosmetic coating demands. Depending on the size of such classical continuous coaters, there are ranges of throughputs from below 100 to over 1000 kg per hour. These coaters are used when manufacturing bulk dietary supplement or some over-the-counter products. The coating quality and the RSD provided are acceptable for products where the coatings tend to be cosmetic or of a simple technical nature [5].

Recent modifications to the pans and processing method have improved the throughput and quality that these machines can provide. For instance, air flow was improved by using perforated pans, as have air shoes that introduce the drying air more efficiently. For some applications, the number of nozzles has been increased to better distribute the coating solution. Suppliers have also experimented with increasing the product load in the pan to boost throughput. This is done by using barriers or locks on each side of the pan that increases the product bed depth. The deeper product bed, however, can decrease tablet exposure to the sprayed solution and thereby adversely affect coating uniformity. The number of spray guns was increased to improve the consistency of the spray [6].

The start-up and shutdown phases of this type of coater create a challenge. During these phases, the balances between the different coating processes vary substantially in the zones. As a result, compromises must be made, and the coater will generate waste material until the process enters a steady and

FIGURE 9.10   A Driaconti-T works with multiple chambers.

reliable process state. The same situation happens at the end of the process. The longer the operation runs, however, the less significant these start-up and shutdown phases become compared to the overall campaign.

4.4 A hybrid: Driaconti-T multichambered continuous coater

A hybrid continuous coater, the Driaconti-T (Fig. 9.10) by Driam, resolves several problems found in batch and traditional continuous systems.

It uses the extended tubular pan of a traditional continuous coater but uses gates between zones to create chambers that allow processing to be controlled similar to a batch coater. Processing, however, occurs simultaneously in all chambers and the type of processing can be different in each chamber. Tablet flow, dosed material, and media are fully controlled in each individual chamber, and processing time can be extended according to the recipe of different products. As a result, the hybrid coater is better at handling complex coatings, big weight gains, and a broad range of coating applications. It is a very flexible continuous coating approach.

The Driaconti-T's segmented pan enables the user to dedicate individual coating processes within a chamber because each has its own spray nozzles and drying functions (Fig. 9.11). The tablets are coated, as a small portion of product passes from one chamber to the next.

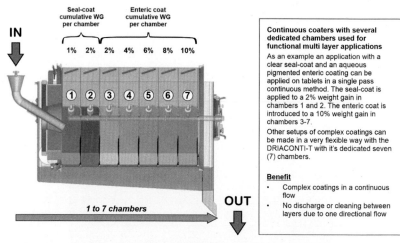

FIGURE 9.11   A Driaconti-T continuous coater with several dedicated chambers allows multi-layer applications.

In operation, all the chambers are separated by gates. When it is time for one portion of tablets to move to the next chamber, all the chamber gates open, which converts the drum into a helix. At this stage, one turn of the pan transfers the entire product load from each chamber to the next. The transfer takes place simultaneously across all the chambers and the product moves from one to the next in a controlled way.

As in standard batch coaters, the entire process in each sectioned chamber is fully controlled as the recipe requires. Thus, the process within each chamber can be customized and dedicated to a specific process step, perhaps applying a different solution or using drying air with different properties. The flexibility of the Driaconti-T's technique allows the application of complex coatings, including enteric, multilayer, and other functional coatings.

The process is time-controlled, so the coating can be prolonged as desired, which allows for large weight gains. Naturally, a higher weight gain requires more processing time and thus reduces throughput. However, tests have shown that using the segmented pan and exercising full control over the process keep the RSD relatively low [3] and result in great process efficiency and superior appearance of the product [7,8].

The Driaconti-T continuous coater is capable of simultaneously applying various film coatings within a single pass continuous coating process [9]. When comparing such a process of applying multiple coatings to a batch coating operation, required intermediate cleanings when switching the coating media are no longer needed.

4.5 Overall comparison

Criteria	Batch coaters	Cascade batch coater	Classic continuous coater	Hybrid continuous coater
Process type	Batch coating	Batch coating	Continuous	Semi-continuous
Cosmetic and/or functional coating applications	Capable of cosmetic and functional coatings	Capable of cosmetic and functional coatings	Capable for cosmetic coatings and basic functional coatings with shorter residence time	Capable of cosmetic and functional coatings. Multilayer complex coatings possible in a single pass
Method of increasing throughput	Based on the working volume of the pan	Based on multiplying the small coating batches	Extend operation time	Extend operation time
Level of controlled process[a]	Control of an entire batch. Center of product bed and outer product flow area are exposed differently	Very frequent presentation of the individual tablets to the spray cone with a high level of controlled processing	Variability of the presentation of the tablet to the spray cone is higher due to product flow	A controlled process using dedicated chambers where individual steps take place
Dedicated zones/sections	When used for complex coatings with different coatings, the batch may need to be discharged when changing coating materials	When used for complex coatings with different coatings, the batch may need to be discharged when changing coating materials	Due to the intermediate tangential mixing in between the overlapping zones, the option for complex coating is limited	Dedicated chambers for individual processes. Ideal for complex and enteric coatings
Start-up and shutdown	No issues; no waste	No issues; no waste	Likely generates some waste until a steady flow is achieved	No issues; no waste

Recipe flexibility	Yes	Yes	Designed to a specific range of products	Yes
Handling in the event of error	The entire batch is impacted for examination	An entire small batch load is impacted for examination	A section including a safety section is to be examined	The content of one chamber is marked and discharged for examination

[a]The level of control describes the capability of processing an individual tablet and is defined by relative standard deviation.

4.6 Considerations for production and other aspects

Converting an existing batch to a continuous OSD production line has a big impact on the manufacturing facility and the way in which the production is managed. This transition requires a change of production philosophy. A significant level of process knowledge and more detailed and careful planning are required for successful implementation. New technologies available in the market are making it easier to implement continuous coating into manufacturing facilities.

There are two main examples of supporting technologies which are driving the growth in continuous film coating projects.

5. Controls and process analytical technology

PAT is helping to drive the development of CM and specifically continuous coating. PAT tools, which entail the sensors and software packages to run the systems, have become very important as they help to investigate, understand, and optimize continuous coating processes.

PAT systems include the use of probes, MSR technology, data collecting software, and statistical analysis programs. Note that the system will not work without the previously mentioned three components of process, formulation, and platform. Typically, various suppliers of hardware and service providers for software and integration are needed for PAT implementation.

5.1 Simulation and modeling of the process

Simulation and modeling of a process is to evaluate and predict the impact of various factors on the coating process performance for a given product. The model is an idealized mathematical data object of a manufacturing process of a specific product.

As an example, a simulation of the kinematics of tablet movement might be beneficial to better understand the overall product motion in a process. When generating a data model to predict the kinematics of certain tablets, the model can become sophisticated, as various parameters have an impact on the simulation. Different sizes and shapes of the tablets, as well as the type of coating materials, its viscosity, surface tension, solid content, and stickiness may lead to varying results.

A computer model of tablet mixing during the coating process can reveal how the overall movement of all the tablets around a center line of the product bed takes place [4,6]. Simulations like this have predicted that a shallow product bed helps achieve a fast and uniform coating with an even distribution of the spray solution on the surface of all the tablets resulting in a low RSD.

Coating tests with tablets in a lab environment are used to prove the results of the computer models. The use of models aids in adjusting processes, running new processes, and predicting results. The data gathered from such coating modeling can be stored and used in future projects resulting in decreased R&D costs and reduced development time. Some examples can be found in Refs. [10,11] .

6. Conclusions

CM is accepted, supported by the FDA, and increasingly established in the pharmaceutical industry. Film coating, as a continuous process, has many benefits if implemented to create a truly continuous end-to-end OSD manufacturing line.

The needs for continuous film coater solutions vary whether it is a cosmetic application that requires a thin shell or a more complex functional coating that likely requires a thicker shell. The continuous coating process is challenging because it must be synchronized with the flow of upstream and downstream processes while achieving the targeted weight gain with the desired properties.

Implementing continuous coating requires genuine process know-how, as well as close working relationships between the pharmaceutical manufacturer, the supplier of the coating materials, and the machine builder, with possible assistance from experts in sensing and PAT technology.

No coating platform prevails over another in every case, as each tablet coating application has unique requirements. Each successful installation of a continuous film coating line must address the needs, goals, expectations, and production strategy of the manufacturer.

References

[1]   Woodcock J. Modernizing pharmaceutical manufacturing — continuous manufacturing as a key enabler. In: MIT-CMAC international symposium on continuous manufacturing of pharmaceuticals, Cambridge, MA; 2014.

[2]   S. F. A. M., Executive Office of the President: National Science and Technology Council. Advanced manufacturing: a snapshot of priority technology areas across the federal goverment. 2016.

[3]   Cunningham C, Birkmire A. Application of a developmental, high productivity film coating in the GEA ConsiGma coater. AAPS 2015 2015.

[4]   Chris N, Cunningham C, Rajabi-Siahboomi A. Evaluation of film coating weight uniformaty, tablet prograssion and tablet transit times in a high throughput continuous coating process. AAPS 2015 2015.

[5]   Neely C, Cunningham C, Rajabi-Siahboomi. Evaluation of film coating weight uniformity, tablet progression and tablet transit time in high through put coating process. 2015.

[6]   Cunningham C, Nuneviller III F, Venczel C, Vilotte F. Evaluation of recent advances in continuous film coating technology in reducing or eliminating potential losses. AAPS 2018 2018.

[7]   Cunningham C, Crönlein J, Nohynek O. Evaluation of a continuous-cycled film coater in applying a high-solids coating formulation. Tablets and Capsules October, 2015.

[8]   Bulletin CT. ontinuous coating performance with Opadrey QX.

[9]   Cunningham C, Krönlein J, Nohynek O, Rajabi-Siahboomi A. Simultaneous application of a two-part delayed release coating in a single pass continunous coating process. 2018.

[10] Suzzi D, Toschkoff GRS, Machold D, Fraser SD, Glasser BJ, G KJ. DEM simulation of continue. Chem Eng Sci 2012;69:107−21.

[11] Boehling P, Toschkoff G, Dreu R, Just S, Kleinbudde P, Funke A, Rehbaum H, Khinast J. Comparison of video analysis and simulation of a drum coating process. Eur J Pharm Sci 2017:72−81.

Further reading

[1]   Chatterjee PS. FDA perspective on continuous manufacturing. In: IFPAC annual meeting, Baltimore MD, USA; 2012.

Chapter 10

Role of process analytical technology in continuous manufacturing

Joseph Medendorp[1], Andrés D. Román-Ospino[2] and Savitha Panikar[3]

[1]Vertex Pharmaceuticals, Boston, MA, United States; [2]Rutgers University, Piscataway, NJ, United States; [3]Drug Product Continuous Manufacturing, Hovione, LLC., East Windsor, NJ, United States

1. Introduction/background

Process analytical technology (PAT) systems share the same purpose in continuous manufacturing (CM) operations as their batch mode counterparts, namely to design, analyze, and control pharmaceutical operations with the goal of ensuring final product quality [1]. PAT may be taken to mean direct spectroscopic measurements, parametric data used for monitoring or controlling the process, or a combination of the two. However, for the purposes of this chapter, PAT will primarily refer to spectroscopic applications. Due to the fundamental differences between continuous and traditional batch manufacturing, namely the continuous charging and discharging of materials in each unit operation, the increased need for PAT in the world of continuous processing is apparent. In order to properly characterize a process for information such as residence time distribution (RTD), the impact of material attributes on flow and the impact of process parameter change on critical quality attributes (CQAs); the level of characterization required is much greater than what can be obtained in practice using offline sample characterization. Online spectroscopic tools provide information that offline analysis is not capable of providing. For example, a specific chemical signature for the analyte(s) of interest can be collected with spectroscopic methods in real time, which can be used quantitatively for content measurements or qualitatively to detect the impact of process changes at a high frequency. Changes to bulk physical properties may also impact spectroscopic measurements, allowing for identification/quantification of the impact of process changes on material attributes and powder flow characteristics in real time, properties that would normally be

How to Design and Implement Powder-to-Tablet Continuous Manufacturing Systems
https://doi.org/10.1016/B978-0-12-813479-5.00005-7
201

unavailable via offline traditional sampling and testing methods. Removing product from the process stream for offline testing is slow, invasive, disruptive to normal operations, and less representative than the online/in-line alternatives. Offline testing is also not practical for use as part of the integrated control strategy required for effective continuous operations; thus, online PAT methods are desirable for process control purposes as well.

There is a wide range of applications where PAT may be and has been used in CM in the last decade. For example, near-infrared (NIR) can be used at various stages of a continuous process to monitor powder blends or core tablets for active pharmaceutical ingredient (API) or moisture content. NIR and Raman can be used to monitor film coat thickness or physical form in core or coated tablets. Both techniques can be used to measure incoming material identification (ID) for excipients or APIs and to confirm material ID for coated or core tablets at the end of the manufacturing process. Laser diffraction methods can be used to measure particle sizes in granulated products. A review article published by Fonteyne et al. [2] describes various PAT sensors and their use in CM applications including continuous material transfer, continuous blending, spray drying, roller compaction, twin-screw granulation, and direct compression. Fonteyne et al. also describe the use of NIR, Raman, and particle size analysis for the assessment of a continuous granulation system [3]. Transmission NIR has been demonstrated for use in monitoring continuously flowing powder for API homogeneity, which effectively provided information about the impact of the sample presentation on partial least squares (PLS) model sensitivity and allowed for an optimized powder flow and sampling design [4]. The research article by Fonteyne et al. [5] demonstrated the combined use of NIR, Raman, and a photometric imaging technique for particle size and surface characterization for the prediction of granules moisture content, tap and bulk density, and material flowability in wet granulated product. Ward et al. presented an NIR application where blend potency is determined in the feed frame of a tablet press, immediately before the compression step [6], which was an easier and faster method of assessing tablet potency than an offline alternative. Järvinen et al. presented a similar application with an NIR sensor in the tablet press, extending the work to include the assessment of both blends and compressed core tablets [7]. Laser diffraction has been utilized for continuous real-time particle size measurements for spray drying operations [8] and could also be used to measure particle sizes of granulated products as an input into dissolution models.

In addition to its utility for process monitoring and control, PAT can also be used to inform batch disposition decisions in tableting operations. PAT-based in-process control (IPC) measurements can be used to develop real-time release testing (RTRt) calculations, allowing for immediate product release upon completion of manufacturing.

2. Method development and life cycle considerations for PAT in CM

Partial least squares regression (PLSR) is the most widely used technique for real-time analysis in the pharmaceutical industry, particularly due to the different applications that can be performed by extracting the latent variables in spectroscopic data. NIR and Raman spectra contain both physical and chemical information, so depending on the application in process monitoring, data pretreatments are required to estimate composition, hardness, dissolution, or bulk density [9—13]. In preparation of PLS models for composition, pre-processing is required because the physical properties in the data are pre-dominant in nondestructive techniques where no preparation of the sample is required, or even possible. Before data acquisition can be performed, standards need to be prepared by modifying the composition of the analyte. PLS looks for covariance in the calibration set and reference values matrix. To generate robust calibration models, correlation in the composition needs to be mini-mized [13,14]. Changes in composition around the target formulation in addition to particle size variation lead to an enormous calibration set to be prepared if a full factorial design is considered. Therefore, more economical experimental design strategies, using design-of-experiment (DOE) methods, need to be implemented to capture the information into a robust calibration model using the lowest possible number of blends. An important aspect is the use of reference values to accompany the spectroscopic measurements because the latter do not measure the attribute directly but rather measure some property of that attribute. The attribute itself is measured using more primary or direct techniques such as the commonly used gravimetric, UV-Vis, HPLC, and GC/LC-MS methods to which these attribute properties need to be tagged to. This means that one has to be very confident about the reference values to avoid mistagging.

Once calibration blends are prepared, the next step is the proper acquisition of spectral data. Particulate systems (pharmaceutical powder blends) are intrinsically heterogeneous and the composition in blends is dependent on the scale of scrutiny. One spectrum for each blend may not be representative of the composition in the entire batch. The reduction of batch dimensionality is required to appropriately collect representative measurements in spectral data. This can be achieved by ensuring the blends (3D sample) are moved as a horizontal bed (1D) or as a vertical fall (1D) (Figs. 10.1—10.3) [15,16]. This process allows representativeness while at the same time connecting two unit operations.

The life cycle of PAT methods is often the focal point of numerous pre-sentations and publications and typically consists of model development, model validation, and model maintenance [17,18]. When model maintenance is required for a model in production, PAT scientists return to some level of model development and revalidation prior to resuming routine production, and

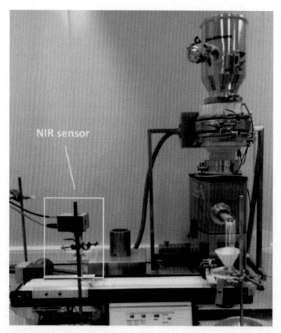

FIGURE 10.1 Conveyor belt setup; arrows indicate direction of powder flow.

1) FT NIR
2) Gravimetric feeder and controller
3) Vibratory feeder

FIGURE 10.2 Vibratory feeder setup.

thus, these life cycle steps form a loop where each step of the loop must be visited and revisited throughout the method life cycle. Special consideration is required for the long-term management of PAT methods as compared with traditional analytical methods due to the high sensitivity of spectroscopy

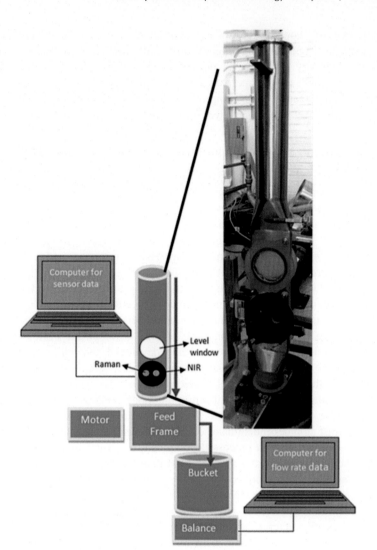

FIGURE 10.3 Chute setup.

methods to process factors and material attributes, particularly in CM processes. Factors such as new material lots or suppliers, changes in process
parameters, intentional or unintentional changes to sample presentation as a
result of material attributes such as particle size and bulk density, each has the
ability to impact PAT measurements, while these same factors are less likely to
impact traditional lab methods. With a formal life cycle management plan,
including routine parallel testing and model maintenance, a spectroscopy-
based method can be evaluated proactively and updated when necessary to

incorporate the new sources of variation that commonly arise in global pharmaceutical operations. While a thorough discussion of the impact of the current regulatory landscape is out of scope for this chapter, it is worth mentioning that as long as spectroscopic PAT applications for CM are still new and rare, maintenance of post-approval submissions and regulatory commitments are still operationally challenging for global pharmaceutical companies.

2.1 Instrument, sampling, reference values, multivariate analysis, sensitivity

Similar to batch manufacturing, PAT model development for in-process monitoring, IPC, and RTRt should all be done concurrently with process development for CM operations. Process development using Quality by Design (QbD) principles offers the widest range of process parameters across all unit operations that may be expected or possible in routine production [19]. In addition, process development using QbD principles is likely to include a number of API and excipient lots, each with ranges of material attributes, in particular when those material attributes are expected to impact a CQA. PAT may not be useful from a control strategy perspective for a number of manufacturing runs at the beginning of process development while data are first being collected and models are being developed and refined, but these runs can serve as the source of spectra and physical samples for subsequent reference method testing. Model development in this context includes data collection, data selection for inclusion in the model, selection of pretreatment and modeling options, model calibration, and testing the performance of the model against an independent data set. Model development is most effectively done in conjunction with process development to ensure that PAT methods operate across the entire design space and that the data necessary for model development have been collected and are available from the process development experiments. Most aspects of model development have been described extensively in a number of guidance documents [17,20]; however, certain development-related topics will be expanded upon here specifically for CM. In particular, model development requires additional considerations beyond development of a stand-alone PAT method, particularly around types of models, sample composition, and process parameter design space ranges.

To take regular samples, a robust in-line monitoring system linked to the integrated control strategy through feedback and feedforward controllers is necessary. Traditional sampling and offline testing, as typically done for a batch process, are challenging in a continuous system, as it only represents a point in time, not the complete "batch," given the continuous flow nature of the system. Even periodic stratified sampling as used for offline testing can be less representative than the in-line alternative, which samples the stream more frequently. For this reason, for CM, in-line analytical testing methods are preferable. While many of the regulatory expectations related to quality for CM are the same as those for batch processing, in continuous processing, sampling considerations will differ from batch and the rationale for testing a continuous batch must be reconciled against the traditional paradigm. A target sampling frequency should be determined based on the combination of process capability and risk. For IPC measurements collected from a unit operation with a high process capability and low probability of affecting a CQA, the target sampling frequency may be lower than for a unit operation with a low process capability and a direct impact on a CQA. Similarly, if an IPC is a required input for a RTRt model, the target sampling rate may be as high as one sample per minute or one sample per kilogram of manufactured material, depending on how a unit is defined in the control strategy. Through a set of designed process shifts or a lag analysis of historical data, target sampling frequency may also be established by assessing the magnitude by which a process could possibly drift between subsequent measurements. For example, if a PAT sensor is unavailable for consecutive measurements of API blend content due to a probe system suitability check, but the process has been shown incapable of drifting beyond 1% API even after 10 consecutive measurements, the target sampling frequency may be established based on the balance of historical performance and intended use.

Instruments should be capable of measuring at scan rates commensurate with operations. If the production line has the capability of operating from 10 to 30 kg/h, PAT sensors must be capable of measuring at an appropriate rate across that range of line rates. In combination with the sampling depth and the sampling rate, it should also be possible to scan a targeted amount of analyte, for example, the equivalent of a unit dose. In order to prevent sampling errors, segregation or nonrepresentative sampling material should not be subsampled for reference method analysis. The reference method should match the analytical sample size interrogated by the PAT method and should be consumed in its entirety for the reference method analysis. In continuous operations for powder measurements, samples should be drawn directly from the product stream and the use of a sample thief in a static powder bed should be avoided. Esbensen et al. demonstrates the importance of proper sampling using PAT for a continuous mixing system [15].

The reference method should be developed and validated thoroughly with the same rigor of the PAT method, such that it is proven to be accurate and precise across the expected range of content and process design space parameters. Content methods, for example, must be capable of extracting 100% of the active ingredients regardless of the material attributes and process parameters used in the manufacturing. Even minor extraction inefficiencies can cause problems when exercising stringent acceptance criteria during parallel testing. Lab-based moisture methods also require additional sample handling instructions. With the PAT sensor installed directly in the product stream or at line for tablet analysis, there is no opportunity for moisture ingress or egress between the times when the measurement is needed versus when it is collected. However, after hours of storage and transport to the receiving lab and sample preparation for reference method testing, it still must be guaranteed that the reference method sees "the same sample" as was presented to the PAT. Dissolution reference methods must be developed across the entire desired manufacturing range, so the release rate can be properly modeled by the online model. When the reference method exhibits sensitivities to incoming material attributes, process parameters, or tablet properties, the PAT method must match the performance of the reference method for those same attributes.

In the context of PAT methods for CM, it is important to note that there are a variety of models one could choose to employ in continuous operations depending on the desired control strategy. For the purposes of this work, model types can be classified into three categories: (1) spectroscopic models with a direct analytical output (e.g., NIR for the measurement of moisture or Raman for the measurement of crystalline form), (2) process models using measured process parameters for the prediction of an analytical output (e.g., feeder mass flow plus tablet hardness for the prediction of dissolution, or fluid bed dryer temperature and drying time as a prediction for powder blend moisture), or (3) a combination of process parameters and measured analytical attributes for the prediction of a final CQA (e.g., tablet weight and tablet hardness plus API content by NIR for the prediction of dissolution). Depending on which model is used in the control strategy and whether it is used for monitoring as a formal IPC measurement or as an RTRt test, development, validation, and long-term life cycle maintenance will require a different level of complexity.

2.2 Sensor location and placement for calibration model building

This section explores possible sampling options for sensor-based characterization of powders during dynamic flow conditions. The intention is to subject powders used for calibration models through the same processing conditions as experienced during a manufacturing process. An essential benefit is the "1D" sampling configuration, which allows for an estimation of the total sampling error and is representative of the bulk powder. Sampling errors, in the form of spectral artifacts, due to sample presentation to the sensor, can greatly influence the prediction of an attribute. In a continuous process, unlike a batch process, where there is a constant generation of material "output" at varying particle sizes and density, the PAT setup needs to be modified to account for this gradient. Inserting a sensor probe orthogonal to the path of a vertical powder flow is one example of how this gradient is factored during data acquisition. Depending on the type of powder monitored, the attribute measured, and the sensor used, there are several options for placing a sensor at the exit of an operation:

- A horizontal but slanted metallic tray placed at the exit which collects the falling powder, during its downward gravitational flow. A suitable "window," preferably sapphire glass, is embedded on the tray allowing for a sensor to be placed on the underside of the tray.

- Conveyor belt: One of the more successful setups involves the use of a conveyor belt for the powder to be transported from an outlet to a receptacle. The sensor is perched over the moving powder bed at a height equivalent to the sensor's optimum focal distance. It is useful to have an overhead scraper that brushes off the excess powder to set a uniform height and result in a smooth surface of the bed. The speed of the belt is adjusted depending on the throughput of the process (Fig. 10.1).
- Vibratory feeder setup: This setup is similar to the conveyor belt with the difference that instead of a moving belt, there is a stainless steel tray that vibrates at different intensities causing powders to move horizontally. Commonly, the sensor is perched over the tray at the target focal length to enable the acquisition of spectra (Fig. 10.2).

Vertical chute setup: An offline chute setup shown in Fig. 10.3 replicates the conditions of powder flowing in a manufacturing line. Apart from the powder being in a dynamic state, this setup also ensures that the powders undergo similar processing conditions of shear and electrostatics experienced during manufacturing [21]. The setup shown in Fig. 10.3 consists of a stainless steel pipe, 10 cm in diameter, partially flattened to form a rectangular interface, an analytical balance, and a feed frame or a rotary valve at its outlet to control the flow out of exit end of the chute. To achieve the in-line monitoring of the powder material, two pairs of acrylic windows are affixed to the rectangular part of the chute. One pair of windows aids in monitoring the level of powder, while the second pair of windows can be customized to integrate PAT sensors.

2.2.1 Sampling volume

Once a PAT sensor has been successfully implemented in the correct location and a robust method has been developed, the question that arises is how much powder is sampled by each sensor during one acquisition. The following equation estimates the mass sampled:

$$m = \frac{V_1}{V_2} t \, v \, \dot{m}$$

where m is the powder mass sampled per acquisition, V_1 is the powder volume illuminated by the sensor, V_2 is the powder volume flowing through the sampling interface setup, t is the acquisition time, and $\dot{m}$ is the powder flow rate. The ratio $\frac{V_1}{V_2}$ represents the fraction of the powder flowing at the interface that is sampled by the sensor.

2.3 PAT method validation overview in CM

Model validation includes demonstration of the typical ICH requirements for method validation such as linearity, accuracy, precision, specificity, and robustness. It consists of the formal demonstration of model performance on a data set specifically designed to probe the requirements of ICH Q2(R1) validation. For example, linearity is demonstrated over a range of 70%—130% of the target or expected concentration. Accuracy is demonstrated as a series of measurements at multiple concentrations across the relevant concentration range. Precision testing is generally performed as replicate measurements at the target concentration. For PAT models based on principal component regression (PCR) or PLS, specificity can be demonstrated by comparison of loading vectors to pure component spectra of the API(s) or analyte(s) of interest. Specificity is also proven when models successfully predict the analyte or property of interest as other chemical and physical changes occur over the design space. Validation always requires predicting a set of unknown samples that have not been used to inform development or calibration of the PAT model. As such, when predicting against a new set of samples, a successful prediction provides a high level of confidence that the response is due to the property of interest.

Some evaluation of robustness should also be considered at the time of PAT method development and validation. By nature of combining process development with PAT method development, model robustness is a desired byproduct of including multiple excipient and API lots across the desired process manufacturing ranges, across time where temperatures and humidity in manufacturing suites vary, and across time spans where instruments are recalibrated during routine PM schedules. All of these factors contribute to the long-term stability of a PAT model. More intentional robustness studies can also be included around parameters such as acquisition rate, number of accumulations, and exposure time, as well as the number of PAT instruments used to collect the calibration data. Some of these robustness studies can also be leveraged to support post-approval regulatory submissions and reduce the filing type and associated time for review and approval.

2.4 Maintenance overview

There are multiple tiers of assessing the health of a PAT model, all of which must be employed to successfully maintain a model for long term. Monitoring trends in predictions and model diagnostics within and across batches can provide a clear and direct indication when the model requires attention. In addition, it is imperative to define the appropriate frequency and testing plan for a formal periodic model assessment. This provides the PAT scientist the opportunity to ensure that the method remains sensitive to the property or analyte of interest through a head-to-head comparison between the PAT method and reference method. Specific acceptance criteria will not be provided in this chapter; however, as a general rule, criteria can be based on known method capability and/or the original method validation criteria. Acceptance criteria should be defined in advance of executing parallel testing. In addition to the periodic model assessments, special cause factors should also be considered and used to trigger model assessments based on their expected level of impact. For example, new excipient/API suppliers and process changes should be evaluated for their impact to PAT methods prior to implementation in commercial production. While these guidelines are similar in traditional manufacturing and in CM processes that rely on spectroscopic PAT as an integral part of the IPC and batch disposition strategies, the factors that affect the life cycle of PAT models are even more impactful.

3. PAT in a CM commercial control strategy

Effective use of PAT methods in a pharmaceutical manufacturing environment requires a clear picture of the intended use of each PAT sensor and where it fits into the overall control strategy: which attributes are being measured and controlled, and whether that should be used to monitor the process, control the process by driving manufacturing targets or driving material diversion decisions, inform a real-time release calculation and a batch disposition decision, or some combination of these options. These decisions guide the level of complexity required for every aspect of PAT implementation. For CM applications, the physical interface of the sensors to the process and the presentation of the sample to the PAT is likely to be online or in-line, where the PAT is either arranged in a sidestream configuration or directly in-line with the product flow. Whether the PAT is used for monitoring or control and diversion of nonconforming material, this should determine the level of integration required between the PAT and the process control software. The intended use of the PAT also sets the expectations for the method development process, validation requirements, amount and type of supporting information required for regulatory submissions, and global regulatory post-approval commitments.

In order to most clearly realize the benefits of PAT in pharmaceutical manufacturing, it is useful to understand the most commonly cited hurdles to implementation. In the authors' collective experience, the factors most likely to deter a company from complete implementation of PAT are cost, lack of expertise, upfront time required before the investment benefits are realized, operational complexity (process equipment design, process integration, software integration, midrun cleaning to prevent probe fouling, etc.), technical aspects of model maintenance (e.g., process and material factors that influence PAT methods throughout the commercial life cycle), regulatory aspects of model maintenance (i.e., novelty in regulatory submissions can translate to more questions and scrutiny for the same price of approval), and finally, the implications that PAT methods have on the global supply chain (i.e., different regional regulations, import testing requirements stifle the benefits of PAT in other regions, different time scales for regional approvals which requires manufacturing with outdated PAT methods to accommodate slower-to-approve regions). Before the widespread use of PAT in pharmaceutical operations can be expected, each of these barriers must be removed or reduced. Particularly in

CM operations, which rely on PAT-based control strategies, these factors inhibit industry-wide PAT implementation. Munson et al. [22] described in 2006 the reasons to adopt PAT and the reasons that some companies hesitated, and largely, to date those reasons have remained unchanged. However, more than a decade ago, CM was not part of this assessment as it is now. As CM becomes more desirable and more attainable, comfort with PAT as a business need should continue to increase.

4. Case studies

This section contains some selected case studies to illustrate how to bring together previously discussed concepts such as the attribute to be monitored, sensor placement, and calibration model building for some individual continuous operations that constitute a CM platform. The examples include feeding, blending, and granulation operations along with a section on using PAT for predicting dissolution performance to enable RTRt. Lastly, a section on offline imaging techniques to understand internal tablet structure is also presented. The latter is a fitting conclusion to the use of PAT as it serves to link tablet production with tablet performance.

4.1 Continuous blending

A CM setup consisting of three levels was used for the blending set of experiments. The top level was composed of feeders, dependent on the number of raw materials in the formulation. The second level included a mill, a continuous blender with the powder falling inside the blender through a hopper. The third level had dual functions, one is to control a level of powder sufficient to collect spectral data for each specific flow rate and the second one is to adapt the NIR. The probe consisted in the NIR source connected to the spectrometer through fiber optic.

The presented case study shows the continuous blending of a typical pharmaceutical blend. Blend homogeneity was performed in real time every 6s by NIR spectroscopy. Calibration models were constructed and loaded in OPUS 7.5 from Bruker. This system allowed the construction of calibration models as well as real-time measurements within the same software. There are several commercial software packages available for data processing and analysis such as Unscrambler X and Process Pulse for real-time measurements from Camo, Inc. SIMCA from Sartorius also offers method development and SIMCA On-Line for analysis in real time. Fig. 10.4 shows the measurement of API concentration post blending using a chute interface as shown in Fig. 10.3. The target concentrations from feeding was 68.8% w/w of API, and the total duration was 50 min. During this time, 502 spectra were acquired, with average predicted value of 68.7% w/w. There was small fluctuation in API

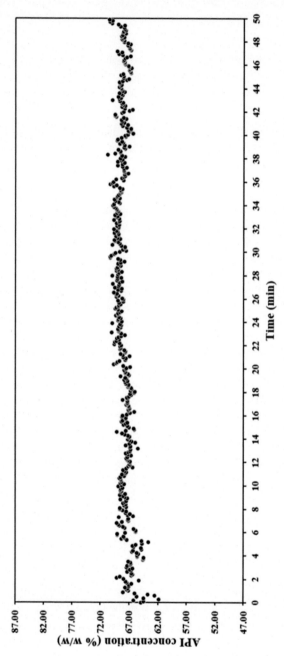

FIGURE 10.4 Active pharmaceutical ingredient concentration monitoring in a continuous blending process.

TABLE 10.1 Figures of merit of predicted values in a continuous blending process.

Target concentration	Spectra acquired	Average	Standard deviation	RSD
68.8	502	68.7	1.41	2.06

concentration during the first 5 min stabilized later and the overall RSD value was 2.06% demonstrating good performance of the system (Table 10.1).

4.2 Granulation

For any powder processing operation, the location and placement of the sensor is very critical, more so for wet granulation because the attributes that are monitored could sometimes be time-sensitive. For instance, if water content within the resulting granules is being quantified, then it is essential to place the sensor close to the exit of the granulator to reduce moisture loss before data acquisition. In addition, as moist granules can sometimes flow less freely than powders, there are chances that the granules could adhere to the sensor causing probe fouling and measurements from a stagnant slug of powder.

The continuous granulation setup consisted of a commercial, twin-screw processor with a sample inlet at one end and product outlet at the other end. The processor has an intricate but interchangeable arrangement of elements— paddles and screws. Depending on the arrangement of these elements, they allow for forward flow and backmixing while simultaneously enabling the kneading of powders at high shear. An AccuRate Schenck loss-in-weight (LIW) single-screw feeder was used for providing a consistent powder flow into the granulator. A peristaltic pump injected liquid, in a drip manner, tangentially through a liquid port located at the side of the processor. To monitor the output rate of the resulting granules, a scale was placed under the granulator's outlet valve. For a granulation process, porosity, liquid content, and granule size distribution are some of the most critical product attributes. Of these attributes, the liquid content, specifically water, can be challenging to monitor due to the constant evaporation that occurs. The rate of moisture loss is usually high in the initial few minutes of a material being exposed to the atmosphere, after which the rate decreases and hence the prompt measurement of the moist granules is imperative.

For this specific study, a preblended mixture of two ingredients was fed through the feeder at 16 kg/h, and water was the liquid used in the formulation (41% w/w). A loss on drying (LOD) technique was used as the reference method for determining the amount of water in the sample. The optimal drying temperature and duration were determined for achieving a near-total loss of moisture with only 2% of bound moisture retained by the sample.

The process parameters that were varied for this continuous granulation operation were L/S ratio, paddle revolutions per minute (RPM), and throughput. Once the operation was determined to be running at a steady state at all the conditions of the DOE, the next step was to integrate PAT in-line.

For the purpose of this study and keeping in mind the attribute that was monitored is water, the conveyor belt setup was found to be the most suitable for presenting the sample to the NIR after trials with other setups failed. The sample moved in a continuous horizontal manner on the belt and the NIR device was placed at the optimal focal length over the moving bed. The belt speed was adjusted such that a continuous bed with a reasonable height was obtained. After spectral acquisition, portions of the sample illuminated by the NIR were scooped aside, at regular intervals, for reference LOD measurements. Spectra were collected for the calibration set at varying water percentages in continuous operation and analyzed in Unscrambler X. Baseline correction using standard normal variate was carried out to remove differences caused due to physical characteristics of the samples. Following this, Savitzky–Golay transformation using second derivative with a 9-point smoothing was performed, causing the spectra to have reduced noise and enhanced peaks. The next step after the pretreatments was to run a principal component analysis (PCA) to observe clustering of like samples. Following PCA, a PLS model was used to cross-validate the model obtained from PCA. The PLSR model develops a linear regression model and determines the fit between the prediction from the NIR data and the LOD reference measurements in terms of the R-square.

For the prediction set, spectral data were collected for samples containing the target water content sample and fit into the model. The final PCA model chosen was used to project the spectra collected from the prediction set onto the calibration set. Such a projection is useful to observe where a sample data set falls, in the new dimensional space, with respect to the calibration data set (Fig. 10.5).

The PLS model predicted the amount of water in the prediction sample set where the NIR spectra were the predictor (X) variables and the LOD values of water were the regressor (Y) variables resulting in an RMSE of 0.13 and R-square of 0.99. From the results obtained, individual error percentages between the LOD values and those predicted by the model were calculated. Low error percentages indicate closeness of the predicted values to the reference LOD values (Table 10.2). Overall, it can be concluded that both the precision and accuracy of the model were excellent for the purpose of predicting water content in granules.

As a next step, the Process Pulse software from Camo was used for the real-time prediction of water content in the granules, at a frequency interval predetermined by the user, during the continuous granulation step. Fig. 10.6 shows that changes in the water content of the granules, in the form of both large and small step changes, were well captured by the model, demonstrating its sensitivity.

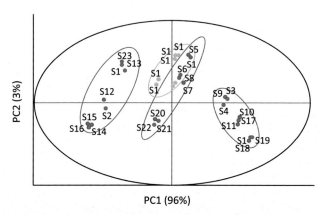

FIGURE 10.5 Principal component analysis (PCA) projection of prediction samples (green) on the calibration samples (blue).

TABLE 10.2 Water percentage predicted from near-infrared spectra compared to reference Loss on drying values.

Sample	Predicted water percentage (%)	Reference water percentage (%)	Error percentage (%)
1	41.80	41.39	0.99
2	41.90		1.23
3	41.88		1.18
4	41.23	41.40	0.41
5	41.34		0.14
6	41.27		0.31
7	41.39	41.73	0.81
8	41.35		0.91
9	41.37		0.86

4.3 Residence time distribution determination in feeders and blenders

4.3.1 Feeders

Feeders are the first set of equipment in manufacturing that initiate the flow of powders and set the concentration of the ingredients. Any perturbation to the feeders can tend to significantly affect the powder uniformity and be the initiation point for generating out-of-specification (OOS) tablets. The powder

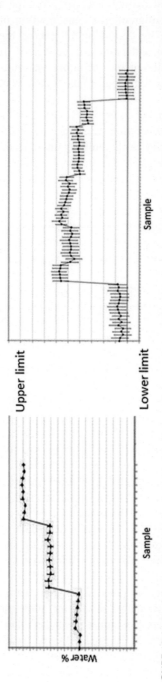

FIGURE 10.6 Real-time monitoring of water content during a continuous wet granulation process demonstrating the capability of observing both large and small step changes.

mass flow of an ingredient exiting a feeder gives the amount fed which is used for determining that ingredient's concentration in a formulation. The RTD within a feeder can be important for tracing the raw materials in a continuous process and for understanding the feeder's response with respect to changes in the properties of the incoming materials. This also helps to answer a key question: how long does it take for a new material to replace the old material?

A tracer experiment is most commonly used method to determine the RTD of a material within a unit operation. A chemically distinguishable ingredient, albeit with similar properties to the bulk, is introduced at the inlet of a unit operation as a pulse (a step change in the concentration of a measurable ingredient can also be made). The tracer concentration is then measured at the outlet of a unit operation, and outlet concentration profile of the tracer provides the RTD of system at those conditions. NIR spectroscopy has been widely used for characterizing the RTD of unit processes. For such applications, the tracer is chosen such that it has an absorbance in the NIR region of the electromagnetic spectrum and preferably peaks that are distinct from that of the bulk material.

To quantify the amount of powder at a given time instance exiting a unit operation, NIR calibration models are developed using varying amounts of the tracer. Prediction performance of the calibration model is evaluated using statistical parameters such as R-square, bias, standard error of prediction, and root mean square error of prediction. The final calibration model once selected is used to determine changes in the predicted concentration of the tracer during RTD trials.

This example discusses the measurement of RTD for a loss-in-weight feeder. After the calibration models were built, the setup for carrying out the RTD experiments was readied. The setup consists of a conveyor belt arrangement as shown in Fig. 10.7. The setup provided an ideal sampling channel for the NIR measurement of flowing powder. To ensure accurate and repetitive NIR spectral acquisitions, three aspects were emphasized: first, the time delay between the exit of the feeder and the NIR location was measured for every experiment. The final RTD results were corrected using this time difference. Second, the NIR probe was mounted at a fixed height, which equaled its focal length, over the conveyor belt. Third, the belt speed was adjusted for every feeder flow rate to allow for a uniform and a level powder bed to be created.

As an example, the effect of powder flow rate (kg/hr) on the shape of the RTD curve was evaluated via the pulse addition of the tracer (Fig. 10.8A). The tracer concentration was kept constant at 5% in the blends (w/w) and flow rates were varied at three levels: 5, 10, and 20 kg/h. Another method of tracer addition investigated was the step response method at the exit of the feeder at the same tracer concentration of 5% and flow rate of 10 kg/h (Fig. 10.8B).

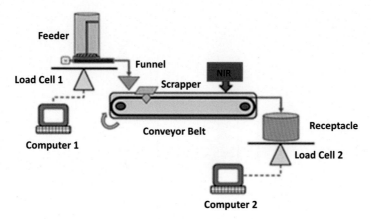

FIGURE 10.7   Feeder residence time distribution setup utilizing a conveyor belt and overhead placement of an NIR.

4.3.2 Blenders

RTD of a continuous blender is a highly desired characterization that leads to process understanding and enables traceability in a CM process [23]. The calibration samples were constructed spanning the tracer concentration range of 0%−9% w/w in API_1 and in steps of 3% w/w. These samples were prepared in a V-blender and the chute interface for vertically flowing powders was used to integrate the PAT tool. In this case, an NIR probe enabling in-line spectral acquisition was used (Fig. 10.9). The experimental setup was similar to that presented in Section 4.1 (Fig. 10.10). Blender RPM and flow rate are the main operational conditions that affect the RTD of the system and were thus varied as part of the study. The process started with feeders and blender running for a particular configuration to achieve a steady state prior to initiating the sensor data collection.

For this study, an API (API_1) was used as the bulk material and a tracer (API_2) with similar flow properties was used to estimate the RTD.

The correct PAT interface, in the case of vertically falling blends, is critical to ensure consistent measurements; thus, a setup similar to the actual RTD experiments was used for the calibration blends too. Calibration blends flowing through the chute get illuminated by the NIR in the flattened rectangular region, identical to the setup for spectral acquisition in the actual experiments. This process of 1D sampling provides the entire batch the same opportunity to be analyzed [15]. Once the calibration data sets have been acquired, the next step is the construction of the models for which a PLS regression method was used.

During the RTD experiments, a preweighed amount of tracer API_2 is added instantaneously into the blender through the hopper entrance following which the concentration of the tracer at the outlet of the blender is constantly

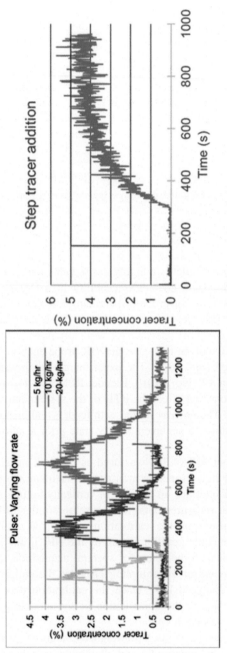

FIGURE 10.8   Tracer concentration exiting a feeder: (A) Pulse addition, (B) step addition.

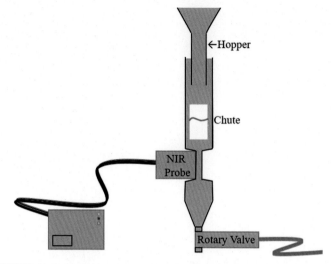

FIGURE 10.9   Experimental setup for spectral acquisition of calibration blends.

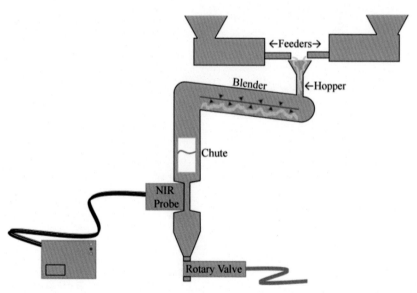

FIGURE 10.10   Experimental setup for determining residence time distribution in a continuous blender by NIR.

monitored. The experiment is considered complete when the entire tracer has washed out.

Bruker's OPUS software was used to perform real-time measurements of the tracer composition using a preconstructed PLS method. Fig. 10.11 shows

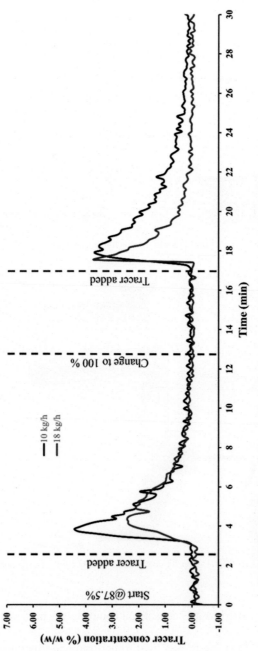

FIGURE 10.11   Residence time distribution in a continuous blender at two blender speeds and two line flow rates.

the RTD of the aforementioned experiments at two blender speeds, 87.5% and 100% of the maximum possible blender RPM, and two line flow rates for each blender speed during the run.

As perceived from Fig. 10.11, the RTD experiments at the same flow rates for different blender speeds were performed consecutively. Once the blender reached steady state, the tracer was pulsed and the RTD curve was obtained. The tracer concentration on the online monitor returning to a 0% (w/w) value was indicative of the entire tracer being flushed out. The blender speed was then changed to the next desired set point, and the system was allowed to reach steady state again. The system was pulsed with the tracer and the experiment was repeated. Fig. 10.11 shows the RTD profiles at different conditions.

4.4 Tablets: dissolution alternatives

In addition to the stated goals of enhanced process understanding and process control through the use of PAT in CM operations, one of the highest aims for a commercial pharmaceutical organization is RTRt. With the right analytical tools in place, knowledge of the incoming material attributes, and access to manufacturing data throughout the process, all the information necessary to assess a batch against the regulatory specifications is known and available. For typical CQAs such as tablet assay, content uniformity, moisture content, physical form, and ID, direct analytical measurements can be made through the manufacturing process with spectroscopic tools. For example, IPC final blend API content could be used to inform an RTR assay or content uniformity calculation. Assuming the tablet coating operation is demonstrated to have no impact on tablet moisture content, NIR in final blend could be used to measure and report moisture content. Physical form and ID could be measured directly by NIR or Raman at various stages throughout the manufacturing run as well. The most difficult test to replace on a typical drug product specification list is dissolution due to the complexity of the laboratory test and the fundamental understanding of the chemistry and mechanisms required to accurately predict in vitro drug product dissolution behavior [24]. As tablet dissolution can be impacted by formulation and manufacturing history (e.g., API content, particle size, polymer content, superdisintegrant moisture content or cross-linking, tablet weight/thickness/hardness) and equally by in vitro method specifics (e.g., pH, buffer capacity, paddle speed, temperature, surfactant concentration), the development of a robust surrogate method for dissolution prediction has proved to be nontrivial.

As mentioned previously, PAT may be taken to mean direct spectroscopic measurements, parametric data used for monitoring or controlling the process, or a combination of the two. For the prediction of dissolution, possibilities exist where a prediction can be made from direct analytical measurement, process parameter inputs, and/or material attributes. In CM, these input variables will vary over the course of a batch due to process variation, thus resulting in different dissolution profiles and consequently different predictions. It may be useful to define a segmented approach where dissolution predictions are made for segments of the batch. For example, in a 12-h run, a discreet prediction could be made for each hour of manufacturing. Thus, the final result would contain 12 values, thus meeting the USP Stage 2 dissolution criteria (USP <711>).

The dissolution prediction output may be a percent dissolved at a single specification time point, or the output may be the predicted coefficients in a model fit, thus allowing for the prediction of the entire dissolution curve. For the latter approach, the actual dissolution profiles must be first fit to a suitable model. A number of model fit options can be explored [25]. Once the best model fit is determined, model coefficients can be regressed to build a predictive model for the dissolution values. It is important to develop this relationship across the desired manufacturing ranges and design spaces to ensure that the fit model appropriately tracks with the reference dissolution profiles in all cases. Using this approach of predicting model coefficients, the percent of dissolved active ingredient at any discrete time point can be easily predicted for product release purposes. This approach contains more information and is more robust than the single time point approach.

4.5 Chemical imaging: offline uniformity, API distribution

Aims of CM include reducing waste, minimizing blend segregation observed in traditional batch manufacturing, and more importantly reducing OOS products. Achieving the latter involves the seamless combination of multiple facets: a thorough understanding of the individual unit operations, PAT sensors efficiently monitoring attributes at the critical locations, and the control systems' ability to initiate the necessary corrective actions. In addition, "good" product should also have the desired performance characteristics for it to pass the standard acceptance tests. Although the understanding of CM in the pharmaceutical industry has rapidly increased in the past decade, there are still known knowledge gaps, particularly regarding the relationship between material properties of ingredients, processing conditions, and resulting product attributes. This is because the tablet itself (or drug product, in general) is still a black box. Understanding the impact of material properties and processing conditions on the tablet microstructure can unravel this black box, enabling quality and performance being built into product by design. Elucidating the product microstructure requires knowledge of the spatial distribution of the different ingredients, which can be achieved by chemical imaging.

Spectroscopic imaging offers a unique advantage over traditional electron microscopy counterparts with respect to sample preparation and ease of operation. Among these, there are the nondestructive methods such as terahertz (THz) imaging with light sources strong enough to penetrate tablets and determine coating thickness [26] but weak enough to prevent damage to samples. THz imaging is gaining traction as an in-line tool in pharmaceutical manufacturing for determining the heterogeneity and integrity of tablet cores and determining coating quality [27]. Although imaging techniques are inherently time-consuming due to the raster scanning method, improved hardware has brought on a generation of devices that can rapidly collect high contrast NIR images of tablets for assessing the blend homogeneity in the final dosage [28], process monitoring [29,30], and functional problems [31,32].

Raman imaging offers superior chemical specificity which is advantageous for multicomponent formulations that result in peaks overlaid over one another. A newly developed Raman system (enter product details) facilitates 3D mapping of samples by physically shaving off layers from solid samples by a grater that is incorporated within the device. The instrument alternates between successive scanning and shaving steps to create multiple 2D maps which are stacked together to create a 3D image of the solid sample. In CM, the individual raw material properties and the process variables affect not only the composition of the tablets produced but also the distribution of the ingredients within the tablet microstructure. Both the composition and the ingredient distribution can singularly or in conjunction have adverse effects on the hardness and dissolution profiles of tablets.

As an example, APIs that are inherently cohesive in nature, and which manufacture steps, often appear as agglomerates in tablets. The automated Raman imaging instrument allows for visualization of these API agglomerates within the tablets (Fig. 10.12). The images can be depicted with respect to the different ingredients of a formulation wherein each color represents an ingredient (Fig. 10.12A) or the image can be of only a single ingredient colored with respect to its varying particle size (Fig. 10.12B). Tablets illustrated in Fig. 10.12 were manufactured in a semi-continuous fashion, where the feeding and continuous blending operations were replaced by a large 36 ft^3 V-blender. The blended powders were then compressed in a Kikusui Libra 2 tablet press.

In another example, the difference in blend homogeneity between batch mixing and continuous mixing of materials has also been illustrated. The CM route in this investigation was a wet granulation process in which the ingredients were premixed, either in batch or continuous blenders, before being granulated and then compacted to form tablets. The tubular continuous blender resulted in a homogenous distribution of all the ingredients, particularly the API as opposed to the tote bin blender (Fig. 10.13).

5. Conclusions

This chapter is intended to provide a high-level guideline with case studies involving some commonly used unit operations to illustrate how PAT, with emphasis on spectroscopic techniques, can be implemented in a manufacturing

(A)   **(B)**

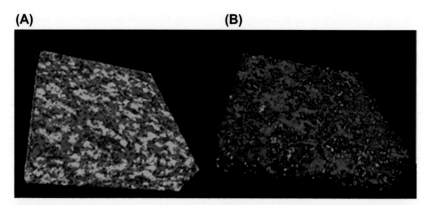

FIGURE 10.12 Three-dimensional Raman images of a subset of a tablet colored with respect to (A) the ingredients of a formulation: red—API, blue—excipient 1, green—excipient 2, yellow—lubricant; (B) API particle size: green—particles <50 microns, red—particles 50–250 microns, blue—particles >250 microns. *API*, active pharmaceutical ingredient.

(A)   **(B)**

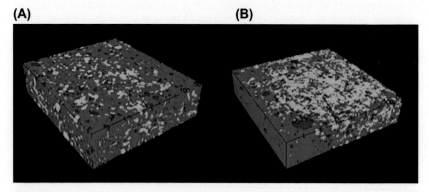

FIGURE 10.13 Three-dimensional Raman images of a subset of a tablet colored with respect to the ingredients of a formulation: red—active pharmaceutical ingredient, blue—excipient 1, green—excipient 2, yellow—lubricant. The ingredients were mixed in (A) a continuous blender and (B) a batch blender.

setting. The use of PAT does not stop at the development and validation of models but extends to include maintenance strategies over the life cycle of a product. This could serve well for those who plan on building CM rigs, modifying existing platforms and even for those who simply need a protocol as a reference. That said, the latter would include instrument, equipment, and even data analysis vendors who wish to customize their product with an application and/or customer in mind. Although PAT is synonymous with in-line measurement of material attributes, the term is also inclusive of offline, at-line, and online assessment of material quality. The approach of ensuring quality has progressed from using reactive to proactive strategies; with the understanding of the product's structure, its desired quality can now be designed to meet specifications.

References

[1] FDA. Guidance for industry PAT — a framework for innovative pharmaceutical develop-
     ment, manufacturing, and quality assurance. 2004. Available from: https://www.fda.gov/
     downloads/drugs/guidances/ucm070305.pdf. Accessed August 2017.

[2] Fonteyne M, Vercruysse J, De Leersnyder F, Van Snick B, Vervaet C, Remon JP, et al.
     Process analytical technology for continuous manufacturing of solid-dosage forms. TrAC
     Trends Anal Chem 2015;67(Suppl. C):159−66.

[3] Fonteyne M, Vercruysse J, Diaz DC, Gildemyn D, Vervaet C, Remon JP, et al. Real-time
     assessment of critical quality attributes of a continuous granulation process. Pharm Dev
     Technol 2013;18(1):85−97.

[4] Alam MA, Shi Z, Drennen 3rd JK, Anderson CA. In-line monitoring and optimization of
     powder flow in a simulated continuous process using transmission near infrared spectros-
     copy. Int J Pharm 2017;526(1−2):199−208.

[5] Fonteyne M, Soares S, Vercruysse J, Peeters E, Burggraeve A, Vervaet C, et al. Prediction of
     quality attributes of continuously produced granules using complementary pat tools. Eur J
     Pharm Biopharm 2012;82(2):429−36.

[6] Ward HW, Blackwood DO, Polizzi M, Clarke H. Monitoring blend potency in a tablet press
     feed frame using near infrared spectroscopy. J Pharm Biomed Anal 2013;80(Suppl.
     C):18−23.

[7] Järvinen K, Hoehe W, Järvinen M, Poutiainen S, Juuti M, Borchert S. In-line monitoring of
     the drug content of powder mixtures and tablets by near-infrared spectroscopy during the
     continuous direct compression tableting process. Eur J Pharm Sci 2013;48(4):680−8.

[8] Medendorp J, Bric J, Connelly G, Tolton K, Warman M. Development and beyond: strategy
     for long-term maintenance of an online laser diffraction particle size method in a spray
     drying manufacturing process. J Pharm Biomed Anal 2015;112(Suppl. C):79−84.

[9] Román-Ospino AD, Singh R, Ierapetritou M, Ramachandran R, Méndez R, Ortega-
     Zuñiga C, et al. Near infrared spectroscopic calibration models for real time monitoring
     of powder density. Int J Pharm 2016;512(1):61−74.

[10] Hattori Y, Otsuka M. NIR spectroscopic study of the dissolution process in pharmaceutical
     tablets. Vib Spectrosc 2011;57(2):275−81.

[11] Hernandez E, Pawar P, Keyvan G, Wang Y, Velez N, Callegari G, et al. Prediction of
     dissolution profiles by non-destructive near infrared spectroscopy in tablets subjected to
     different levels of strain. J Pharm Biomed Anal 2016;117(Suppl. C):568−76.

[12] Blanco M, Alcalá M. Content uniformity and tablet hardness testing of intact pharmaceu-
     tical tablets by near infrared spectroscopy: a contribution to process analytical technologies.
     Analytica Chimica Acta 2006;557(1):353−9.

[13] Barnes RJ, Dhanoa MS, Susan JL. Standard normal variate transformation and de-trending
     of near-infrared diffuse reflectance spectra. Appl Spectrosc 1989;43(5):772−7.

[14] Small GW. Chemometrics and near-infrared spectroscopy: avoiding the pitfalls. TrAC
     Trends Anal Chem 2006;25(11):1057−66.

[15] Esbensen KH, Román-Ospino AD, Sanchez A, Romañach RJ. Adequacy and verifiability of
     pharmaceutical mixtures and dose units by variographic analysis (Theory of Sampling) — a
     call for a regulatory paradigm shift. Int J Pharm 2016;499(1):156−74.

[16] Esbensen KH, Paasch-Mortensen P. Process sampling: theory of sampling — the missing
     link in process analytical technologies (PAT). Process analytical technology. John Wiley &
     Sons, Ltd; 2010. p. 37−80.

[17] EMEA. Guideline on the use of near infrared spectroscopy by the pharmaceutical industry and the data requirements for new submissions and variations. 2014. Available from: http://www.ema.europa.eu/docs/en_GB/document_library/Scientific_guideline/2014/06/WC500167967.pdf.

[18] Lichtig M. Lifecycle management of process analytical technology procedures. 2015. Available from: http://www.infoscience.com/JPAC/ManScDB/JPACDBEntries/1426003537.pdf.

[19] Morton M. A quality-by-design (QbD) approach to quantitative. 2011. Available from: http://www.americanpharmaceuticalreview.com/Featured-Articles/36924-A-Quality-by-Design-QbD-Approach-to-Quantitative-Near-Infrared-Continuous-Pharmaceutical-Manufacturing/.

[20] ASTM. ASTM E1790-04(2016)e1. Standard Practice for Near Infrared Qualitative Analysis. West Conshohocken, PA: ASTM International; 2016. Available from: https://doi.org/10.1520/E1790-04R16E01.

[21] Alam MA, Shi Z, Drennen JK, Anderson CA. In-line monitoring and optimization of powder flow in a simulated continuous process using transmission near infrared spectroscopy. Int J Pharm 2017;526:199–208.

[22] James M, Stanfield CF, Bir G. A review of process analytical technology (PAT) in the U.S.pharmaceutical industry. Curr Pharm Anal 2006;2(4):405–14.

[23] Engisch W, Muzzio F. Using residence time distributions (RTDs) to address the traceability of raw materials in continuous pharmaceutical manufacturing. J Pharm Innov 2016;11:64–81.

[24] Shanley A. Moving toward real-time release testing. Pharm Technol 2017;41(7).

[25] Costa P, Sousa Lobo JM. Modeling and comparison of dissolution profiles. Eur J Pharm Sci 2001;13(2):123–33.

[26] Zeitler JA, Shen Y, Baker C, Taday PF, Pepper M, Rades T. Analysis of coating structures and interfaces in solid oral dosage forms by three dimensional terahertz pulsed imaging. J Pharm Sci 2007;96(2):330–40.

[27] Niwa M, Hiraishi Y, Terada K. Evaluation of coating properties of enteric-coated tablets using terahertz pulsed imaging. Pharm Res 2014;31(8):2140–51.

[28] El-Hagrasy AS, Morris HR, D'Amico F, Lodder RA, Drennen 3rd JK. Near-infrared spectroscopy and imaging for the monitoring of powder blend homogeneity. J Pharm Sci 2001;90(9):1298–307.

[29] Clarke F. Extracting process-related information from pharmaceutical dosage forms using near infrared microscopy. Vib Spectrosc 2004;34(1):25–35.

[30] Gowen AA, O'Donnell CP, Cullen PJ, Bell SEJ. Recent applications of chemical imaging to pharmaceutical process monitoring and quality control. Eur J Pharm Biopharm 2008;69(1):10–22.

[31] Clarke F, editor. NIR microscopy: utilization from research through to full-scale manufacturing. European Conference on Near Infrared Spectroscopy; 2003.

[32] Lewis EN, John EC, Fiona C. A near infrared view of pharmaceutical formulation analysis. NIR News 2001;12(3):16–8.

Chapter 11

Developing process models of an open-loop integrated system

Nirupaplava Metta[1,2] and Marianthi Ierapetritou[3]

[1]*Automation Products Group, Applied Materials, Logan, UT, United States;* [2]*Department of Chemical and Biochemical Engineering, Rutgers University, Piscataway, NJ, United States;* [3]*Department of Chemical and Biomolecular Engineering, University of Delaware, Newark, DE, United States*

1. Introduction

Process modeling plays a crucial role in the advancement of continuous pharmaceutical manufacturing. In addition to the potential to act as surrogates for experiments, process modeling also enhances understanding of process variability. Through process systems engineering tools, the optimal design and operating conditions can be attained, details of which are discussed in Chapter 13 [Process Optimization]. Ultimately, advanced control strategies can be applied that help achieve consistent product quality.

This chapter provides an overview of modeling approaches generally used for process model development. Attempt has been made to give relevant equations where applicable and to explain the modeling strategies. However, readers are referred to corresponding published work where more details on the model development can be found. In the following sections, approaches to develop process models for various unit operations in pharmaceutical manufacturing are discussed. Specifically, models for loss-in-weight (LIW) feeder, blender, wet granulator, fluidized bed dryer, roller compactor, co-mill, and tablet press are discussed. In Section 13.9, a brief discussion on the integration of these models to develop a flowsheet model via various routes of continuous manufacturing is presented.

2. Loss-in-weight feeder

Although variability in feeding performance can propagate downstream and affect the final product quality, dynamic models published for feeding

How to Design and Implement Powder-to-Tablet Continuous Manufacturing Systems
https://doi.org/10.1016/B978-0-12-813479-5.00004-5

equipment are limited when compared to other unit operations. In order to get stable and constant flow, LIW feeders are operated under gravimetric mode where the flow rate is controlled by adjusting the screw speed inside the feeder. Wang et al. [1] used multivariate analysis to correlate feeding performance to powder flow properties. The feeding performance was quantified using relative standard deviation of mass flow rate from the feeder, which correlated to the powder flow properties using partial least squares regression. This approach is particularly useful if the new material to be fed is expensive or is available in limited quantities. Boukouvala et al. [2] developed data-driven models from experimental data collected using LIW feeder provided by a Gericke feeder. Screw speed, screw size, screw configuration, and powder flow index were the significant feeder variables identified from experimental data using an Analysis of Variance (ANOVA) method. Modeling approaches such as kriging and response surface methodology were used to correlate these variables to feeder flow rate standard deviation, which is a feeder performance metric. Such approaches are valuable, as a mechanistic understanding of powder behavior inside the screw feeders is still lacking.

Wang et al. [3] used a semi-empirical equation to dynamically model the mass flow rate out of the feeder $F_{out}(t)$ using Eq. (11.1):

$$\dot{F}_{out}(t) = ff(t)\omega(t) \tag{11.1}$$

where $\omega(t)$ is the screw speed; $ff(t)$ is the feed factor, defined as the maximum mass of powder fitting in a screw flight and is expressed in Eq. (11.2):

$$ff(t) = \rho_{effective}(t) V_{Screw\ Pitch} \tag{11.2}$$

where $\rho_{effective}$ is the effective density of materials in the screw pitch with a volume $V_{Screw\ Pitch}$. ff was found to be dependent on the amount of material in the hopper, as the effective density of powder entering the screws is expected to change due to change in pressure exerted by static head of the material above. This relationship was determined to be following a pseudo first order as expressed in Eq. (11.3):

$$ff(W(t)) = ff_{level}^{sat} - e^{-\beta W(t)}\left(ff_{level}^{sat} - ff_{level}^{min}\right) \tag{11.3}$$

where ff_{level}^{sat} is the saturated feed factor; ff_{level}^{min} is the minimum feed factor; and β is the feed factor exponential decay constant. These parameters are regressed from experimental data and can be related to material properties such as bulk density, permeability, compressibility, and cohesion. This is extensively covered in Chapter 3 [Loss-in-weight Feeding] and the readers are referred to this for further details.

3. Continuous blender

Powder blending is a crucial unit operation in pharmaceutical industry as the individual components in a formulation are effectively mixed in this unit, thus impacting the content uniformity in the final drug product. This section reviews various models that have been used to model continuous powder blenders. Discussion has been restricted to modeling tubular blenders, the most popular form of powder blenders that have been used in continuous processing of pharmaceuticals.

Numerous modeling techniques have been used to simulate the powder behavior in a continuous blender. The relative standard deviation (RSD) of the concentration of the active ingredient at the blender exit is typically used as a blender performance metric and is defined in Eqs. (11.4) and (11.5) [4]:

$$RSD = \frac{\sigma}{\overline{\overline{C}}} = \frac{\text{standard deviation}}{\text{average concentration}} \tag{11.4}$$

$$\sigma = \sqrt{\frac{\sum_{i=1}^{N}\left(C_i - \overline{C}\right)^2}{N - 1}} \tag{11.5}$$

where N is the number of samples collected and C_i is the concentration of sample i.

Several models in the literature for mixing processes used discrete element method (DEM) to simulate the mixing and segregation behavior [5,6]. DEM models are used to capture the effects of blender geometry, material properties, and operating conditions [4,7] on the RSD. While mechanistic models such as DEM are computationally expensive, the discrete element−reduced order modeling methodology has opened avenues to capture mechanistic effects in a unit modeling framework. In this methodology, the distributed parameter information from DEM simulations, such as velocity profiles, is efficiently represented using a reduced order model. The lower dimensional velocity profiles are then correlated to blender performance predictors such as RSD using partial least squares modeling. Sen et al. [8] coupled population balance models (PBMs) with DEM models to predict the concentration of active pharmaceutical ingredient (API) at the blender exit. Velocity profiles obtained from a periodic section of the DEM blender model was used to define the change in the number of particles with time in the PBM equation. More details on PBM are discussed further in this chapter in Sections 13.5 and 13.7.

In addition to RSD, the residence time distribution (RTD) of particles in the blender is also of interest, and as discussed in Chapter 4 [Continous Powder Mixing and Lubrication], it dictates the amount of time a material takes to exit the blender, thus enabling material traceability. In Wang et al. [3], modeled a continuous blender as continuous stirred tank reactors (CSTRs) in series, which was used to characterize the RTD. The CSTR in series model uses multiple ideal stirred tanks in series to simulate mixing of powders along the length of the blender. A delay time τ_{delay} is used, which represents the time particles take to convectively move through the blender in axial direction. The blender RTD is then represented as given in Eq. (11.6):

$$E(t) = \text{Unit Step}\left[\tau - \tau_{\text{delay}}\right] \frac{\left(t - \tau_{\text{delay}}\right)^{n-1} e^{\left(-\frac{t-\tau_{\text{delay}}}{\bar{\tau}}\right)}}{(n-1)!\bar{\tau}^n} \tag{11.6}$$

where n is the number of tanks and $\bar{\tau}$ is the mean residence time of one tank in the model. The blender's mean residence time then is $\tau_{\text{blender}} = \tau_{\text{delay}} + n\bar{\tau}$.

The mass balance in the blender is modeled assuming the holdup in the blender reaches a steady state asymptotically, i.e., it follows a first order. The flow rate out of the system, which is used as an input in the downstream unit, can thus be modeled as given in Eqs. (11.7) and (11.8):

$$\bar{\tau}\frac{dM(t)}{dt} + M(t) = M_{ss} \tag{11.7}$$

$$\frac{dM(t)}{dt} = F_{\text{in}}^{\text{total}} - F_{\text{out}}^{\text{total}} \tag{11.8}$$

where the parameters n, τ, and M_{ss} depend on inlet flow rate and blade speed and are estimated from experimental data.

4. Roller compactor

Roller compaction is an intrinsically continuous operation, and thus its adoption and use in a continuous process is relatively straightforward. Moreover, as the unit itself and its operation is intrinsically continuous, models developed to describe a roller compaction operation in a batch paradigm can be directly leveraged. The most popular model for a roller compaction was developed by Johanson in 1965 [9]. The model can predict nip angle and the resulting ribbon bulk density. Various assumptions relating to powder properties and compaction conditions are made in the model which has propelled research to further improve this model. Johanson assumed that the powder is isotropic, frictional, cohesive, and compressible and obeys Jenike—Shield yield criterion. The powder's relative density is assumed to be related to applied stress via a power law relationship that is fitted to experiments. The model uses additional powder properties such as effective angle of internal friction and powder roll friction angle that are assumed constant. Powder flow through the roller compactor is modeled as one-dimensional. In the upstream "slip" region, speed of powder is less than roll speed, and in the downstream "no-slip" region, speed is assumed equal to roll speed. Transition between regions occurs at "nip" angle α, which is calculated by equating powder stress gradients in the two regions. The Johanson's model is used to predict nip angle as well as the final ribbon relative density at the minimum gap and the acting force on the rolls.

Hsu et al. [10] developed a dynamic model and extended capabilities of the Johanson model to predict the stress and ribbon density profiles through the use of an additional material balance equation to model the roll gap change. Johanson model assumes that the gap width does not change. However, in the common design of a roller compactor, the screw speed, roll speed, and roll pressure affect ribbon density as well as gap width.

The applied hydraulic roll pressure applied, P_h^{rc}, on a roll designed to resist the roll-separating force is given by Eq. (11.9):

$$P_h^{rc}(t) = \frac{W^{rol}}{A^{rol}} \frac{\sigma_{out}(t)R^{rol}}{1+\sin\delta} \int\limits_0^\alpha \left[\frac{h_0(t)}{R^{rol}\left(1+\dfrac{h_0(t)}{R^{rol}}-\cos\theta\right)\cos\Theta} \right]^{K^{rc}} \cos\theta d\theta$$

(11.9)

where δ is the powder effective angle of friction, W^{rol} is the roll width, R^{rol} is the roll radius, A^{rol} is the compact surface area, ω^{rc} is the angular velocity of the rolls, and u_{in} is the linear velocity of the feed. An empirical model is used to describe the compression behavior of the material in the "no-slip" region as given in Eq. (11.10):

$$\sigma_{out}(t) = C_1^{rc}\left(\rho_{out}^{rib}(t)\right)^{K^{rc}}$$

(11.10)

where C_1^{rc} and K^{rc} are constants to be estimated from experiments.

To model the change in $h_0(t)$, i.e., the half roll gap at time t, material balance equation was used by Hsu et al. [10] as given in Eq. (11.11):

$$\frac{d}{dt}\left(\frac{h_0(t)}{R^{rol}}\right) = \frac{\omega^{rc}\left[\rho_{bulk_{in}}(t)\cos\Theta_{in}\left(1+\dfrac{h_0(t)}{R^{rol}}-\cos\Theta_{in}\right)\left(\dfrac{u_{in(t)}}{\omega^{rc}(t)R^{rol}}\right)-\dfrac{\rho_{out}^{rib}(t)h_0(t)}{R^{rol}}\right]}{\int_0^{\Theta_{in}}\rho(\Theta)\cos(\Theta)d\Theta}$$

(11.11)

where $\Theta_{in} = \pi/2 - 0.5\left(\pi - \sin^{-1}\dfrac{\sin\varphi}{\sin\delta} - \varphi\right)$, where φ is the angle of surface friction. The powder bulk density entering the roller compacter is denoted by $\rho_{bulk_{in}}$. To solve the denominator of Eq. (11.11), the density profile $\rho(\Theta)$ needs to be solved for each timestep. In the nip region, $\rho(\Theta)$ can be calculated using Eq. (11.10). In the slip region, the density is assumed to be the same as the inlet density, as the stress is relatively small, thus allowing an analytical solution of the equation. Boukouvala et al. [11] used the above given set of equations in a dry granulation flowsheet model and demonstrated the capability to predict changes in roll ribbon density and roll gap with changes in input process parameters.

Recent work published by Liu et al. [12] provided strengths and limitations of the Johanson's model which is summarized as follows. The predicted nip angle from Johanson's theory was noted to be in agreement with experiment results based on work of [13]. However, the predicted and experimental roll force agreed only for roll gaps less than 0.15 mm. Also, a slight change in powder bulk porosity predicted a large change in roll force, which did not agree with experimental observations. From FEM simulations, Muliadi et al. [14] found that the model overpredicts ribbon relative density and roll pressure significantly. This has been attributed to the one-dimensional (1D) powder flow assumption in the Johanson model. Contrary to the assumption, FEM studies found that the powder flow in the "no-slip" region is fastest at the rolls and slowest at the centerline. As the powder speed is assumed equal to the roll speed, the predicted mass flow rate through the unit is greater than what actually occurs, resulting in overprediction of the ribbon density. To address this, Liu and Wassgren [12] used a mass correction factor f_θ as shown in Eq. (11.12):

$$\frac{f_\Theta}{f_0} = 1 + \frac{1 - f_0}{f_0}\left(\frac{\Theta}{\alpha}\right)^n \tag{11.12}$$

The dependence of mass correction factor on position Θ is also considered in this work through the empirical fit. Two fitting parameters, power constant n and correction factor at minimum gap width f_0, are estimated using measurements of roll force and gap width from experiments.

5. Continuous wet granulator

Population balance modeling is the most commonly used approach to model a continuous wet granulation process, when modeling both a twin screw process and a high shear granulator, the two most common forms of continuous wet granulators. A general PBM equation is given in Eq. (11.13).

$$\frac{\partial F(x,t)}{\partial t} + \frac{\partial}{\partial x}\left[F(x,t)\frac{dx(x,t)}{dt}\right] = R_{\text{form}}(x,t) - R_{\text{dep}}(x,t) + \dot{F}_{\text{in}}(x,t) - \dot{F}_{\text{out}}(x,t)$$

$$\tag{11.13}$$

where the number of particles F is the particle density and x is the vector that represents the granule characteristic. The partial differential term with respect to x accounts for property changes due to mechanisms such as layering, liquid addition, or consolidation. R_{form} and R_{dep} are functions representing birth and death rates of particles with property x. $\dot{F}_{\text{in}}$ and $\dot{F}_{\text{out}}$ are the flow rates of particles entering and exiting the granulator, respectively.

A three-dimensional (3D) PBM is widely used to account for distributions in size, liquid content, and porosity simultaneously. Granulation processes in pharmaceutical industry involve multiple solid components, i.e., an API granulated with one or more excipients. In this case, the distribution of API in the granule population is of interest. Nonuniform API distribution is undesired as it can affect the uniformity of the final solid oral dosage form [15].

To model multicomponent granulation systems, a fourth dimension must be added to the 3D PBM as given in Eq. (11.14):

$$\frac{\partial}{\partial t}F(s_1,s_2,l,g,t) + \frac{\partial}{\partial l}\left[F(s_1,s_2,l,g,t)\frac{dl}{dt}\right] + \frac{\partial}{\partial g}\left[F(s_1,s_2,l,g,t)\frac{dg}{dt}\right]$$

$$= R_{\text{nuc}}(s_1,s_2,l,g,t) + R_{\text{agg}}(s_1,s_2,l,g,t) + R_{\text{break}}(s_1,s_2,l,g,t)$$

(11.14)

where l, g indicate the liquid and gas volumes in the granule, respectively. s_1 and s_2 are the solid volumes in the granule of the two different components used, typically API and excipient. The formation and depletion rates are governed by nucleation, aggregation, and breakage processes, which constitute the right hand side of the PBM equation.

There is a lot of work that has been published and is ongoing to accurately model the mechanisms in a granulation process. Specifically, several aggregation and breakage kernels have been developed that are empirical, mechanistic, or a combination of both [16−18]. The terms in the kernels may be estimated based on experimental data or data obtained from mechanistic models such as DEM models. Nucleation process entails formation of nuclei as liquid droplets are added to the system and come in contact with powder particles. In drop controlled regime [19], each droplet forms one nuclei. Rate of nucleation is typically modeled as following zero or first-order reaction [17]. Barrasso and Ramachandran [20] modeled rate of nucleation as R_{nuc} as ratio of rate of liquid added to powder and an assumed droplet volume, $\dot{L}_{\text{in,powder}}/V_{\text{droplet}}$. Here, the fraction of liquid added to the powder, $\dot{L}_{\text{in,powder}}$, is assumed as its volume fraction or the ratio of total powder volume to total volume of granules and powder.

Various aggregation kernels have been used and published in the literature. [18,21] provide a tabulated list of aggregation kernels used. As an example, aggregation kernel proposed by Ref. [22] is given in Eq. (11.15):

$$\beta\left(s_1,s_2,l,g,s_1',s_2',l',g'\right) = \beta_0(V+V')\left(\left(LC+\acute{LC}\right)^\alpha\left(100-\frac{LC+\acute{LC}}{2}\right)^\delta\right)^\alpha$$

(11.15)

Aggregation rate is strongly dependent on liquid binder content and granule size, which explains the kernel structure. In this kernel, the two colliding particles are represented as (s_1,s_2,l,g) and (s_1',s_2',l',g'). V and LC represent total volume and fractional liquid binder content. β_0, α, and δ are the parameters to be estimated from experimental data. For pharmaceutical processes, a composition-dependent aggregation kernel proposed by Matosukas et al. [23] may be used, as distinct solid phases may attract or repel each other. This can be accounted for using a multiplication factor

$\Psi(s_1, s_2, s_1', s_2') = \exp(-a_{ab}(x + x - 2xx'))$, where x is the mass fraction of the first component. Once the kernel is formulated, the rate of aggregation is included in the model through Eqs. (11.16)–(11.18):

$$R_{agg}(s_1, s_2, l, g, t) = R_{agg}^{form}(s_1, s_2, l, g, t) - R_{agg}^{dep}(s_1, s_2, l, g, t) \qquad (11.16)$$

$$R_{agg}^{form}(s_1, s_2, l, g, t) = \frac{1}{2} \int_0^{s_1} \int_0^{s_2} \int_0^l \int_0^g \beta(s_1 - s_1', s_2 - s_2', l - l', g - g', s_1', s_2', l', g')$$

$$F(s_1 - s_1', s_2 - s_2', l - l', g - g', t) F(s_1', s_2', l', g', t) dg' \, dl' \, ds_2' \, ds_1'$$

$$(11.17)$$

$$R_{agg}^{dep}(s_1, s_2, l, g, t) = \frac{1}{2} \int_0^\infty \int_0^\infty \int_0^\infty \int_0^\infty \beta(s_1, s_2, l, g, s_1', s_2', l', g')$$

$$F(s_1', s_2', l', g', t) dg \, dl \, ds_2 \, ds_1 \qquad (11.18)$$

Similarly, breakage mechanism may be also be included in the model through the use of breakage kernels. Breakage kernels as discussed in Chapter 5 [Continuous Dry Granulation] for comminution processes are also applicable for granulation processes.

It is worth noting that the 4D models as described above are computationally expensive to evaluate. This limits its applicability for model-based control or advanced model applications such as sensitivity analysis or flowsheet modeling. For practical purposes, the dimensionality of the model can be reduced and a lumped parameter approach can be used, that is, one or more granule properties are lumped into the remaining distributions. A new equation is used for each lumped parameter to track its evolution with time. For example, if gas volume is taken as the lumped parameter, the 3D reduced model is given by Eq. (11.19) where nucleation and layering effects are not considered.

$$\frac{\partial}{\partial t} F(s_1, s_2, l, t) + \frac{\partial}{\partial l} \left[F(s_1, s_2, l, t) \frac{dl}{dt} \right] = R_{agg} + R_{break} \qquad (11.19)$$

Eq. (11.20) is the gas balance equation where total volume of gas in each bin is given by $G(s_1, s_2, l, t) = g(s_1, s_2, l, t) F(s_1, s_2, l, t)$

$$\frac{\partial}{\partial t} G(s_1, s_2, l, t) = F(s_1, s_2, l, t) \frac{dg}{dt} + R_{agg,gas} + R_{break,gas} \qquad (11.20)$$

Barrasso and Ramachandran [24] provided a detailed account of several model order reduction strategies and compared them to the full 4D model. It was found that the 3D model with gas volume as the lumped parameter showed most promising results in terms of accuracy and computational time, possibly due to low influence of gas phase on aggregation and breakage rates.

Among various continuous granulators available, the twin-screw granulator (TSG) is the most widely used model in continuous manufacturing processes. Barrasso and Ramachandran [20] demonstrated the ability of a multidimensional PBM coupled with a DEM model to qualitatively predict effects of screw design and configuration on granule properties. A compartmental PBM was used where the TSG was represented as four well-mixed axial, spatial compartments in series. Powder and liquid were introduced in the first compartment and the granulated product exited the last compartment. Within each compartment, the residence time of particles was evaluated using DEM simulations. In addition, collision and velocity data gathered from DEM simulations were also used to evaluate mechanistic expressions for aggregation, breakage, and consolidation. Despite its promise, the use of mechanistic models for predictive purposes is still not widely accepted due to the computational expense it carries. Use of surrogate modeling techniques such as artificial neural networks (ANNs), kriging, etc. [25], Metta et al. [26] to represent and predict the high dimensional and computationally expensive mechanistic data, shows tremendous potential.

6. Fluidized bed dryer

Fluidized bed drying has been the widely adopted form of drying for pharmaceuticals, and discussions on modeling of dryers has thus been restricted to fluidized bed drying. It is important to note that some dryers operate in semi-continuous modes Chapter 7 [Continuous Fluidized Bed Processing], and it is advised that the practitioner is cognizant of the mode of operation when adopting the drying models reviewed below.

Drying models with varying extent of detail ranging from empirical models to detailed mechanistic models are published in the literature. A detailed review of the mechanistic drying models available is published in Mortier et al. [27]. Drying of pharmaceutical material is typically described using single particle drying models, where pore structure is ignored and the porous material is treated as a whole. The single particle drying model can be used to predict the drying behavior of a population of granules using PBM. The effect of fluidization of the particles can be thoroughly studied using computational fluid dynamics models where the spatial distribution of the moisture content can be analyzed.

In single particle drying models, evaporation of liquid is described using a diffusion equation. In particular, Mezhericher et al. [28] describe drying of a motionless single porous droplet in a flow of air in two phases. In the first phase, temperature of the water increases and water evaporates from the surface. This can be described using Eq. (11.21).

$$\dot{m}_v = h_D \left(\rho_{v,s} - \rho_{v,\infty} \right) A_d \tag{11.21}$$

where $\dot{m}_v$ is the mass transfer rate, h_D is the mass transfer coefficient, $\rho_{v,s}$ is the partial vapor density over the droplet surface, $\rho_{v,\infty}$ is the partial vapor density in the ambient air, and A_d is the surface area of the droplet.

When the radius of the droplet becomes equal to the radius of the drying particle, the second drying phase starts during which two regions, the wet core and the dry crust, are formed. In this phase, water evaporates inside the particle at the receding interface between the crust and the wet core. The vapor generated over the interface diffuses and forms a thin boundary layer over the particle surface. The drying air then takes away the vapor on the particle surface through convection. The evaporation rate in the second phase is given by a complicated equation that is developed based on a moving evaporating interface. The equations from both drying phases are to be solved simultaneously with ODEs (Ordinary Differential Equations) for decrease in droplet radius, temperature of the droplet, decrease in wet core radius, PDEs (Partial Differential Equations) for temperature profile in the dry crust and wet core. The reader is referred to Mortier et al. [29] for details on the corresponding equations and their solution.

Mortier et al. [30] proposed reduction of the complex single particle drying model to be used in a population balance equation. To reduce the complex drying model, a global sensitivity analysis is performed and the critical operating parameters that most impact the model outputs are chosen. Empirical relationships can then be used to represent the complex drying model as a function of the critical processing parameters. Based on the sensitivity analysis, particle radius was found to be important for the first drying phase and gas temperature for the second drying phase.

For constant ambient conditions, drying of a population of wet granules can be described by Eq. (11.22).

$$\frac{\partial}{\partial t} n(R_w, t) + \frac{\partial}{\partial R_w} \dot{R}_w (R_w, Y) n(R_w, t) = 0 \qquad (11.22)$$

where $n(R_w, t)$ is the number density distribution of particles of wet radius R_w at time t. The growth term $G_r = \dot{R}_w (R_w, Y)$ accounts for the decrease in moisture content. Here, Y represents the ambient conditions in the system such as gas temperature, gas velocity, air humidity, etc.

It is worth paying attention to the empirical and ANN-based models that are published in literature [31] to predict the evolution of average moisture content of granules with time. The empirical models are generally expressed in exponential terms to represent the drying curves. While no mechanistic understanding of the process is incorporated in these models, they may serve as an easy way to predict moisture content evolution.

7. Conical screen mill

Conical screen mills (Comills) are commonly employed in integrated continuous processes to condition granules produced by a dry or a wet granulation process. Comills are also commonly used to delump ingredients for mixing. Modeling a Comill as a comminution unit is discussed first, followed by its use as a delumping unit.

Particle size change due from "comilling" can be simulated using PBM. A review of various PBMs published for a comilling process is discussed in Chapter 5 [Continuous Dry Granulation]. A 1D PBM tracks change in mass of particles of various sizes over time as shown in Eq. (11.23).

$$\frac{dM(w,t)}{dt} = R_{\text{form}}(w,t) - R_{\text{dep}}(w,t) + \dot{M}_{\text{in}}(w,t) - \dot{M}_{\text{out}}(w,t) \qquad (11.23)$$

where $M(w,t)$ represents mass of particles of volume w at time t, and R_{form} and R_{dep} represent rates of formation and depletion of particles, respectively. $\dot{M}_{\text{in}}$ and $\dot{M}_{\text{out}}$ are the mass flow rates of particles entering and exiting the mill, respectively. The terms R_{form} and R_{dep} entail breakage kernel and breakage distribution function that represent the probability of the breakage event occurring and the distribution of the particles formed once the event occurs. The breakage kernels and breakage distribution functions used vary depending on the extent of information that the practitioner intends to incorporate. Capece et al. [32], Loreti et al. [33], Metta et al. [34] developed and used mechanistically based breakage kernels and distribution functions, whereas Barrasso et al. [35] used an impeller speed−based breakage kernel. Reynolds [36] used a generalized Hill-Ng distribution function, while Barrasso et al. [35] applied a log-normal distribution function. Typically, parameters relating to the breakage kernel and breakage distribution function are calibrated using experimental data generated at various milling conditions.

Another aspect of modeling a comilling process is formulating the mass flow rate of particles exiting the mill as they depend on impeller speed, screen size, and also the distribution of particles inside the mill. In Metta et al. [37], the mass flow rate out of the mill $\dot{M}_{\text{out}}(w,t)$ is formulated using a screen model as given in Eq. (11.24).

$$\dot{M}_{\text{out}}(w,t) = \left(R_{\text{form}}(w,t) - R_{\text{dep}}(w,t) + \gamma d_{\text{in}}(w,t) \right)\left(1 - f_d(w,t)\right) \qquad (11.24)$$

where the feed particle size distribution entering the mill is denoted by d_{in}. A parameter $\Delta = d_{\text{screen}} * \delta$ is used, where δ is referred to as critical screen size ratio and d_{screen} is the screen size. The critical screen size ratio reflects the size limit below while the particle exits the mill instantaneously. If the size of the particle is greater than the screen size, it does not exit the mill. A linear model is used to represent the flow rate of particles of various sizes exiting the mill. In addition, a relationship between the critical screen size ratio δ and the impeller speed v_{imp} is proposed as given in Eq. (11.25).

$$\delta = \varepsilon \left(\frac{v_{\text{imp,min}}}{v_{\text{imp}}} \right)^{\alpha} \tag{11.25}$$

The relationship is reflective of the reduced apparent screen size available for the particle to exit the mill as the impeller speed increases. At higher impeller speeds, the particle motion tangential to the screen leads to a reduction in the apparent screen size that is available for the particle to exit the mill. The parameters γ, ε, and α are estimated from experimental data.

The particle size distribution of the milled granules predicted by PBM, along with data on particle shape, moisture content, etc., can be used to build empirical models in order to predict other bulk properties such as bulk density, tapped density, etc. These bulk properties are required as inputs in the modeling of tablet press which is a downstream unit. Metta et al. [37] used a partial least squares modeling approach, where particle size distribution predicted using a PBM along with residual moisture content of granules, and feed granule properties are used to predict milled product bulk density, tapped density, and friability.

In addition to reducing the size of roller compacted ribbons and wet agglomerates, conical screen mill is also used as a delumping device for cohesive powders. Vanarase et al. [38] studied the various configurations of blender and comill and concluded that a comilling step before mixing provides the optimal strategy for blending cohesive materials. In this application, comilling imparts additional residence time to the powders and the holdup in comill can be modeled as adhering to first-order dynamics [3]. Recent studies have also showed the use of a comill to improve the flowability of cohesive powders by coating them with nanosized particles [39]. Deng et al. [40] developed a DEM model of a conical screen mill as a dry coating device to investigate the effect of impeller speed, feed rate, and screen size on residence time of particles inside the comill.

8. Tablet press

Similar to the roller compaction, the tablet press unit operation is identical in both batch and continuous operations. The compaction process in a tablet press has thus been modeled since decades. Tablet press unit modeling constitutes modeling of powder residence in the feed frame as well as powder compaction behavior after flow from the feed frame into the dies of a rotating turret. The uniformity of powder flow from the feed frame to the die impacts the tablet weight and its potency. Modeling of powder flow in the feed frame through DEM simulations is published and have provided deeper understanding of powder flow patterns, potential particle attrition, overlubrication in the feed frame, and effect on tablet weight variability [41,42]. Mateo-Ortiz and Mendez [43] studied the effect of paddle wheel speed and disc speed on the RTD of powder in the feed frame through laboratory experiments as well as DEM simulations. The experiments conducted showed that higher paddle wheel speeds lead to a lower mean residence time, narrow RTD profiles, and that the RTD profiles are similar to an ideal CSTR. Boukouvala et al. [11] used experimental data published in Mendez et al. [44] and developed response surface models for predicting mean residence time of powder as a function of turret speed and paddle wheel speed.

Compaction of powders has been modeled extensively. Patel et al. [45] provided a review of the various models proposed to characterize powder compressibility. These models relate powder properties such as porosity, volume, density, etc., which are a measure of the state of consolidation of powder, with compacting pressure. Among many such models proposed, the Heckel equation, the Kawakita equation, and the Kuentz–Leuenberger (KL) equation are the most commonly used in the pharmaceutical area. The Heckel equation given in Eq. (11.26) assumes that the reduction in porosity e due to applied pressure P on the powder obeys a first-order relationship.

$$ln\frac{1}{e} = kP + A \tag{11.26}$$

Here, constant k represents plasticity of the material and the constant A is the sum of two densification terms as given by Eq. (11.27):

$$A = ln\frac{1}{e_0} + B \tag{11.27}$$

where the first term is related to initial die filling and B gives densification due to rearrangement of particles.

The Kawakita model, described by Eqs. (11.28) and (11.29), assumes that when subjected to load, the product of pressure and volume terms is constant, as the particles are in equilibrium at all stages of compression.

$$\frac{P}{C_1} = \frac{1}{ab} + \frac{P}{a} \tag{11.28}$$

$$C_1 = \left(\frac{V_0 - V}{V_0}\right) \tag{11.29}$$

Here, V is the volume of compact at pressure P and V_0 is the initial apparent volume of powder. a is the initial porosity and $1/b$ is the plasticity parameter.

The KL equation (Eq. 11.30) was obtained by considering the pressure susceptibility of porosity reduction as a function of powder bed porosity [46].

$$P = \frac{1}{C_2}\left[(\varepsilon - \varepsilon_c) - \varepsilon_c ln\left(\frac{\varepsilon}{\varepsilon_c}\right)\right] \tag{11.30}$$

Here ε_c denotes the critical porosity when the powder attains a state of mechanical rigidity and $1/C_2$ is the plasticity parameter.

All of the parameters can be determined experimentally through regression from force-displacement data. Paul and Sun [47] systematically evaluated the performance of these equations and suggested that the KL equation is superior to Heckel and Kawakita equations applied to powders exhibiting a wide range of mechanical properties. Singh et al. [48] described a detailed model based on Kawakita equation. This model also incorporates prediction of tablet hardness as a function of compression force proposed by Kuentz and Luenberger [46] as given in Eqs. (11.31) and (11.32):

$$H = H_{\max}\left(1 - \exp\left(\rho_r - \rho_{r,cr} + \lambda_H\right)\right) \qquad (11.31)$$

$$\lambda_H = \ln\left(\frac{1 - \rho_r}{1 - \rho_{r,cr}}\right) \qquad (11.32)$$

where ρ_r is the relative density. The parameters $\rho_{r,cr}$ and $H_{\max}$ are to be fitted from experimental data. Escotet-Espinoza et al. [49] estimated these co-efficients from experimental data and developed empirical equations to relate these parameters to the original blend properties.

9. Integration

Developing individual unit operation models enables a superior understanding of each unit operation, the relationship between the material properties, process, and design parameters, and the quality attributes of intermediate materials. However, equally important, it enables the development of a flowsheet model that simulates the continuous manufacturing line. A robust and detailed flowsheet simulation is an approximate representation of the actual plant operation [50]. Dynamic behavior of the entire manufacturing line with respect to disturbances or changes in the input factors can be simulated using a flowsheet model. In addition, flowsheet models can be systematically used to perform sensitivity analysis. Sensitivity analysis is a tool to identify the input factors that are most influential on the output of interest, such as tablet properties. Once the critical input factors are identified, a design space can be determined, within which all the process, product quality, equipment, and production rate constraints are met. Thereafter, the optimal operating conditions in this feasible region that requires least operating and material costs can be determined. Another advantage of development of flowsheet model is enabling the traceability of material, which aids in identifying and discarding potential off-spec material. Lastly, it enables the testing of various control strategies on the process. The entire exercise can be performed in silico, resulting in savings in time, effort, and money from performing laboratory experiments or trials on the manufacturing setup.

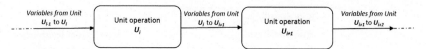

FIGURE 11.1   Schematic showing the flow of information in a generic flowsheet model.

Once the individual unit operation models are developed and the required parameters are estimated, the integrated model is built via connecting the inlet of a unit to the outlet of the preceding unit. Relevant material properties, operating conditions, and individual unit model variables are sent to the following unit. Fig 11.1 is a pictorial representation of information flow between unit operations in a generic flowsheet model.

Process simulators such as gPROMS and ASPEN facilitate the development of flowsheet models. The commercially available simulators aid in process simulation as well as the aforementioned advanced applications of flowsheet models such as design space identification, sensitivity analysis, and process optimization. To further describe flowsheet model development, the required connections and flow of information in a direct compaction, dry granulation, and wet granulation lines are shown in Fig 11.2. In the schematic, ρ, $\dot{m}$, *PSD*, x, z denote density, mass flow rate, particle size distribution of the powder or granules, concentration, and moisture content in the granule, respectively. The tablet weight, hardness, and potency are denoted by w_{tablet}, λ_{tablet}, and x, respectively. It is important to note that connections of the same type are to have the same property set. For example, the text in red, blue, gray, and green corresponds to variables related to powder, granules, ribbons, and tablets, respectively. The variable set for each of these phases is to be maintained consistent throughout the flowsheet in order to avoid simulation errors. The schematic shown in Fig. 11.2 is developed for a two-component system as an example. However, this can be easily extended to a multicomponent system. The readers are to note that the variables listed are shown as an example and the comprehensive list of variables; properties that need to be transferred from a unit model to another will depend on requirements and validity of the unit operation models for the specific manufacturing line in question.

The schematic shown in Fig. 11.3 is an example of an integrated model of a direct compaction line developed in gPROMS. The flowsheet model shown integrates the feeder, blender, feed frame, and tablet press models. The feeder model also includes a refill unit and a controller. The controller manipulates the feeder screw speed in order to control feeder flow rate. *Feeder_GEA001* and *Feeder_GEA002* were set to a flow rate of 20 kg/h and 5 kg/h, respectively. The refill unit intermittently fills the feeder with material when the feeder fill level decreases below a set point. In this example, the refill unit

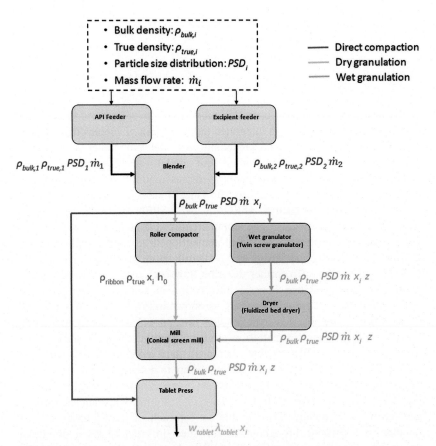

FIGURE 11.2 Schematic of an integrated flowsheet model for direct compaction, dry granulation, and wet granulation routes of pharmaceutical manufacturing.

drops material into the feeder when the fractional feeder fill level decreases to 0.1. This is equivalent to setting a refilling operation when the height of the material in the feeder reaches 10% of the total feeder height.

In order to demonstrate the capability of the flowsheet model to predict dynamic behavior, the properties of the material in the refill unit *Refill_Unit001* were set to be different than the material in the feeder unit *Feeder_GEA001*. Specifically, the bulk density of the material in *Refill_Unit001* is 450 kg/m^3 where the bulk density of the material in *Feeder_GEA001* is 400 kg/m^3. Bulk densities of the material in *Feeder_GEA002* as well as *Refill_Unit002* are set to 450 kg/m^3. Hence, any change seen in bulk density is the result of refilling in *Feeder_GEA001*. The simulation is run for 400 s, when the flow rates are observed to fluctuate around the set point values. As refill occurs when the fractional feeder fill level reaches 0.1, the propagation of the material with a

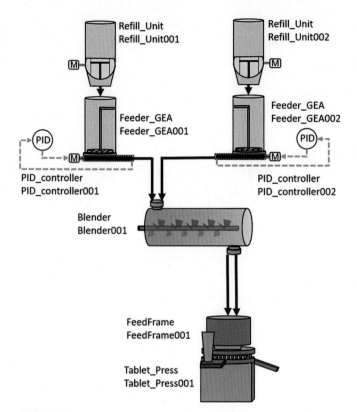

FIGURE 11.3   Schematic of an integrated model developed in gPROMS.

different bulk density and its eventual influence on tablet properties is demonstrated.

Fig. 11.4 shows feeder fill levels for the two feeders *Feeder_GEA001* and *Feeder_GEA002*. It can be seen that feeder refill occurs when the fractional fill level reaches 0.1.

Fig. 11.5 shows feeder flow rates fluctuating around a mean of 20 kg/h and 5 kg/h. A buildup in blender flow rate to the total flow rate of 25 kg/h is also shown in Fig. 11.5. It is also worth noting the larger deviations in feeder flow rates that occur periodically. The larger deviations are a result of feeder refill operations which is also seen experimentally [51].

Fig. 11.6 shows a change in bulk density from the outlet of *Feeder_GEA001* (red line) due to refilling operation from *Refill_Unit001*. An eventual change in bulk density from the outlet of *Blender001* is also shown (dotted line).

Bulk density of the powder blend has an effect on tablet properties. Specifically, for a fixed fill depth, the blend bulk density has an effect on the

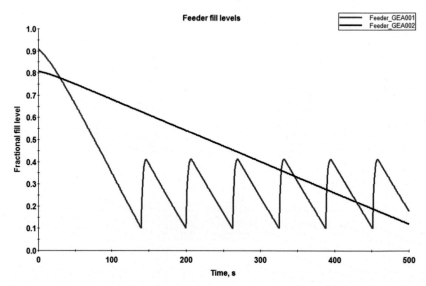

FIGURE 11.4 Fractional fill levels in *Feeder_GEA001* and *Feeder_GEA002*.

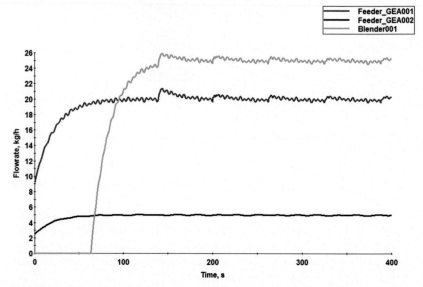

FIGURE 11.5 Flow rates in the feeder and blender units, *Feeder_GEA001*, *Feeder_GEA002*, and *Blender001*.

weight of tablet. Fig. 11.7 shows a change in tablet weight profile which is due to variation in the incoming powder blend bulk density.

With this simple case study, an application of the integrated process model is demonstrated. Using the unit operation models described in this chapter, it is clear how the flowsheet model could be used for applications such as material traceability.

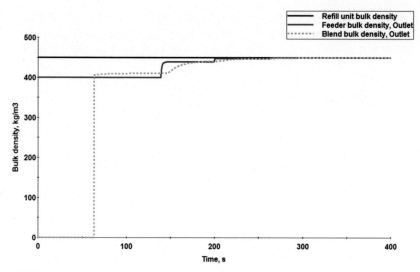

FIGURE 11.6 Bulk density of material in *Refill_Unit001* (black), outlet of *Feeder_GEA001* (red), and outlet of *Blender001* (*dotted line*).

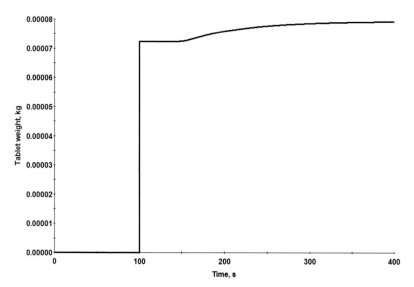

FIGURE 11.7 Plot showing the change in tablet weight as a result of change in blend bulk density.

10. Conclusions

This chapter provided an overview of several modeling approaches used to simulate processes in pharmaceutical manufacturing such as feeding, blending, roller compaction, wet granulation, drying, milling, and tableting. An attempt has been made to discuss modeling strategies ranging from data-driven models to fully mechanistic models. A brief discussion on the utilization of the developed process models to build an integrated model and its potential applications is included. An application of the flowsheet model is also demonstrated through a simple case study.

References

[1] Wang YF, Li TY, Muzzio FJ, Glasser BJ. Predicting feeder performance based on material flow properties. Powder Technol 2017;308:135—48.

[2] Boukouvala F, Muzzio FJ, Ierapetritou MG. Design space of pharmaceutical processes using data-driven-based methods. J Pharm Innov 2010;5(3):119—37.

[3] Wang Z, Escotet-Espinoza MS, Ierapetritou M. Process analysis and optimization of continuous pharmaceutical manufacturing using flowsheet models. Comput Chem Eng 2017;107:77—91.

[4] Rogers A, Ierapetritou MG. Discrete element reduced-order modeling of dynamic particulate systems. AIChE J 2014;60(9):3184—94.

[5] Liu XY, Hu Z, Wu WN, Zhan JS, Herz F, Specht E. DEM study on the surface mixing and whole mixing of granular materials in rotary drums. Powder Technol 2017;315:438—44.

[6] Sarkar A, Wassgren CR. Effect of particle size on flow and mixing in a bladed granular mixer. AIChE J 2015;61(1):46—57.

[7] Boukouvala F, Gao Y, Muzzio F, Ierapetritou MG. Reduced-order discrete element method modeling. Chem Eng Sci 2013;95:12—26.

[8] Sen M, Chaudhury A, Singh R, John J, Ramachandran R. Multi-scale flowsheet simulation of an integrated continuous purification-downstream pharmaceutical manufacturing process. Int J Pharm 2013;445(1—2):29—38.

[9] Johanson JR. A rolling theory for granular solids. J Appl Mech 1965;32(4):842—8.

[10] Hsu SH, Reklaitis GV, Venkatasubramanian V. Modeling and control of roller compaction for pharmaceutical manufacturing. Part I: process dynamics and control framework. J Pharm Innov 2010;5(1—2):14—23.

[11] Boukouvala F, Niotis V, Ramachandran R, Muzzio FJ, Ierapetritou MG. An integrated approach for dynamic flowsheet modeling and sensitivity analysis of a continuous tablet manufacturing process. Comput Chem Eng 2012;42:30—47.

[12] Liu Y, Wassgren C. Modifications to Johanson's roll compaction model for improved relative density predictions. Powder Technol 2016;297:294—302.

[13] Bindhumadhavan G, Seville JPK, Adams N, Greenwood RW, Fitzpatrick S. Roll compaction of a pharmaceutical excipient: experimental validation of rolling theory for granular solids. Chem Eng Sci 2005;60(14):3891—7.

[14] Muliadi AR, Litster JD, Wassgren CR. Modeling the powder roll compaction process: comparison of 2-D finite element method and the rolling theory for granular solids (Johanson's model). Powder Technol 2012;221:90—100.

[15] Oka S, Emady H, Kašpar O, Tokárová V, Muzzio F, Štěpánek F, Ramachandran R. The effects of improper mixing and preferential wetting of active and excipient ingredients on content uniformity in high shear wet granulation. Powder Technol 2015;278:266—77.

[16] Immanuel CD, Doyle FJ. Solution technique for a multi-dimensional population balance model describing granulation processes. Powder Technol 2005;156(2):213—25.

[17] Poon JMH, Immanuel CD, Doyle IIIFJ, Litster JD. A three-dimensional population balance model of granulation with a mechanistic representation of the nucleation and aggregation phenomena. Chem Eng Sci 2008;63(5):1315−29.

[18] Liu LX, Litster JD. Population balance modelling of granulation with a physically based coalescence kernel. Chem Eng Sci 2002;57(12):2183−91.

[19] Hapgood KP, Litster JD, Smith R. Nucleation regime map for liquid bound granules. AIChE J 2003;49(2):350−61.

[20] Barrasso D, Ramachandran R. Qualitative assessment of a multi-scale, compartmental PBM-DEM model of a continuous twin-screw wet granulation process. J Pharm Innov 2016;11(3):231−49.

[21] Cameron IT, Wang FY, Immanuel CD, Stepanek F. Process systems modelling and applications in granulation: a review. Chem Eng Sci 2005;60(14):3723−50.

[22] Madec L, Falk L, Plasari E. Modelling of the agglomeration in suspension process with multidimensional kernels. Powder Technol 2003;130(1):147−53.

[23] Matsoukas T, Kim T, Lee K. Bicomponent aggregation with composition-dependent rates and the approach to well-mixed state. Chem Eng Sci 2009;64(4):787−99.

[24] Barrasso D, Ramachandran R. A comparison of model order reduction techniques for a four-dimensional population balance model describing multi-component wet granulation processes. Chem Eng Sci 2012;80:380−92.

[25] Barrasso D, Tamrakar A, Ramachandran R. Model order reduction of a multi-scale PBM-DEM description of a wet granulation process via ANN. Proc Eng 2015;102:1295−304.

[26] Metta N, Ramachandran R, Marianthi Ierapetritou. A novel adaptive sampling based methodology for feasible region identification of compute intensive models using artificial neural network. AIChE J 2020;67(2). https://aiche.onlinelibrary.wiley.com/doi/abs/10.1002/aic.17095.

[27] Mortier S, De Beer T, Gernaey KV, Remon JP, Vervaet C, Nopens I. Mechanistic modelling of fluidized bed drying processes of wet porous granules: a review. Eur J Pharm Biopharm 2011;79(2):205−25.

[28] Mezhericher M, Levy A, Borde I. Theoretical drying model of single droplets containing insoluble or dissolved solids. Dry Technol 2007;25(4−6):1025−32.

[29] Mortier S, De Beer T, Gernaey KV, Vercruysse J, Fonteyne M, Remon JP, Vervaet C, Nopens I. Mechanistic modelling of the drying behaviour of single pharmaceutical granules. Eur J Pharm Biopharm 2012;80(3):682−9.

[30] Mortier S, Van Daele T, Gernaey KV, De Beer T, Nopens I. Reduction of a single granule drying model: an essential step in preparation of a population balance model with a continuous growth term. AIChE J 2013;59(4):1127−38.

[31] Aghbashlo M, Hosseinpour S, Mujumdar AS. Application of artificial neural networks (ANNs) in drying technology: a comprehensive review. Dry Technol 2015;33(12):1397−462.

[32] Capece M, Bilgili E, Dave RN. Formulation of a physically motivated specific breakage rate parameter for ball milling via the discrete element method. AIChE J 2014;60(7):2404−15.

[33] Loreti S, Wu CY, Reynolds G, Mirtic A, Seville J. DEM-PBM modeling of impact dominated ribbon milling. AIChE J 2017;63(9):3692−705.

[34] Metta N, Ierapetritou M, Ramachandran R. A multiscale DEM-PBM approach for a continuous comilling process using a mechanistically developed breakage kernel. Chem Eng Sci 2018;178:211−21.

[35] Barrasso D, Oka S, Muliadi A, Litster JD, Wassgren C, Ramachandran R. Population balance model validation and predictionof CQAs for continuous milling processes: toward QbDin pharmaceutical drug product manufacturing. J Pharm Innov 2013;8(3):147−62.

[36] Reynolds GK. Modelling of pharmaceutical granule size reduction in a conical screen mill. Chem Eng J 2010;164(2−3):383−92.

[37] Metta N, Verstraeten M, Ghijs M, Kumar A, Schafer E, Singh R, De Beer T, Nopens I, Cappuyns P, Van Assche I, Ierapetritou M, Ramachandran R. Model development and prediction of particle size distribution, density and friability of a comilling operation in a continuous pharmaceutical manufacturing process. Int J Pharm 2018;549(1):271−82.

[38] Vanarase AU, Osorio JG, Muzzio FJ. Effects of powder flow properties and shear environment on the performance of continuous mixing of pharmaceutical powders. Powder Technol 2013;246:63−72.

[39] Han X, Jallo L, To D, Ghoroi C, Dave R. Passivation of high-surface-energy sites of milled ibuprofen crystals via dry coating for reduced cohesion and improved flowability. J Pharmaceut Sci 2013;102(7):2282−96.

[40] Deng XL, Scicolone J, Han X, Dave RN. Discrete element method simulation of a conical screen mill: a continuous dry coating device. Chem Eng Sci 2015;125:58−74.

[41] Ketterhagen WR. Simulation of powder flow in a lab-scale tablet press feed frame: effects of design and operating parameters on measures of tablet quality. Powder Technol 2015;275:361−74.

[42] Mateo-Ortiz D, Mendez R. Microdynamic analysis of particle flow in a confined space using DEM: the feed frame case. Adv Powder Technol 2016;27(4):1597−606.

[43] Mateo-Ortiz D, Mendez R. Relationship between residence time distribution and forces applied by paddles on powder attrition during the die filling process. Powder Technol 2015;278:111−7.

[44] Mendez R, Muzzio F, Velazquez C. Study of the effects of feed frames on powder blend properties during the filling of tablet press dies. Powder Technol 2010;200(3):105−16.

[45] Patel S, Kaushal AM, Bansal AK. Effect of particle size and compression force on compaction behavior and derived mathematical parameters of compressibility. Pharmaceut Res 2007;24(1):111−24.

[46] Kuentz M, Leuenberger H. A new model for the hardness of a compacted particle system, applied to tablets of pharmaceutical polymers. Powder Technol 2000;111(1−2):145−53.

[47] Paul S, Sun CC. The suitability of common compressibility equations for characterizing plasticity of diverse powders. Int J Pharm 2017;532(1):124−30.

[48] Singh R, Gernaey KV, Gani R. ICAS-PAT: a software for design, analysis and validation of PAT systems. Comput Chem Eng 2010;34(7):1108−36.

[49] Escotet-Espinoza MS, Vadodaria S, Singh R, Muzzio FJ, Ierapetritou MG. Modeling the effects of material properties on tablet compaction: a building block for controlling both batch and continuous pharmaceutical manufacturing processes. Int J Pharm 2018;543(1):274−87.

[50] Ramachandran R, Arjunan J, Chaudhury A, Ierapetritou MG. Model-based control-loop performance of a continuous direct compaction process. J Pharm Innov 2011;6(4):249−63.

[51] Engisch WE, Muzzio FJ. Feedrate deviations caused by hopper refill of loss-in-weight feeders. Powder Technol 2015;283:389−400.

Chapter 12

Integrated process control

Ravendra Singh and Fernando J. Muzzio

Engineering Research Center for Structured Organic Particulate Systems (C-SOPS), Department of Chemical and Biochemical Engineering, Rutgers, The State University of New Jersey, Piscataway, NJ, United States

1. Introduction

The pharmaceutical process control area has been significantly investigated in the past few years. Singh et al. [1] suggested a monitoring and feedback control system for a batch tablet manufacturing process. Singh et al. [2] designed a feedback control system for an integrated roller compaction route of the continuous tablet manufacturing process. A detailed review of feedback control for a fluid bed granulation process has been performed by Ref. [3] and has been further discussed by Ref. [4] along with control aspects for efficient operation of a high shear mixer. Sanders et al. [5] have performed extensive feedback control studies using proportional integral derivative (PID) and model predictive control (MPC) methods on an experimentally validated fluidized bed granulation model. Singh et al. [6] developed an MPC system for a direct compaction continuous tablet manufacturing process. Singh et al. [7] also implemented an MPC-based feedback control system into a direct compaction tablet manufacturing process. The performance of the implemented feedback control system utilizing process analytical technology (PAT) tools was demonstrated [8]. Singh et al. [9] also proposed a combined feedforward/feedback control system for a tablet compaction process. An advanced hybrid MPC-PID—based combined feedforward/feedback control system has also been developed [10]. The advanced MPC has been implemented into a tablet press [11]. Furthermore, a moving horizon based real-time optimization (MH-RTO) technique has also been integrated on top of the advanced MPC system [12].

One of the main advantages of continuous manufacturing (CM) is that it can enable process-wide real-time quality control, thus paving the way for Quality by Control (QbC) and real-time release (RTR). Enabling a line with such capabilities requires a significant amount of effort and resources and needs to be carefully considered. Thus, a general strategy on the desired depth of sensing and control should be agreed upon prior to its implementation into

How to Design and Implement Powder-to-Tablet Continuous Manufacturing Systems
https://doi.org/10.1016/B978-0-12-813479-5.00011-2
251

the process. Characterization of individual unit operations and open-loop experiments on the entire process can reveal complex, dynamic relationships between critical process parameters (CPPs) and critical quality attributes (CQAs) of the product. However, the required level and control implementation, as may become defined by regulation in the near future, is likely to be based on a risk-based analysis of the product and the process. Low-dose formulations with highly potent active pharmaceutical ingredients (APIs) will likely require a higher degree of sensing and control. Similarly, a higher degree of sensing and control will be needed for the development of a RTR strategy, if that is desirable.

However, in the author's opinion, this diversity is likely only a temporary situation. As it becomes clear to regulators worldwide that real-time quality assurance via QbC is attainable, such capability is likely to become a widespread expectation [13]. Thus, this chapter attempts to set up a basic blueprint for implementing such systems. A model-based design of continuous pharmaceutical tablet manufacturing process has been reviewed. A closed-loop process flowsheet model of direct compaction continuous pharmaceutical tablet manufacturing process has been provided, along with guidelines on implementing a control system on a CM process. Lastly, the performance of a process operating under closed-loop control has been evaluated.

2. Design of the control architecture

A direct compaction tablet manufacturing process has been considered here for illustration of the design of the control architecture. The direct compaction process is shown in Fig. 12.1. The equipment present in the process includes three gravimetric, loss-in-weight feeders, with capability to add to more feeders if necessary. Following the feeders, a co-mill is integrated for delumping the powders and creating contact between the components. The lubricant is introduced in the system, through a lubricant feeder, after the co-mill in order to prevent overlubrication of the formulation in the co-mill. All these streams are then connected to a continuous blender to create a homogeneous mixture of all ingredients. The exit stream from the blender is fed to a rotary tablet press via a feed frame. The powder blend fills a die, which is subsequently compressed in order to create a tablet. The control system is also shown in Fig. 12.1.

When a continuous process is integrated with a real-time distributed control system (DCS) in a manner that real-time measurements are used to actively control process parameters and achieve desirable process or product performance, it is called a "closed-loop process." A properly implemented closed-loop process consistently ensures the achievement of desired, predefined product quality. Under closed-loop operation, the raw and intermediate critical material attributes (CMAs), the CPPs, and the final product CQAs are measured in real time. These values are used to take real-time corrective

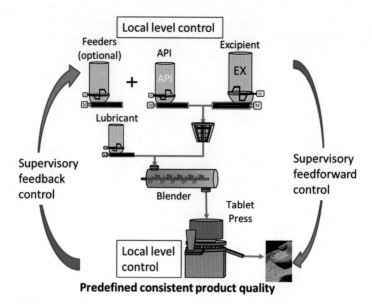

FIGURE 12.1 Illustration of closed-loop continuous tablet manufacturing process.

actions using feedback and/or feedforward controllers (FFCs). A controller can be defined as a mathematical equation or algorithm able to calculate the desirable quantitative actions needed to achieve the control goals.

Importantly, closed-loop control is *not* new to the pharmaceutical industry, as many unit operations have, and have had for decades, "in-process control" ("IPC") capabilities. For example, force and speed controllers in most tablet presses have been used for many years; however, these local controllers only act on individual process units and concern themselves primarily with mechanical parameters (as opposed to quality attributes). In contrast, a distributed control system acts on a unit operation to modify the outcome of another unit operation and to ensure efficient operation of the entire system.

The concept of a closed-loop process is illustrated in Fig. 12.1. If there are no variations in the raw material properties and if there are no process disturbances, then one may be able to manufacture the product with the desired quality by running a steady state process. The assumption that such an operation is possible is, in fact, one of the underlying assumptions, often inaccurate, of classic "process validation" methods. In this open-loop operational scenario, the product quality can be fitted, but cannot be guaranteed. In practice, there are always variations in raw material properties as well as in process disturbances. In some cases, these variations do not affect product quality significantly; however, in other cases, it is very difficult to achieve the desired product quality for open-loop operation, and one must take corrective action to achieve consistent quality.

In contrast, in closed-loop operation, accurate sensors for real-time monitoring of critical process variables are placed in the optimal locations, and therefore, appropriate controllers are added so that the critical process variables can be controlled automatically and in real time. The goal is to achieve the desired product quality irrespective of variations in raw material properties and process disturbances. The control system is also useful to manufacture product safely, to satisfy flexible market demands, to reduce manufacturing expenses (e.g., labor cost), and to assure regulatory requirements.

The overall control architecture of a continuous pharmaceutical manufacturing process includes a local level control system and a supervisory control system (see Fig. 12.1). As mentioned, the local level control system (IPC) is unit operation centric, while the supervisory control system governs the overall plant in an integrated manner. The local level controllers are normally built into the unit operations, but their performance needs to be evaluated. A supervisory control system is externally added and its function is more complex; therefore, more attention needs to be paid to its design and evaluation. As mentioned, supervisory control system could be feedback, feedforward, or a combination of both.

The control architecture needs to be designed carefully before implementing it into the manufacturing plant. The design of the control architecture involves the identification of critical control variables, pairing of control variables with suitable actuators, selection of a real-time monitoring tool for each control variable, selection of controllers and tuning of controller parameters, implementation of control loops into a process model, and performance evaluation of the control system [14]. The process flowsheet model is an important tool that can be used to design and perform preliminary tuning of an efficient and robust control architecture. The different steps required for designing a control architecture are described next [15].

The first step is to identify the critical process variables that need to be controlled. The critical control variables can be selected based on sensitivity analysis of CQAs to process variables, combined with process understanding. The process variables that have the most significant impact on CQAs should be controlled. The variables that are not self-regulating must be kept within the equipment and operating constraints. If a specific variable or variables significantly interact with controlled variables, then the former should also be controlled. The set points of control variables can be obtained from a design space analysis. The design space can be generated using feasibility analysis and optimization methods (see Chapter 12).

The next step is to select an actuator for each control variable. The general criterion to select an actuator is that it should have a large effect on the corresponding controlled variable. The controlled variable should be most sensitive to the selected actuator with respect to other actuator candidates. The selected actuator should rapidly affect the controlled variable, and if possible, it should also affect the controlled variable directly, rather than indirectly, and there should be minimum delay time. The relative gain array (RGA) method can be used to pair the controlled variables with corresponding actuators [16]. In the RGA method, the relative gain should be calculated for each pair of controlled variables and actuators, followed by the construction of a RGA matrix. Relative gain is the ratio of open-loop gain and closed-loop gain. The open-loop gain can be obtained by partial differentiation of a control variable function, with respect to an actuator candidate, while assuming that other actuator candidates are constant. This calculation is virtually identical to performing sensitivity analysis. Similarly, the closed-loop gain can be calculated by partial differentiation of the control variable function, with respect to an actuator candidate, while assuming that all other controlled variables are constant. The elements of the RGA matrix are the relative gains. The values in the array describe the relationship between the inputs (actuators) and outputs (control variables). Negative values indicate an unstable relationship, while a value of zero indicates no relationship. A value of one indicates that the specific input variable is the only influence on that output variable. A value between zero and one indicates an interaction among the control loops.

Subsequently, the possibility of cascading the control loops needs to be investigated. In cascade arrangements, the inner and outer loops need to be integrated such that the outer loop provides the set point for the inner loop. The inner loop is called the slave loop while the outer loop is called the master loop. Cascade control loops can improve performance in many instances; for example, when a large time delay is involved and/or when disturbances affect a measurable intermediate that directly affects the controlled variable. The cascade control system, however, is more difficult to tune and requires a larger number of variables to be measured in real time. In a cascade arrangement, the dynamics of the inner (slave) loops should be significantly faster than the outer (master) loop.

It is often desired to couple a feedback control system with feedforward control capabilities. The next step is thus to identify the feedforward control loops [9]. The basic concept of feedforward control is to measure important disturbance variables and take corrective action before the disturbances upset the process. The FFC therefore takes into account, proactively, the known and predictable effect of raw material variability and process disturbances. The precise control of the quality of the pharmaceutical product requires the effective implementation of corrective actions in the process/raw material variability before product quality is influenced. In comparison, feedback control is implemented in response to a measured (rather than predicted) deviation in product quality. While the method is "safer" because actual product quality is measured, feedback control systems are intrinsically less efficient, in that nonconforming product is more likely to be produced.

The ability of combining feedforward and feedback control methods to respond to real-time disturbances throughout the multiple unit operations is one of the key advantages of CM [17]. An illustrative example of combined feedforward/feedback control loops is shown in Fig. 12.2. As shown in the

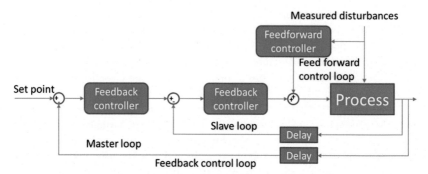

FIGURE 12.2 Combined feedforward/feedback cascade control architecture.

figure, the measured disturbance is the input to an FFC. The output of the FFC has been integrated with the feedback control loop. The FFC is a mathematical model relating input (disturbance) with the output (actuator). The feedforward control loops require additional sensors for real-time monitoring of process disturbances. A cascade feedback loop is also shown in Fig. 12.2. As mentioned, the cascade controller comprises of a slave loop and a master loop. The master loop provides the set point for the slave control loop. Depending on process requirements, the cascading of control loop may or may not be needed.

The next step in the design of the control architecture is the selection of real-time monitoring tools for critical control variables, slave control variables, and feedforward variables. The selection of sensors is important in the design phase of the control architecture because the performance of various sensors could vary from manufacturer to manufacturer. These sensor specifications can have significant impact on control loops performance. Therefore, on changing the sensors, the control loops might need to be retuned. Furthermore, as sensors typically introduce both error and delays, sensor models need to be integrated with the process model before testing the control architecture. Different performance criteria (e.g., accuracy, precision, operating range, response time, resolution, sensitivity, drift) and cost need to be considered before selecting a sensor [1]. Many pharmaceutical processes require the implementation of spectroscopic techniques to monitor different process variables [7] (see also Chapter 12).

After identifying control variables, actuators, and sensors, the decision on the type of controllers to be used for each control loop needs to be made. There are two main classes of controllers available, PID controllers and model predictive controllers. Within the PID structure, P, I, and D can be used individually or in any combination as per requirement.

A PID controller is simpler, easier to implement, easier to use, and in most cases (but not all) works very well. The performance of PID controllers can be substantially limited by process nonlinearity, process dead time, process interactions, and process constraints. A dead time compensator (e.g., Smith predictor) might need to be integrated with PID controller if the process is dead time dominated.

MPC refers to a family of control algorithms that employ an explicit model to predict the future behavior of a given process over an extended prediction horizon [6]. These algorithms are formulated as a performance objective function, which is defined as a combination of set point tracking performance and control effort. This objective function is minimized by computing a profile of controller output "moves" across a control horizon. The first controller output move is implemented, and then, the entire procedure is repeated at the next sampling instance [6]. MPC has proven to be a very effective control strategy and has been widely used in oil refining and chemical manufacturing. There are several advantages of using MPC. For example, MPC is better than PID when handling multivariable control problems, process constraints (e.g., actuator limitations, constraints on controlled variable, system constraints), process delays, system disturbances, equipment (sensor/actuator) failure, and process variable interactions. It is important to note that both PID and MPC need to be tuned using appropriate methods. Of the two, MPC is easier to tune. The method in Ref. [18] can be used for the tuning of a PID controller [18], while an optimization based method (e.g., ITAE [integral of time absolute error]) can be employed to tune MPC [6]. After designing the control architecture, it needs to be implemented into the process model for in silico performance evaluation before implementing it into the plant.

3. Develop integrated model of closed-loop system

Once the control architecture has been selected, it then needs to be implemented into an integrated process flowsheet model of the continuous pharmaceutical manufacturing process. This integration allows the user to obtain an integrated closed-loop model of the system [6,19,2]. The integrated closed-loop process model is required to tune the controller parameters and evaluate the performance of the control architecture. gPROMS, a dynamic flowsheet software developed by PSE, has been used extensively in previous work as a simulation platform to demonstrate the development of an integrated model of a closed-loop continuous pharmaceutical manufacturing process. The control architecture designed in the previous step, together with the integrated open-loop process flowsheet model, is the starting point for this development step.

To integrate the controller inputs/outputs with the predictions of the integrated flowsheet model, the first task is to create input/output control ports in the unit operation models, wherever needed. The purpose of the input/output control ports is to transmit the information between the process model and the controllers. An output port is the location where a sensor needs to be integrated and an input port is the location where an actuator needs to be placed. In the case of local level control, the input/output ports will be in same unit operation model, while in case of supervisory control, the input/output ports could be in different unit operation model.

The second task is to add the controllers into the flowsheet model and to create feedback loops by connecting the input/output ports with the controllers input/output. The integration of a control loop with the process model is illustrated in Fig. 12.3. As shown in the figure, the sensor outlet is connected with the controller inlet, and the controller outlet is connected with the actuator inlet. Within the controller, there is a port to integrate external set point signals obtained from the master controller (e.g., when using cascade control loops). When using a model to simulate the action of the control architecture, the sensor input signal comes from the process model, while the actuator output signal goes back to the process model. This integration resembles the actual implementation of control loops in a real-world processing plant.

The third step is to provide the controller parameters and constraints. In the case of a PID controller, gain, reset time, and rate are the tuning parameters. The controller parameters can be tuned using either a heuristic-based method (e.g., Ziegler and Nichols) or an optimization-based method [20]. In the case of an optimization-based method (e.g., ITAE), an objective function needs to be minimized using the optimization routine of gPROMS, and the controller parameters that give the minimum error need to be identified.

In addition to tuning controller parameters, there are other control parameters that need to be specified appropriately to achieve the ideal controller performance. These parameters are minimum and maximum limits of controller inputs (control variables), minimum and maximum limits of

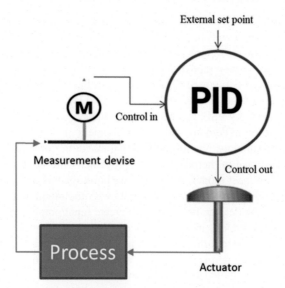

FIGURE 12.3   Integration of a control loop with the process model.

controller outputs (actuators), bias (controller offset used for smooth switching), and rate limits [6].

The closed-loop process flowsheet model of a direct compaction process is illustrated in Fig. 12.4. As shown in the figure, each feeder has the local level control to achieve the desired flow rate via manipulating the screw rotational speed. There is a ratio controller that dictates the set point for each feeder. The ratio controller is also useful when a change in line throughput is needed. In such a situation, the operator just needs to provide the "new total flow rate" to the ratio controller, and this will automatically change the feeder settings. The drug concentration needs to be controlled at the blender outlet via a feedback control loop. This control loop provides the ratio set point for the ratio controller. Note that when this control loop is not used, then the ratio set point needs to be provided manually.

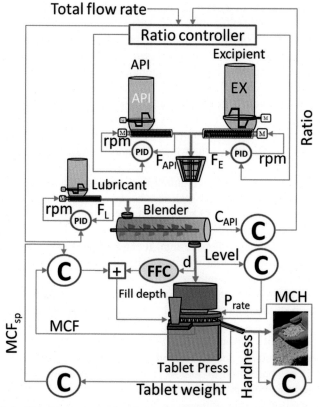

FIGURE 12.4 Closed-loop process flowsheet model. *API*, active pharmaceutical ingredients; *C*, controller; *d*, density; *E*, excipient; *F*, flow rate; *FFC*, feedforward controller; *L*, lubricant; *MCF*, main compression force; *MCH*, main compression height; *P*, production; *rpm*, revolution per minute; *sp*, set point.

A transfer pipe (chute) connects the blender outlet with tablet press inlet. The powder level needs to be controlled in this transfer pipe because of multiple reasons. Firstly, it ensures a consistent powder level in front of the PAT sensors placed in the chute. Second, it keeps the balance between line inputs and outputs in order to avoid overflow or underflow into the tablet press. Finally, a consistent powder level is often needed for a consistent tablet quality, as in many situations, the level in the chute directly affects powder density into the feed frame of the tablet press.

In the tablet press, the tablet weight is controlled through a cascade control arrangement using one master loop and one slave loop. The master loop is used to control the tablet weight that provides the set point to slave controller, which has been designed to control the main compression force (MCF) by manipulating the fill depth. A feedforward control loop has also been added. The real time measured powder bulk density is the input to the FFC that actuates the fill depth [21]. An FFC has been added here to take proactive action to mitigate the effects of variations in powder bulk density. Subsequently, the tablet hardness is controlled by manipulating the punch displacement. Note that, there are several other alternatives to control the tablet press as described in Bhaskar et al. [11]. Closed-loop process flowsheet models of continuous pharmaceutical manufacturing via wet granulation and roller compaction have been developed as well [19,2].

The off-spec products need to be rejected in real time to avoid mixing the "good" tablets with "bad" tablets. Most of the tablet presses commercially available have the inbuilt capability to reject tablets that fail to achieve the desired MCF. To the author's knowledge, currently, none of the commercially available CM processes have inbuilt rejection systems for off-spec tablets that fail to satisfy the drug potency, required by regulators. Therefore, a residence time distribution (RTD) based control system was developed to reject the off-spec tables in real time [22,23]. RTD, by definition, is the probability distribution of time that solid or fluid materials stay inside one or more unit operations in a continuous flow system. It can be used to characterize mixing and flow behavior of material within a unit operation. RTDs describe how a material travels through the unit operations of a continuous process, including the presence of stagnant (or semi-stagnant) regions, bypassing, etc. A method for prediction of tablet potency based on process parameters and RTDs will enable the in-line rejection of tablets that do not meet quality standards. A major complexity surrounding the determination of the RTD in the tablet press is the difficulty to measure blend composition in real time in the feed frame. RTD has been used to address the lack of a real-time tablet rejection system based on concentration inadequacies.

The closed-loop performance evaluation of the continuous tablet manufacturing process has been reported elsewhere [9,6,19,2]. In Ref. [2], the design of the control architecture required by a continuous tablet manufacturing process via roller compaction has been described. The control architecture was implemented into an integrated process flowsheet model and controller parameters were tuned. The performance of the control system was evaluated for set point tracking and disturbance rejection. In Ref. [6], a hybrid MPC-PID control architecture was developed for a direct compaction process. MPC was implemented in MATLAB and PID was implemented with the model simulated in gPROMS. Matlab and gPROMS communicated via gOMATLAB tool of gPROMS. The design, implementation, and evaluation of control architecture for continuous tablet manufacturing via wet granulation has been presented in Ref. [19]. Additional feedforward/feedback control architecture has also been designed, implemented, and evaluated [9].

4. Implementation and verification of the control framework

The developed control architecture has been implemented in the authors' pilot plant facility situated at C-SOPS, Rutgers University for experimental verification. A snapshot of the plant is shown in Fig. 12.5. The plant spans three levels to take advantage of gravity for material flow purposes. The top level is devoted to powder feeding and storage, while the middle level is dedicated to delumping and blending, and the bottom floor is used for compaction. Each level spans an area of 10×10 feet. Gravimetric feeders placed on the top floor are used to feed the API(s) and excipients. The API and excipient streams then go through a co-mill placed on the second floor, upstream of the blender. The

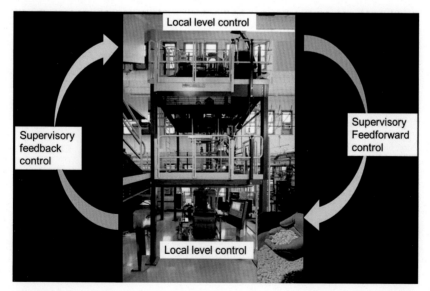

FIGURE 12.5 Continuous pharmaceutical tablet manufacturing pilot plant with an integrated control systems.

lubricant feeder is also placed on the second floor, after the mill but before the blender. A connecting pipe then transfers the output of the blender to the tablet press. In the connecting pipe, there is an interface to integrate PAT sensors. Through the transfer pipe, the blend is fed to the tablet press via a rotary feed frame to make tablets. This plant is modular in nature, thus, enabling the use of equipment in different combinations specific to the required experiments.

Next, we discuss the implementation of the designed control architecture into the physical line. A detailed, step-by-step procedure to implement and verify the control architecture in an actual plant has been previously presented by Ref. [7]. The key challenge is to enable communication between the real-time process monitoring sensors, their software, the data management tool, the control platform, and the plant actuators. Each component of the sensing and control framework communicates through its own language. Discussed below is a high-level introduction to the effort required to facilitate these "multi-lingual" components to communicate accurately and efficiently, thereby enabling closed-loop control.

The first step is to integrate all the required control hardware, software, and sensors with the pilot plant so that the plant can be run through a centralized control platform, and all the data from unit operations as well as from external sensors can be collected in a data historian. Fig. 12.6 provides an overview of the integration of control hardware, software, and sensors with the plant. The different unit operations of the plant have been integrated with two control platforms, DeltaV (Emerson) and PCS7 (Siemens). Note that only one control platform is needed at a time to operate the plant and therefore a switch has

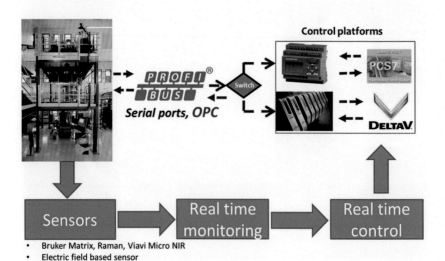

FIGURE 12.6 Integration of control hardware and software with a continuous pharmaceutical tablet manufacturing plant.

been placed to select the platform to be used for a particular run. The feeders are integrated with the control platform via Profibus. The co-mill and the blender are integrated with the control platform via serial port connections. The tablet press has been integrated with the control platform via OLE process control (OPC). Through these integration, the pilot plant can be run from a centralized PC and the process operational data can be collected in real time.

There are some inbuilt sensors in each unit operation to monitor different variables in real time. Feeder weight, powder flow rate, and feeder screw rotational speed are monitored at the feeding operation using inbuilt sensors. Impeller speed is monitored at the Comil operation. Blender impeller RPM is monitored at the blending operation. Tablet thickness, compression forces, ejection force, and other operating parameters (e.g., fill depth, feed frame speed, turret speed) are monitored at the tableting operation using inbuilt sensors. Some external sensors are also integrated with the plant for real-time process monitoring and control. Near-infrared (NIR) (Bruker Matrix, Viavi) and Raman (Kaiser) spectrometers are used for drug concentration monitoring in the blend. An electric field—based sensor Triflex (Fluidwell) is used for real-time monitoring of powder level. A check master module (FETTE) is used for monitoring of tablet weight, hardness, and thickness. A catch scale—based method, which works on the principle of gain in weight, has been also developed for real-time monitoring of average tablet weight [11].

The physical sensor measures a material attribute (e.g., blend composition) or a process parameter (e.g., compression force) and creates a representative signal. In the case of spectroscopic sensors, the generated signal is multivariate in nature and thus needs to be regressed to an ordinary number. The regressed signal is sent to data management software. The data management tool communicates this signal to the control system, typically through an OPC protocol, in addition to performing several other functions. The control system assimilates this signal and decides on the necessary corrective actions. The process hardware or the actuators receive directions from the control system and implement the corrective action to the plant to bring the CQAs/CPPs back to its set point. We will briefly discuss each of these communication steps, considerations during implementation of each step, and commonly available platform technologies to set up each communication junction.

At the beginning of the control loop is the *sensing probe*. The probe senses the material and creates a representative signal. This signal is typically stored in the sensor's proprietary software. For some probes, the signal could be a simple number. For NIR and some other spectroscopy-based sensors, the signal is multivariate in nature. In such cases, the multivariate signal must be converted to one (or a few) ordinary number. This is because most control algorithms are designed to assimilate ordinary numerical or binary numbers as input. Multivariate signals are converted to ordinary digits in the *online prediction tool* with the help of previously developed calibration models (see Chapter 12). The chemometric models are not developed in the online prediction tool but have to be developed using commercially available, *multivariate data analysis software*, such as Unscrambler X by Camo or Simca P+ by Umetrics. These models are then imported to the online prediction tool. The online prediction tool receives the multivariate signal from the sensor and converts it to an ordinary number using the imported calibration model. The predicted value of the CQA/CPP is communicated to a *PAT data management tool*. A schematic illustrating the flow of communication is shown in Fig. 12.7.

The PAT data management tool receives the predicted value of the CQA. This tool is designed for systematic data collection and storage. Moreover, the tool has an OPC communication protocol and thus can communicate with the control platform, which too is OPC-compliant. The data management tool allows data to be stored, protected, and plotted. It allows for alarms to be created. synTQ from Optimal Industrial Automation Limited and SiPAT from Siemens are common commercially available PAT data management platforms. Process Pulse II (Camo) can be also used for real-time prediction as well as communicating the data to control platform via OPC.

The control system receives data from the PAT data management toolbox via an OPC protocol. The control system is the component of the control framework which takes the input signal and decides on the necessary corrective action based on a previously developed algorithm. It communicates this action to the actuators, which make physical changes to ensure the CQA/CPP returns to its set point. The control system has hardware and software components. The software component is tasked with data reception, analysis,

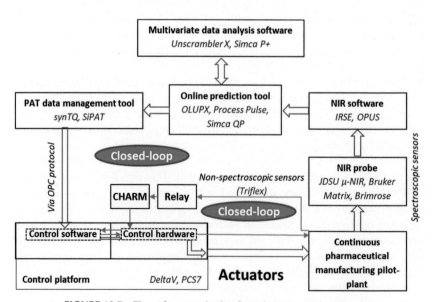

FIGURE 12.7 Flow of communication from the sensor to the actuator.

and decision-making. The hardware component is tasked with communicating with the actuators. This is typically done via a standard industrial communication protocols like Fieldbus or Ethernet. The flow of information is illustrated in Fig. 12.7. DeltaV from Emerson Process Management and PCS7 by Siemens are commercially available control platforms, which are particularly suitable for continuous pharmaceutical processes.

The implementation of the control loop in case of a nonspectroscopic sensor is relatively easier. The information and data flow for this case is also shown in Fig. 12.7. A nonspectroscopic sensor can be directly integrated with the control platform as shown in Fig. 12.7. A sensor that generates a standard 5−20 mA signal has been considered to demonstrate the concept of control system implementation. As shown in the figure, the sensor is integrated with the control panel at the relay through serial ports. From the relay, the signal is transmitted to a charm and from the charm to the controller. From the controller block (placed in the control panel), the signal is transmitted to the control platform where the control loop is implemented. In the control platform, the 5−20 mA signal is converted into a relevant variable to be monitored and controlled. From the control platform, the signal goes to the plant using the standard communication system depending on the unit operation where the signal needs to go (OPC, serial ports, Profibus).

Once the implementation of the control system is complete, the next step is to verify it in order to ensure that all its components are performing as expected. As an illustrative example, Singh et al. [8] performed a verification exercise on a ratio controller for two feeders. The example is described in brief here. Two feeders, one feeding semi-fine acetaminophen (APAP) (Mallinckrodt) and the other feeding silicified microcrystalline cellulose (Prosolv, JRS Pharma), were linked by a ratio controller. The feeders fed into a co-mill, which then feeds into a continuous blender. The blender output was discharged into a chute, on which a Viavi NIR microspectrometer was mounted. The NIR spectrometer monitored the APAP concentration. A previously developed calibration model in Unscrambler X was imported to Process Pulse I. OLUPX, an online prediction engine, was integrated with Process Pulse I. The APAP concentration was communicated to the DeltaV control system via a MATLAB OPC toolbox. An MPC algorithm analyzed the input and communicated the corrective action to the actuators, the feeder screws. The results were previously reported elsewhere [7,8]. Similarly, other control loops, previously discussed, have also been verified [11,24].

The operating and control user interface for a direct compaction continuous tablet manufacturing pilot plant is shown in Fig. 12.8. As shown in the figure, four feeders (one API feeder, two excipient feeders, one lubricant feeder), a co-mill, a blender, and a tablet press have been connected with the control platform. Note that each unit operation can also be run individually, if deemed necessary. Operational parameters (e.g., set points and actual values) have been displayed in real time for each unit operation.

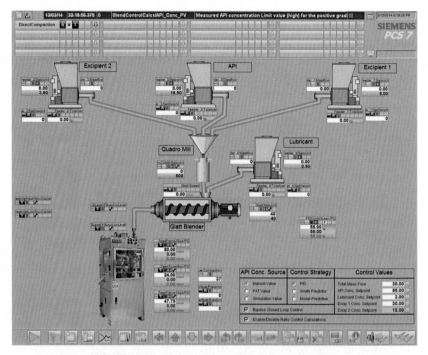

FIGURE 12.8 Plant operating and control user interface.

5. Characterize and verify closed-loop performance

The next step is to characterize and verify closed-loop performance of the entire process. This involves changing certain process parameters and observing the response of system to this change. This allows the experimenter to quantify the sensitivity of the system to process inputs in the presence of the control system, whose objective is to ensure that CQAs do not deviate from their prescribed set point. Such characterization increases process knowledge, facilitating quality to be built into the product by design. The exercise also confirms that the highest quality product is produced at the optimum process parameters.

All steps up to this point involved the implementation of the individual control loops and their verification. However, when running the entire process, all control loops operate simultaneously, which may create unanticipated interactions that must be understood and conflicts that might require resolution. Thus, a thorough characterization of the effect of process parameters on the CQAs of the final product is necessary. Moreover, running the process under the supervision of the control system allows one to understand the system performance not only at the set point but also around the set point. For instance, consider a scenario where the effect of the blender speed on the final product is being tested. At a certain design point of the DOE, the blender speed is set at a given value, for example, 100 rpm. When this design point was tested, the blender speed, in the absence of a control system, always stayed constant at 100 rpm. With the control system enabled, the blender speed may vary depending upon the CQAs that are linked to the speed. This allows for the population of the response surface near the blender speed of 100 rpm.

Interaction between control loops, as previously mentioned, may also create unprecedented interaction effects. Consider the previous example where the blender speed is set at 100 rpm. Fluctuations in this speed occur if the speed is linked in a control loop. Consider a scenario where the speed increases from 100 to 105 rpm. The slight increase in the speed results in a decrease in the blender holdup. The excess material is ejected from the blender and is added to the chute following the blender. A level sensor detects an increase in the holdup level within the chute. The system, to maintain its set point, might then increase the speed of tableting. An increase in tableting speed could result in a change in properties of the final product (e.g., hardness and dissolution properties of tablets), triggering new control actions. The process, if allowed to continue evolving, might then become unstable. Thus, it is highly recommended that DOE performed in open loop should be repeated in the presence of the control system. It enables one to characterize the effect of control loop interaction on the CQAs of the process stream and the final product.

Lastly, the long-term stability of the system, in the presence of the control system, should also be tested as a part of this step. As discussed in the chapter on regulatory compliance (see Chapter 16), this is increasingly becoming a standard expectation of regulators. This helps to ensure that the process does not drift with time, and in case it does, the control system can detect this and take corrective action. It also ensures no physical changes to the system occur as a result of extended operation, for example, sticking of powder material to surfaces over time, choking of hoppers, coating of blender and mill blades with material, etc.

The step is designed to enhance the experimenter's understanding of the system and enhance process knowledge. It helps to understand sources of variability and establish control over the variability. It also helps to establish relationships between CMAs, CPPs, and CQAs. Lastly, it facilitates process optimization to ensure the highest quality product. This step is the key enabler for successful implementation of Quality by Design [25].

6. Conclusions

A distributed control system is an essential part of a modern, continuous pharmaceutical tablet manufacturing process. Integration of sensing and control functions and implementation of capabilities for real-time quality assurance is a large undertaking. The closed-loop process flowsheet model helps to design, tune, and evaluate the control system prior to its implementation into the manufacturing setup. Using the CM process at Rutgers University as an example, steps to automate a process and optimize resources in order to establish consistency in operation and data management have been presented. Implementation of a real-time monitoring system in conjunction with a distributed control system has been discussed. The real-time control of continuous pharmaceutical manufacturing is a significant advancement and is considered essential for RTR as well as patient safety.

Acknowledgment

This work is supported by the National Science Foundation Engineering Research Center on Structured Organic Particulate Systems, through Grant NSF-ECC 0540855 and U.S. Food and Drug Administration (FDA).

References

[1] Singh R, Gernaey KV, Gani R. An ontological knowledge-based system for the selection of process monitoring and analysis tools. Comput Chem Eng 2010;34(7):1137−54.

[2] Singh R, Ierapetritou M, Ramachandran R. An engineering study on the enhanced control and operation of continuous manufacturing of pharmaceutical tablets via roller compaction. Int J Pharm 2012;438(1−2):307−26.

[3] Burggraeve A, Tavares da Silva A, Van den Kerkhof T, Hellings M, Vervaet C, Remon JP, Vander Heyden Y, Beer TD. Development of a fluid bed granulation process control strategy based on real-time process and product measurements. Talanta 2012;100:293−302.

[4] Bardin M, Knight PC, Seville JPK. On control of particle size distribution in granulation using high-shear mixers. Powder Technol 2004;140(3):169−75.

[5] Sanders CFW, Hounslow MJ, Doyle III FJ. Identification of models for control of wet granulation. Powder Technol 2009;188(3):255−63.

[6] Singh R, Ierapetritou M, Ramachandran R. System-wide hybrid MPC−PID control of a continuous pharmaceutical tablet manufacturing process via direct compaction. Eur J Pharm Biopharm 2013;85(3, Part B):1164−82.

[7] Singh R, Sahay A, Fernando M, Ierapetritou M, Ramachandran R. A systematic framework for onsite design and implementation of a control system in a continuous tablet manufacturing process. Comput Chem Eng 2014;66:186−200.

[8] Singh R, Sahay A, Karry KM, Muzzio F, Ierapetritou M, Ramachandran R. Implementation of an advanced hybrid MPC−PID control system using PAT tools into a direct compaction continuous pharmaceutical tablet manufacturing pilot plant. Int J Pharm 2014;473(1−2):38−54.

[9] Singh R, Muzzio F, Ierapetritou M, Ramachandran R. A combined feed-forward/feed-back control system for a QbD based continuous tablet manufacturing process. Processes 2015;3:339−56.

[10] Haas NT, Ierapetritou M, Singh R. Advanced model predictive feedforward/feedback control of a tablet press. J Pharm Innov 2017;12(2):10−123. https://doi.org/10.1007/s12247-017-9276-y.

[11] Bhaskar A, Barros FN, Singh R. Development and implementation of an advanced model predictive control system into continuous pharmaceutical tablet compaction process. Int J Pharm 2017;534(1−2):159−78. https://doi.org/10.1016/j.ijpharm.2017.10.003.

[12] Singh R, Sen M, Ierapetritou M, Ramachandran R. Integrated moving horizon based dynamic real time optimization and hybrid MPC-PID control of a direct compaction continuous tablet manufacturing process. J Pharm Innov 2015;10(3):233−53.

[13] Lee SL, O'Connor TF, Yang X, Cruz CN, Chatterjee S, Madurawe RD, Moore CMV, Yu LX, Woodcock J. Modernizing pharmaceutical manufacturing: from batch to continuous production. J Pharm Innov 2015;10(3):191−9.

[14] Singh R, Gernaey KV, Gani R. Model-based computer-aided framework for design of process monitoring and analysis systems. Comput Chem Eng 2009;33(1):22−42.

[15] Singh R. Model-based computer-aided framework for design of process monitoring and analysis systems. In: Chemical and biochemical engineering. Denmark: Technical University of Denmark; 2009. p. 296.

[16] Bristol E. On a new measure of interaction for multivariable process control. IEEE Trans Automat Contr 1966;11(1):133−4.

[17] Myerson AS, Krumme M, Nasr M, Thomas H, Braatz RD. Control systems engineering in continuous pharmaceutical manufacturing. J Pharmacol Sci 2015;104:832−9.

[18] Ziegler JG, Nichols B. Optimum settings for automatic controllers. Trans A.S.M.E. 1942;64:759−65.

[19] Singh R, Barrasso D, Chaudhury A, Sen M, Ierapetritou M, Ramachandran R. Closed-loop feedback control of a continuous pharmaceutical tablet manufacturing process via wet granulation. J Pharm Innov 2014;9:16−37.

[20] Seborg DE, Edgar TF, Mellichamp DA. Process dynamics and control. 2nd ed. John Wiley & Sons, Inc; 2004.

[21] Singh R, Román-Ospino AD, Romañach RJ, Ierapetritou M, Ramachandran R. Real time monitoring of powder blend bulk density for coupled feed-forward/feed-back control of a continuous direct compaction tablet manufacturing process. Int J Pharm 2015;495(1):612−25.

[22] Singh R. Systematic framework for implementation of RTD based control system into continuous pharmaceutical manufacturing pilot-plant. Pharma 2017;(34):43−6.

[23] Bhaskar A, Singh R. Residence time distribution (RTD) based control system for continuous pharmaceutical manufacturing process. J Pharm Innov 2018;14:316−31. https://doi.org/10.1007/s12247-018-9356-7.

[24] Singh R. A novel continuous pharmaceutical manufacturing pilot-plant: advanced model predictive control. Pharma 2017;(28):58−62.

[25] Yu LX. Pharmaceutical quality by design: product and process development, understanding, and control. Pharmaceut Res 2008;25(4):781−91.

Chapter 13

Applications of optimization in the pharmaceutical process development

Zilong Wang[1], Marianthi Ierapetritou[2]

[1]*Manufacturing Intelligence, Global Technology and Engineering, Pfizer Global Supply, Pfizer Inc., Peapack, NJ, United States;* [2]*Department of Chemical and Biomolecular Engineering, University of Delaware, Newark, DE, United States*

1. Introduction

There is an increasing number of collaborative efforts in improving pharmaceutical process development and manufacturing, jointly from academia, regulatory agencies, and the industry. Such a trend is stimulated by a few factors. From the economic perspective, it has been well acknowledged by the pharmaceutical industry that efficient manufacturing processes need to be developed and adopted in order to produce qualified products and maximize profits within a drug's patent life. From the regulatory perspective, it is the vision for Food and Drug Administration (FDA) to have a highly efficient, agile, flexible pharmaceutical sector that can consistently provide drugs of high quality [1]. To achieve the goal of improving pharmaceutical product quality, FDA had a number of initiatives, including process analytical technology [2], the Quality by Design initiative [3], and more recently, the creation of an Emerging Technology Team designed to facilitate adoption of new technologies (e.g., continuous manufacturing, 3D printing technologies [4], etc.) in the improvement of pharmaceutical product quality [5].

In order to embrace the benefits of emerging technologies in pharmaceutical manufacturing, it is critical to have in-depth process knowledge. Process modeling tools have become increasingly important in gaining insights into processes and assisting risk assessment via prediction based on process data [6,7]. For a review on the recent development in the modeling of pharmaceutical processes, the interested readers are referred to Ref. [8] and Chapter 2 in Ref. [9]. Commonly used modeling approaches include first-principle models (e.g., discrete element models [DEM] [10], finite element models [FEM] [11]), population balance models [12], phenomenological models [9],

How to Design and Implement Powder-to-Tablet Continuous Manufacturing Systems
https://doi.org/10.1016/B978-0-12-813479-5.00012-4

and reduced order models (e.g., response surface models [13], artificial neural network [ANN] [14], latent variable methods [15], etc.). A well-developed process model is a powerful tool to predict process dynamics [16], characterize design space (Chapter 6 in Ref. [17]), investigate critical process parameters (CPP) [18], facilitate process control [19], and perform process optimization [20].

Based on process models, mathematical optimization approaches have long been used in other industries to improve process performance, with a variety of applications covering process design, operations, and control. A general classification of the mathematical optimization problem is given in Ref. [21], where the general formulation of an optimization problem is as follows:

$$\min Z = f(x, y)$$

such that

$$h(x, y) = 0 \qquad (13.1)$$

$$g(x, y) \leq 0$$

$$x \in X, y \in \{0, 1\}^m$$

where $f(x, y)$ is the objective function (e.g., cost); $h(x, y)$ are the equality constraints describing the process systems (e.g., mass balance); $g(x, y)$ are the inequality constraints defining the process constraints (e.g., specifications on product qualities); x represents continuous variables; and y denotes discrete variables. Problem (1) corresponds to a mixed-integer problem (MIP). When no discrete variables exist in the system, Problem (1) is reduced to a nonlinear program (NLP) when any of the functions involves nonlinearities, or a linear program when all the functions are linear.

In pharmaceutical process development, mathematical optimization approaches have been implemented to improve product formulations, drug delivery systems, and manufacturing processes. This chapter aims to provide an overview of the applications of optimization in the pharmaceutical process development and briefly introduce the mathematical tools (e.g., data-driven models, optimization algorithms) that are generally used, with a particular focus on continuous manufacturing systems.

2. Optimization objectives in pharmaceutical process development

In process optimization, the first question that should to be considered is what needs to be optimized. The answer to this question determines the objective function. Depending on the nature of the study, there are a number of objective functions that have been used when optimizing a pharmaceutical process. In this section, we first introduce some commonly adopted objective functions for single-objective optimization and then discuss the cases when we need to consider multiple objectives simultaneously.

2.1 Single-objective optimization

One common choice for the objective function is to optimize a critical quality attribute (CQA) of a drug product. For example, Velásco-Mejía et al. [22] determined the optimal operating conditions for a drug crystallization process that provided the highest crystal density. Monteagudo et al. [23] developed an optimized formulation to obtain a pharmaceutical product with best taste-masking efficiency. Chavez et al. [24] optimized the formulation of a pharmaceutical tablet product to maximize the joint probability for five CQAs to meet a minimum satisfactory level of quality. Pal et al. [13] determined the optimal formulation that generates a desired drug release profile.

The optimization of process performance and efficiency has been investigated in various pharmaceutical processes. Based on a thermodynamic model that can predict phase equilibria of multicomponent systems, Sheikholeslamzadeh et al. [25] investigated the optimal operating conditions that maximized the crystallization yields for a batch cooling−antisolvent crystallization process. Zhang and Huang [26] calculated the optimal operating conditions that led to the largest chemical oxygen demand (COD) of a pharmaceutical wastewater treatment concentration in an active pharmaceutical ingredient (API) synthesis process for a benzazepine class of heterocyclic compound (a weight loss drug).

Economic objectives are often used in process optimization as they are critical to decision-making. Jolliffe and Gerogiorgis [27] formulated an NLP optimization problem of a conceptual upstream continuous process for the production and purification of ibuprofen, with the objective of minimizing the total cost consisting of capital and time-discounted operating expenditure. Abejón et al. [28] minimized the total costs for a separation process using multistage membrane cascades, with the operation variables as independent variables and the product specifications as constraints. More closely related to the topic of this book, Boukouvala and Ierapetritou [20] formulated a constrained optimization problem to minimize the total cost of a continuous direct compaction (CDC) process while meeting product quality requirements. A similar optimization problem of the CDC process was solved in Ref. [18].

Recently, there has been growing interest in improving the flexibility of operating pharmaceutical manufacturing processes. Flexibility is a quantitative measure of the capability of a process to remain feasible in the presence of process uncertainties (e.g., variations in the flow rate), which was initially proposed by Grossmann and Morari [29]. Grossmann et al. [30] provided an overview of recent advances in quantifying flexibility of chemical processes. The concept of flexibility has also been introduced to pharmaceutical processes and its applications in the context of characterizing the design space. Rogers and Ierapetritou [31] developed a surrogate-based approach to investigate both the steady state and the dynamically changing design space of a roller compaction process. Rogers and Ierapetritou [32] further evaluated the flexibility of the roller compaction process by computing the stochastic flexibility index, with the uncertainties in process inputs being described by an arbitrary probability distribution. Adi and Laxmidewi [33] computed the volumetric flexibility index to evaluate the operational flexibility of a separation process using membranes.

The analysis of environmental impacts of pharmaceutical processes has attracted increasing attention over the past few years. A systematic approach is via Life Cycle Assessment (LCA), which evaluates the environmental impacts throughout a product's life cycle, covering raw material acquisition, production and usage, and waste disposal [34]. Proper integration of environmental considerations in pharmaceutical processes can make significant contribution to building a more sustainable and environmental benign pharmaceutical process. Ott et al. [35] presented a holistic life cycle—based process optimization and intensification for an API production process of an anticancer drug, which investigated the main bottlenecks of the process and made recommendations for optimization strategies by accounting for a number of partly contradictory environmental effects. Jolliffe and Gerogiorgis [27] evaluated the environment impact of an API process by using the environmental factor (E-factor), which was defined as the total mass of waste generated per unit mass of product. Ott et al. [36] performed an environmental assessment of different pathways of the production of rufinamide, based on both simplified metrics (e.g., process mass intensity, cumulative energy demand) and a holistic LCA investigation. The results indicated the potential environmental benefits of switching from a multistep batch process to a continuous flow process for the production of rufinamide.

2.2 Multiobjective optimization

Many pharmaceutical processes have more than one objective to be optimized. For example, when considering finished product drug content, it is usually needed to simultaneously control the mean values (to be as close to targets as possible) and the variance (to be as small as possible). This multiobjective problem has been formulated into a robust design problem, with applications demonstrated in Refs. [37,38]. In addition, it has also been investigated to achieve target product qualities and meanwhile reduce the operational costs for a pharmaceutical granulation process [39]. For continuous pharmaceutical processes, multiobjective optimization has been used to improve a crystallization process of Paracetamol [40] and to estimate kinetic parameters of a continuous plug flow antisolvent crystallizer [41]. Ardakani and Wulff [42] gave an overview of a wide range of methods for multiobjective optimization problems. Below, we first show the general form of a multiobjective optimization problem and then provide a review on the mathematical approaches that are widely adopted in pharmaceutical processes. Interested readers are referred to Ref. [43] for more mathematical details.

A multiobjective optimization has the following general form

$$\min\{f_1(\mathbf{x}), f_2(\mathbf{x}), \ldots, f_k(\mathbf{x})\}$$
$$\text{such that } \mathbf{x} \in S,$$

(13.2)

which has k (≥ 2) ≥ 2) ≥ 2) conflicting objective functions $f_i : \mathbb{R}^n \to \mathbb{R}$ that need to be minimized simultaneously. The decision variable $\mathbf{x}$ belong to a nonempty feasible region $S \subset \mathbb{R}^n$. Objective vectors are denoted as $\mathbf{z} = \mathbf{f}(\mathbf{x}) = (f_1(\mathbf{x}), f_2(\mathbf{x}), \ldots, f_k(\mathbf{x}))^T$. In the multiobjective optimization, objective vectors are regarded as optimal (i.e., Pareto optimal) if none of their components can be improved without deterioration to at least one of the other components [43].

The most direct way of solving a multiobjective optimization problem is to find the Pareto curve consisting of multiple Pareto optimal solutions. Abejón et al. [28] used the Pareto curves to interrelate two variables (i.e., the product purity and the process yield) for a continuous organic solvent nanofiltration process. Brunet et al. [44] formulated MIP model for the process design of the production of penicillin V, which aims to determine the optimal operating conditions of the pharmaceutical plant (continuous variables) and the plant topology (integer variables) that optimize simultaneously the profitability of the process and the associated environmental impact.

Another way to solve a multiobjective optimization problem is by maximizing a desirability function. Derringer [45] defined the desirability as a weighted geometric mean of transformed objectives f_i (scaled to the range between 0 and 1). The advantage of using the desirability is that it can use a single measure to characterize the overall performance. However, the resulted optimal solution is highly sensitive to the assigned weights. Uttekar and Chaudhari [46] selected the optimal formulation that maximizes the desirability function in order to achieve the targeted particle size distribution for budesonide (a drug used for the treatment of asthma), which was produced by using the amphiphilic crystallization process. Sato et al. [47] calculated the optimal conditions of a crystallization process that satisfy both the required amount of residual solvent and the particle size D50 based on the desirability. Chakraborty et al. [48] optimized the formulation of a fast-dissolving pharmaceutical wafer containing loratadine (a drug to treat allergies), with the desirability as the objective function which involves four CQAs of the drug. Kermet-Said and Moulai-Mostefa [49] applied electrocoagulation to a pharmaceutical wastewater treatment process and investigated the optimal operating conditions to maximize COD removal and turbidity removal.

Alternatively, multiobjective optimization problems can also be solved by the goal programming approach. It requires the decision-maker to specify a goal point and finds a feasible solution that is as close to the goal as possible [42]. Nha et al. [37] developed a lexicographical dynamic goal programming to account for the dynamic nature of pharmaceutical quality characteristics and utilized it in testing for in vitro bioequivalence (considering two time-dependent responses: gelation kinetics and drug release rates) of a generic drug. Li et al. [50] proposed a priority-based optimization scheme, which incorporates goal programming methods, modified desirability functions, and higher-order response surface models, to address the multiresponse pharmaceutical formulation optimization problem.

3. Applications of data-driven models in optimization

In order to formulate the optimization problem, data-driven models have been widely used for pharmaceutical process optimization. Such models investigate the system input—output relationship only on the basis of system data, without requiring any explicit knowledge of the physical behavior of the system [51]. In some research areas, the data-driven model is also known as "surrogate model" [52], "metamodel" [53], "reduced-order model" [54], or "response surface" [55]. For pharmaceutical processes, data-driven models are a useful tool to enhance fundamental understanding when first-principle models are not available, due to, for example, a lack of knowledge on the mechanical and physiochemical properties of raw materials [56]. They are also useful in cases where first-principle simulations (e.g., DEM, FEM) can be performed, but they are usually too computationally expensive to be directly applied in the process optimization or design settings [53]. In such cases, data-driven models can be used as a computational efficient approximation to the expensive simulation and contribute to the definition and solution of an optimization problem.

In this section, we first briefly review the sampling plans for data-driven models and then discuss a variety of modeling techniques that have been adopted in pharmaceutical processes, which is followed by model validation methods. Finally, we will demonstrate how data-driven models can support process optimization.

3.1 Sampling plans

In order to extract the most meaningful process information within a limited sampling budget, we need an organized sampling plan to determine how to sample the input space. The selection of a specific sampling plan depends on what kind of experiments are being conducted—whether it's a physical experiment or a computational experiment—and on what data-driven modeling technique is being used.

For pharmaceutical process development, it is mainly dependent on the physical experiments conducted to gain process knowledge. In such cases, sampling plans are chosen by using "Design of Experiment" (DoE) methods. The main idea is to plan experiments in order to minimize the effects of random errors [53]. Singh et al. [57] presented an extensive review on classical DoE sampling plans. Widely used designs include (fractional) factorial designs [13,48,58], central composite designs [26,38,46], mixture designs [37],

Box—Behnken designs [59,60], and Plackett—Burman designs [61,62]. These experimental plans usually focus on sampling around the input space boundaries with a few sample points at the center of the input space.

When computer experiments are to be conducted, as most of simulations are deterministic, the goal of an experimental plan is focused on reducing systematic errors rather than random errors. For this type of experiments, Sacks et al. [63] claimed that a good design should fill out the entire input space rather than only concentrate on the boundaries. Commonly used "space-filling" designs include Latin hypercube designs (LHD) [64,65], Hammersley sequence sampling [66], orthogonal arrays [67], and uniform designs [68]. An extensive discussion of modern designs for computer experiments is provided in Ref. [69].

3.2 Building a data-driven model

There is a variety of data-driven models that have been developed and used in different engineering fields. Below, we discuss four types of models that are widely used in modeling and optimizing pharmaceutical processes. These models can be applied when the process output to be modeled can be assumed to be continuous and smooth. These assumptions are generally valid for most engineering processes. Forrester and Keane [70] commented that when the continuity assumption is not guaranteed, the process can then be modeled with multiple data-driven models which are patched together at discontinuities. Such an exception will not be considered in this chapter.

3.3 Response surface methodology

Response surface methodology (RSM) was first proposed by Box and Wilson [71] to improve chemical manufacturing processes. An RSM model approximated the process input—output relationship with a low-degree polynomial model [72] in the following form:

$$y = f'(x)\beta + \varepsilon \tag{13.3}$$

where $x = (x_1, x_2, ..., x_k)'$; $f(x)$ if a vector function consisting of powers and cross-products of powers of $x_1, x_2, ..., x_k$ up to a degree d ($\geq 1) \geq 1) \geq 1$); β is a vector of p model parameters; ε is a random experimental error with zero mean. Two commonly used RSM models are the first-degree model

$$y = \beta_0 + \sum_{i=1}^{k} \beta_i x_i + \varepsilon \tag{13.4}$$

and second-degree model

$$y = \beta_0 + \sum_{i=1}^{k} \beta_i x_i + \sum_{i=1}^{k}\sum_{i<j} \beta_{ij} x_i x_j + \sum_{i=1}^{k} \beta_{ii} x_i^2 + \varepsilon \tag{13.5}$$

The values of β can be estimated using analytical expressions (i.e., ordinary least squares estimators) [72,73].

In order to evaluate the significance of model parameters of a specific RSM model, analysis of variance (ANOVA) can be carried out, together with Student's t-test [57]. Only significant model parameters should be retained in the final model. In addition, to choose the best RSM model and prevent the danger of overfitting, Singh et al. [57] suggested the use of several metrics to evaluate the model fitting, including R^2, R^2_{adj}, predicted residual sum of squares, and Q^2.

RSM is a regression technique. An RSM model is usually constructed using experimental designs from DoE theory [72]. The applications of RSM in modeling pharmaceutical processes can be found in a variety of studies [13,37,38,49,74]. Forrester and Keane [70] stated that RSM models are appropriate for problems with low dimensions, uni- or low modality, where physical experiments were conducted. For deterministic computer experiments, however, Jones [55] demonstrated that RSM models could be insufficient to capture the shape of the function and thus may not be able to identify an optimal solution in optimization settings.

3.4 Partial least squares

Partial least squares (PLS) regression, also termed as projection to latent structures, is a multivariate regression approach [75,76]. Assuming the data have been mean-centered and scaled to unit variance, PLS projects the process inputs $\mathbf{X}$ (of size N-by-K) and process outputs $\mathbf{Y}$ (of size N-by-M) to the common latent space of A latent variables, with the following model structure:

$$\mathbf{X} = \mathbf{TP'} + \mathbf{E_X} \tag{13.6}$$

$$\mathbf{Y} = \mathbf{TQ'} + \mathbf{E_Y} \tag{13.7}$$

$$\mathbf{T} = \mathbf{XW} \tag{13.8}$$

where $\mathbf{T}$ (of size N-by-A) is the score matrix; $\mathbf{P}$ (of size K-by-A) and $\mathbf{Q}$ (of size M-by-A) are loading matrices; $\mathbf{E_X}$ and $\mathbf{E_Y}$ represent residuals; and $\mathbf{W}$ (of size K-by-A) is the weight matrix.

Different algorithms can be applied to calculate the PLS model, such as nonlinear iterative partial least squares algorithm [75] and the expectation maximization algorithm [77]. The number of latent variables A is determined by cross-validation (CV), which is performed by dividing the data in a number of groups and then developing different models in parallel using reduced data with one of the groups removed [76]. Among different CV methods, Shao [78] suggested not to use the leave-one-out (LOO) approach. Details on CV will be further discussed in Section 3.3.

By collecting data based on the DoE theory, the PLS approach is a powerful tool for analyzing data with high dimensions, noise, and strong collinearities in both **X** and **Y** [76]. This modeling technique seeks to approximate effectively both of the data tables **X** and **Y** as well as maximize the correlation between **X** and **Y** [79]. In cases when there exist strong collinearities in **X** and/or **Y** is only sensitive to a reduced set of input combinations, PLS can sufficiently reduce the problem dimensions and simplify the problem without losing much information. Recently, advances have been made to extend PLS to the modeling of nonlinear systems [80]. Pharmaceutical process applications involving PLS techniques can be found in Refs. [23,39,81].

3.5 Artificial neural network

ANN is a biologically inspired modeling technique that simulates the human brain's way of processing information. An ANN model is formed by numbers of single units (known as processing elements or neurons), which are connected with coefficients (weights) and constitute the neural structure [82]. As the building component, a neuron passes a sum of weighted inputs to a transfer function (e.g., sigmoidal function) and yields the output. The versatility of ANN comes from the various ways that neurons can be connected in a network. A vast number of ANNs have been developed and are surveyed in Ref. [83], among which the most widely used type of ANN is the back-propagation (BP) network [84]. A BP network has a multilayer perception architecture: (1) an input layer of nodes representing the process inputs; (2) an output layer of nodes for the process outputs; and (3) one or more hidden layers containing nodes to capture the nonlinear relationship. Practically, the BP network with one hidden layer is sufficient to approximate most functions [85].

For a BP network, the number of nodes in the input and output layers is determined by the dimension of process inputs and outputs, respectively. However, it is challenging to determine the number of nodes in the hidden layer. When the number is too large, training the model requires a high computational cost and also raises the risk of overfitting. On the other hand, if there are too few hidden nodes, the model may not be able to accurately represent input—output relationship [86]. In most applications, the number of hidden nodes is still determined by trial and error, while some practical rules have been applied in Refs. [86,87].

The model coefficients (weights) of an ANN model are calculated by using a training algorithm. The most commonly used algorithm is the feedforward error-backpropagation learning algorithm (BP algorithm) [88]. Each iteration of the BP algorithm involves two steps. A forward propagation is performed to pass information from the input layer through the hidden layer to the output layer. Then a BP of error stage is used to calculate the error of each layer of nodes sequentially from the output layer to the input layer. Based on the calculation, the weights are adjusted to reduce the error by using the gradient descent method. This BP algorithm is performed iteratively until it reaches a certain prespecified level of accuracy.

As a useful method of modeling nonlinearities, the ANN model and its different variants have been implemented both as a regression technique and an interpolation technique [89]. Applications of ANN in pharmaceutical processes are described in Refs. [22,90−92].

3.6 Kriging

Kriging is a widely used interpolation method which is named after the South African mining engineer Krige [93]. In different fields, Kriging is also known as stochastic process model [94] or Gaussian process model [95]. The ordinary Kriging represents a process output with the following model:

$$y(x) = \beta + \varepsilon(x), \tag{13.9}$$

where β is a model parameter which represents the surface mean, $\varepsilon(x)$ is the realization of a stationary Gaussian random field $(\mathbb{R}^d \to \mathbb{R})$: $\varepsilon(x) \sim \text{Normal}(0, \sigma^2)$. It is assumed that $\varepsilon(x)$ is spatially correlated: if two points x and x' are spatially close to each other, then $\varepsilon(x)$ and $\varepsilon(x')$ will tend to be similar. The spatial correlation can be modeled with various correlation functions. The mostly used is the Gaussian function:

$$\text{Corr}[\varepsilon(x), \varepsilon(x')] = \exp\left[-\sum_{h=1}^{d} \theta_h |x_h - x'_h|^2\right], \quad (\theta_h \geq 0) \tag{13.10}$$

With the Kriging model, the best linear unbiased predictor at x^* can be expressed as

$$\hat{y}(x^*) = \beta + r'R^{-1}(y - 1\beta) \tag{13.11}$$

where $y = (y^{(1)}, \ldots, y^{(n)})'$ is the n-vector of observed function values; R is the n-by-n matrix with the (i,j) entry being $\text{Corr}[\varepsilon(x^{(i)}), \varepsilon(x^{(j)})]$; 1 is an n-vector of ones; r is an n-vector with the i-th entry being $\text{Corr}[\varepsilon(x^*), \varepsilon(x^{(i)})]$.

In addition, the mean squared error of the predictor can be derived as follows:

$$\hat{s}^2(x^*) = \sigma^2\left[1 - r'R^{-1}r + \frac{(1 - 1'R^{-1}r)^2}{1'R^{-1}1}\right] \tag{13.12}$$

The model parameters β, $\theta_h, \ldots, \theta_d, \sigma^2$ are obtained by maximizing a likelihood function [94].

Kriging is mostly used to approximate computer simulations with data sampled from a space-filling design (e.g., LHD). On the other hand, Kriging can be modified by introducing a "nugget" factor in order to approximate a stochastic simulation with a homogeneous noise level [96]. Recently, Ankenman et al. [97] developed a stochastic Kriging to deal with stochastic simulations with heterogeneous noises. For the pharmaceutical processes, the Kriging has been applied to model steady state pharmaceutical processes with missing data [56], dynamic pharmaceutical processes [98], and a continuous blending process [74].

3.7 Model validation

As a critical step of the model development, model validation is defined as the process of determining the degree of accuracy to which a model can represent the real world within the intended use of the model [99]. Based on the calculation of quantitative validation measures to assess the model fidelity, the user can decide whether it is needed to conduct more experiments to increase the model accuracy.

CV is a widely used approach to estimate the error of a constructed model [100]. Starting with a dataset of N samples, $S\{X, Y\}$, with a p-fold CV, this original dataset is divided (randomly) roughly equally into p different subsets: $S\{X, Y\} = S1\{X1, Y1\}, ..., Sp\{Xp, Yp\}$. Then, the model is fitted p times, each time with one of the subset removed from the training set, and this left-out subset is used to calculate the error. As a variation of the p-fold CV, the leave-k-out approach considers $\binom{N}{k}$ subsets, each with k elements left out from the training set. The special case when k = 1 is called LOO CV, which can be computed efficiently [101]. Meckesheimer and Booker [100] recommended that LOO CV was appropriate to estimate the prediction error for low-order RSM and radial basis function models (a special case of ANN). However, for Kriging, it was suggested to choose $k = 0.1N$ or $k = \sqrt{N}$.

The advantage of CV is that it can provide a nearly unbiased estimation of the generalization error (compared to the "split sample" case where the sample data are divided once into a training set and a test set), as every point is used in a test set once and in a training set (k-1) times (for the LOO approach) [102]. The disadvantage, however, is that we need to fit the data-driven model multiple times, which can be computationally expensive. Furthermore, Lin [103] stated that the LOO CV was not a sufficient metric to evaluate the model accuracy. The LOO method actually measures the insensitivity of a model to information loss at the data points, while an "insensitive" model is not necessarily equivalent to an "accurate" one. Therefore, it is suggested to conduct additional experiments for model validation. A vast number of additional validation metrics have been surveyed in Ref. [104].

3.8 Data-driven models in support of optimization needs

In pharmaceutical process development, a data-driven model is used to support process optimization mainly in two ways, which are shown in Fig. 13.1.

The first approach (Fig. 13.1A) is a "sequential approach." Based on an initial dataset, a data-driven model is constructed, which, after being validated (mostly with a CV approach), is optimized using a mathematical programming approach. In some cases, after the optimal solution is found, a new experiment is performed at this optimal point. The idea is to validate whether the predicted optimal value is consistent with the experiment. Note that this "sequential approach" is mostly adopted when physical experiments are performed. Thus, the initial data points are usually sampled by using a DoE sampling plan. The advantage of this approach is its simplicity. However, the efficacy of this approach depends on the assumption that a sufficiently accurate model can be constructed based on a prespecified number of sample points. In cases when the model is not accurate enough (which can be reflected at Step 3 and Step 5), this approach does not give guidance on further sampling directions. In pharmaceutical processes, "sequential approach" is still the dominant method of process optimization, with applications in Refs. [13,22,23,26,39,47,49].

The second approach (Fig. 13.1B) is an "adaptive approach" which features an adaptive sampling stage. Similar to "sequential approach," the "adaptive approach" also starts with an initial dataset and an initial data-driven model.

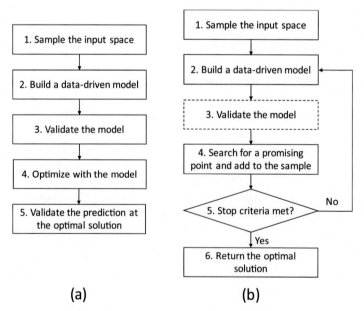

FIGURE 13.1 Two approaches that use a data-driven model to support optimization. (A) sequential approach; (B) adaptive approach.

In some cases, model validation is optional after the initial data-driven model is built because the model accuracy can be further improved in the subsequent adaptive sampling stage. The crucial part of "adaptive approach" is the search criteria that are used in Step 4. Instead of directly optimizing the data-driven model, different mathematical approaches have been developed to guide the search direction, with a balance between finding a better optimal solution and reducing the uncertainty in the data-driven model. Once a new sample point is added, the data-driven model also gets updated. This adaptive sampling (i.e., from Step 4 back to Step 2) is iteratively performed until some stop criteria are met. This approach is also called surrogate-based optimization (SBO) [70], of which details will be further discussed in Section 4. Because "adaptive approach" is mostly used when deterministic computer simulations are available, it usually does not require the validation of the optimal value with the experiment (i.e., Step 5 in Fig. 13.1A). Applications of the "adaptive approach" in pharmaceutical processes can be found in Refs. [20,105,106].

4. Optimization methods in pharmaceutical processes

In previous sections, we have discussed different optimization objectives in pharmaceutical processes and how data-driven models can be used within an optimization setting. One more important aspect that remains to be discussed is how we can find an optimal solution, i.e., the mathematical programming methods. A classification of optimization problems and solution methods has been presented in Ref. [21]. Generally, there are two major approaches to solve an optimization problem: (1) the derivative-based methods and (2) the derivative-free methods. In this section, we review the most popular algorithms under these two categories and refer interested readers to the applications in pharmaceutical process development.

4.1 Derivative-based methods

The derivative-based methods require the derivative information (e.g., gradient, Hessian, etc.) to direct the search to an optimal solution. Such methods are appropriate for problems whose derivative information is reliable and easy to obtain (either provided by users or estimated by computational tools). The advantages of this type of approach are the fast rate of convergence and its capability of dealing with large-scale problems. In this section, we give an overview on the derivative-based optimization approaches and their applications to pharmaceutical processes.

4.2 Successive quadratic programming

Successive (or sequential) quadratic programming (SQP) [107] is a conceptual method that has evolved into numbers of different specific algorithms for constrained NLP problems. The basic idea is to model the NLP at iteration k (with an approximate solution $x^{(k)}$) by a quadratic programming (QP) subproblem, to which the solution is used to direct to the search to the next iteration k+1. The process is iteratively conducted until it converges to a local optimal solution. There are two main reasons for using a QP subproblem: (1) it is relatively easy to solve; and (2) its objective can reflect the nonlinearities of the original problem. For the generation of the QP subproblem, a Hessian matrix of the Lagrangian function needs to be constructed, which can be obtained by either (a) the second derivatives for the objective or constraint functions or (b) positive definite quasi-Newton approximations [21].

SQP methods can only guarantee the convergence to a local optimal solution [107]. The convergence from poor starting points can be promoted by using a line search or a trust-region method [21]. It is found that SQP solvers generally require fewest function evaluations to solve NLPs [108,109]. In addition, SQP methods do not require feasible points at any stage of the process, which is advantageous because it can be usually difficult to find a feasible point in the existence of nonlinear constraints. However, modifications have been made to ensure SQP always remain feasible through the process [110]. A list of SQP-based solvers was presented in Ref. [21]. Aside from the SQP algorithm, a vast number of other gradient-based NLP solvers are also available, which can require more function evaluations than SQP but provide a good performance when interfaced to optimization model platforms (e.g., GAMS [111], AMPL [112]).

Using the SQP method, Sen et al. [113] estimated the parameters of a two-dimensional population balance model (PBM) of a cooling crystallization process based on experimental data. Acevedo et al. [114] calculated the optimal temperature profile for an unseeded batch cooling crystallization system, with the goal of achieving the desired shape and size distribution of crystals subject to a set of process constraints considering temperature range, product yield, and batch time. Yang and Nagy [115] identified the optimal steady state operating profiles of a continuous mixed suspension, mixed product removal (MSMPR) cascade system by maximizing the crystal mean size, which was constrained on temperature, solvent composition, and residence time. Gagnon et al. [116] computed the optimal control strategy based on a phenomenological state-space model of a fluid bed drying (FBD) process. Compared to traditional open-loop FBD operations, the control approach could improve the process by reaching the target particle moisture content while limiting operation problems, including under/overdrying and particles overheating. With the CONOPT solver, Wang and Lakerveld [117] maximized the attainable region of crystal size of a continuous membrane-assisted crystallization (cMAC) process and demonstrated the advantages of cMAC over conventional crystallization processes. In order to find the optimal reactor design of an API synthesis, Emenike et al. [118] used the elementary process functions methodology, which was formulated into a dynamic optimization problem and solved with the CONOPT solver. Furthermore, in references mentioned in previous sections of this chapter, applications of SQP can be found in Refs. [39,44], while other derivative-based NLP solvers have been used in Refs. [18,27,28].

4.3 Derivative-free methods

The derivative-free optimization (DFO) methods find the optimal solution only based on the objective function values (and constraint values) without any derivative information. These methods are successful in cases where derivative information is unreliable or impractical to get (e.g., when the model is expensive or noisy). However, for many algorithms, difficulties still remain for the proof of global convergence. A survey of DFO algorithms for bound-constrained problems was presented in Ref. [119]. Traditionally, most of the DFO methods are only suitable for low-dimensional problems. However, recent efforts have been made to adapt some of the methods to high-dimensional problems; for a review, the reader is encouraged to consult reference [104]. Below, we briefly introduce three types of DFO methods that have been applied to pharmaceutical processes.

4.4 Direct search methods

The Nelder–Mead algorithm [120] involves iteratively building and updating a simplex formed by a set of vertices. (Note that, geometrically, a simplex is a defined as a polyhedron to arbitrary dimensions; a vertex is a corner point of the polyhedron.) At each iteration, it aims to replace the worst vertex by a new vertex and then forms a new simplex. This process is conducted by a series of operations considering the centroid of the current simplex, including reflection, expansion, contraction, and shrink. The convergence of the Nelder–Mean algorithm was investigated in Ref. [121], while further developments can be found in Ref. [122]. Another direct search algorithm is the generalized pattern search (GPS) algorithm [123], which was initially developed for unconstrained problems. It is a generalization of direct search methods including the Hooke and Jeeves method [124] and the coordinate search method [125]. GPS updates the current iterate by sampling at a finite number of points along a suitable set of search directions, with the goal of finding a decrease in the objective function values. It has been extended to bound-constrained problems [126] and linearly constrained problems [127]. GPS was further generalized as Generating Search Set methods by Kolda et al. [128], which were shown to converge to stationary points under mild conditions [123].

Grimard et al. [129] used the Nelder–Mead algorithm to evaluate the parameters for a mathematical model consisting of mass and energy balance equations for a hot melt extrusion process. Besenhard et al. [130] combined

Nelder—Mead with global optimization techniques to estimate the crystal growth model parameters of a PBM for crystallization processes. Based on a DoE, Paul et al. [131] identified and quantified CPP (i.e., sodium chloride concentration, pH value for elution) of a multimodal ion exchange step for the purification of biopharmaceuticals. The optimal values of the CPP were calculated to maximize the purity and recovery of this purification process. Zou et al. [132] fitted a Korsmeyer-Peppas model with dissolution data, which was used to describe the kinetics of an in vitro drug release process using nanoparticle formulations. Xi et al. [133] computed the optimal values for three categories of design variables (related to device, particles, and patients) to maximize the efficiency of an electric-guided drug delivery system for the treatment of rhinosinusitis. Moudjari et al. [134] estimated the values of interaction parameters of thermodynamic models for the solubility prediction of pharmaceutical compounds in various solvents. In earlier mentioned references from Section 2, direct search algorithms were used in Refs. [49,50].

4.5 Genetic algorithms

Genetic algorithms (GAs) [135], sometimes known as evolutionary algorithms, are a family of population-based heuristic search algorithms that mimic the mechanistic of natural selection and reproduction processes. A GA starts with a randomly sampled initial population of chromosomes (i.e., initial generation), which are basically sample points with variables represented by binary strings. Then, the structures of the chromosomes are evaluated and reproductive opportunities can be allocated in a way so that those with a better solution have higher chances to reproduce. At each iteration, the descendant generation is generated successively via a series of operations including selection, recombination, and mutation. GA can be categorized as a stochastic global search algorithm [119]. It usually requires the objective function to be fast to evaluate [136] and is only suitable for low-dimensional problems [137].

For a continuous crystallization process modeled with the PBM approach, Ridder et al. [41] used a GA technique to calculate the Pareto optimal solutions of a multiobjective optimization problem, which simultaneously maximized the average crystal size and minimized the coefficient of variation. Zaki et al. [138] used the ANN approach to model the fabrication process of bupropion HCl-loaded agar nanospheres which were used for sustained drug release. Based on this model, the GA approach was applied to optimize the process, namely, to minimize the particle size, release efficiency, and maximize loading efficiency, etc. Allmendinger et al. [139] formulated a constrained optimization problem to improve the performance (considering process cost, time, and product waste) of a chromatography purification process. Four types of GA methods were applied to identify the optimal equipment sizing strategies of this biopharmaceutical process. Rostamizadeh et al. [140] identified the optimal process parameters for fabricating a type of nanoparticles (used for oral insulin delivery) in order to achieve its best performance with respect to six performance measures. Kalkhorana et al. [141] applied the GA approach to estimate the parameters of a drug release model, which was developed to predict the drug diffusion rate from a hydrogel-based drug delivery system. Wang et al. [142] used the GA approach to find the optimal formulation of Doxy inclusion complex (a type of broad-spectrum antibiotic drug) that could lead to optimum inclusion efficiency and stability in the aqueous solution. For those references that were mentioned earlier in Section 2, case studies applying GA can be found in Refs. [22,39].

4.6 Surrogate-based optimization methods

SBO methods treat the original problem as a black-box process. Accordingly, a surrogate model is built as a fast approximation to this black-box process and then guides the search direction to the next sample point. The general optimization framework is shown in Fig. 13.1B, and its description is mentioned in Section 3.4. Depending on the choice of surrogate models, there are mainly two SBO algorithms: (1) Kriging-based approach (sometimes known as Bayesian optimization) and (2) radial basis function (RBF) based approach. The seminal work of the Kriging-based approach is the efficient global optimization (EGO) algorithm by Jones [94], which used an expected improvement function as the infill criteria to search for the next sample point. Jones [55] presented a survey of various infill criteria that could be applied with the Kriging-based methods. The classical EGO algorithm can be seen as a greedy search approach. Recently, it has been combined with dynamic programming which can account for the remaining number of evaluations [143,144]. This look-ahead approach finds the optimal strategy by maximizing a long-term reward and is shown to be more effective when a limited sampling budget is available. In terms of the RBF-based methods, Gutmann [145] proposed an RBF-based approach which used a bumpiness measure to search for the next sample point. This approach has very similar characteristics with the one-stage approach for the Kriging-based method that was discussed in Ref. [55]. Regis and Shoemaker [146] developed an RBF-based method for black-box constrained optimization problems. Regis [147] further proposed an RBF-based method for high-dimensional constrained optimization problems.

The number of applications of SBO methods in pharmaceutical processes is relatively small compared to other optimization algorithms. Luna and Martínez [148] used the Bayesian optimization approach to maximize biomass growth based on a hybrid cybernetic model which was used to simulate the animal cell metabolism of bioreactors. Based on a reduced model for a perfusion bioreactor, Mehrian et al. [149] applied the Bayesian optimization approach to find the optimal medium refreshment regime that would result in the maximized neotissue growth kinetics in the bioreactor. Boukouvala and Ierapetritou [20] and Wang et al. [106] have developed Kriging-based methods for the optimization of CDC pharmaceutical manufacturing processes.

5. A case study of the optimization of a continuous direct compaction process

In order to demonstrate how an optimization algorithm can be used to improve the performance of a pharmaceutical process, we use a CDC process as an example. This example was reported in Ref. [106].

The flowsheet model of this CDC process has been developed on the basis of the plant installed at the Center for Structured Organic Particulate Systems (C-SOPS) at Rutgers University. The process flowsheet is shown in Fig. 13.2. Two feeders are installed at the top to continuously feed raw materials (an API and an excipient) to the system. The API and excipient go through a co-mill where big chunks of materials are delumped. After the co-mill, the mixtures, together with a lubricant, are transferred to a blender where materials are continuously mixed. The blends are finally sent to a tablet press. Details of the mathematical models were given in Ref. [18].

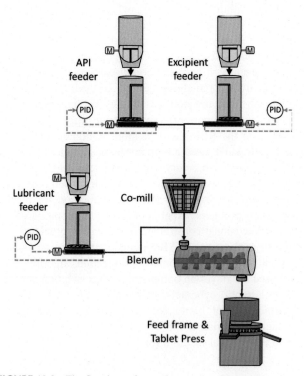

FIGURE 13.2 The flowsheet of a continuous direct compaction process.

In Ref. [106], the process optimization problem was formulated as follows. Find the optimal operation conditions that result in the minimal total costs subject to a constraint that no wasted products should be generated once the process arrives at its steady state. In this problem, the total costs include material cost, utility cost, and waste cost. The considered decision variables include "API flow rate set point" and "API refilling strategy." These two operation conditions are chosen because they were found to be most influential to the process variabilities [18].

This flowsheet model has a relatively high computational cost: each simulation run takes CPU time roughly between 40 and 60 s. Therefore, a more efficient way is to use the SBO approach to find the optimal solution. Wang et al. [18] developed a Kriging-based adaptive sampling approach to solve this problem. After a total number of 100 sample points, a near-optimal solution was identified. The results indicate that a relatively low API flow rate set point and a high frequency of API refilling are desired. This is expected because such settings are beneficial to reduce process variations based on experimental studies [150].

To further see the benefits of running the process at the returned near-optimal operation conditions, we use the flowsheet model to simulation the first 1000 s of the process. The simulation results are shown in Fig. 13.3. From the results, we can see that the profiles (denoted with solid lines) for product qualities (i.e., API concentration, tablet weight, tablet hardness) are within specified lower and upper bounds (denoted with dashed lines) once the steady state is achieved. The only wasted products are generated in the starting stage when the tablet weight is not yet at its steady state. This is reflected in the bottom right figure, where "1" indicates that wasted products are made; "0" for in-spec products. Therefore, by running the process at the identified near-optimal solution, we can achieve a reduced total cost while satisfying the specified process constraint.

6. Discussion and future perspectives

In this chapter, we have discussed recent developments of optimization in pharmaceutical processes. A review is first provided on various objectives that are mostly considered in different pharmaceutical processes, such as API processes, wastewater treatment process, downstream tableting processes, etc. Furthermore, we introduced four types of data-driven models that are commonly used under two optimization frameworks for pharmaceutical processes, including RSM, PLS regression, ANN, and Kriging. We also included an overview on several optimization algorithms that are widely adopted to solve an optimization problem, which may or may not require the derivative information. Lastly, a case study for optimizing CDC process was provided.

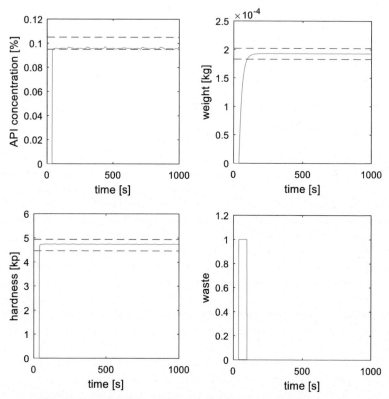

FIGURE 13.3 Simulation results of continuous direct compaction process.

Compared to traditional chemical and petrochemical processes, the application of optimization tools in the pharmaceutical process development is still in a primary stage. Challenges remain to improve the mechanistic understanding and models' predicting capabilities of the pharmaceutical processes. From this respective, hybrid models (e.g. Ref. [151]) which combine both the mechanistic knowledge from first-principle models and the efficiency from reduced order models can be a highly promising tool to be used in the optimization settings. In addition, the flowsheet modeling approach, acting as a representation of the whole integrated process, is also gaining an increasing attention from the research community. However, due to its high model complexity, it also raises the challenge of optimizing for large-scale systems, which can involve black-box model components and may potentially be computationally expensive to compute. A promising approach to address such difficulties is via the use of SBO methods. In addition, for pharmaceutical manufacturing processes, the variability of product qualities is usually a critical aspect that requires attention. Such variabilities can be modeled by introducing a random error term to the simulation (i.e., a stochastic simulation). To address the challenges of optimizing with black-box stochastic simulations, the simulation optimization methods can be used. This has recently been investigated by Wang and Ierapetritou [152] in the optimization of a CDC flowsheet model.

Acknowledgments

The authors would like to acknowledge financial support from FDA (DHHS - FDA - 1 U01 FD005295-01) as well as National Science Foundation Engineering Research Center on Structured Organic Particulate Systems (NSF-ECC 0540855).

References

[1]   Food U, Administration D. Pharmaceutical CGMPs for the 21st Century—a risk-based approach. 2004. Final report. Rockville, MD.

[2]   Food, Administration D. Guidance for industry: PAT—a framework for innovative pharmaceutical development, manufacturing, and quality assurance. Rockville, MD: DHHS; 2004.

[3]   Lawrence XY. Pharmaceutical quality by design: product and process development, understanding, and control. Pharm Res 2008;25(4):781−91.

[4]   Khaled SA, Burley JC, Alexander MR, Roberts CJ. Desktop 3D printing of controlled release pharmaceutical bilayer tablets. Int J Pharm 2014;461(1):105−11.

[5]   O'Connor TF, Lawrence XY, Lee SL. Emerging technology: a key enabler for modernizing pharmaceutical manufacturing and advancing product quality. Int J Pharm 2016;509(1):492−8.

[6]   Ketterhagen WR, am Ende MT, Hancock BC. Process modeling in the pharmaceutical industry using the discrete element method. J Pharm Sci 2009;98(2):442−70.

[7]   McKenzie P, Kiang S, Tom J, Rubin AE, Futran M. Can pharmaceutical process development become high tech? AIChE J 2006;52(12):3990−4.

[8]   Rogers A, Ierapetritou M. Challenges and opportunities in modeling pharmaceutical manufacturing processes. Comp Chem Eng 2015;81:32−9.

[9]   Kleinebudde P, Khinast J, Rantanen J. Continuous manufacturing of pharmaceuticals, vol. 7703. John Wiley & Sons; 2017.

[10] Zhu H, Zhou Z, Yang R, Yu A. Discrete particle simulation of particulate systems: a review of major applications and findings. Chem Eng Sc 2008;63(23):5728−70.

[11] Wu C-Y, Ruddy O, Bentham A, Hancock B, Best S, Elliott J. Modelling the mechanical behaviour of pharmaceutical powders during compaction. Powder Technol 2005;152(1):107−17.

[12] Chaudhury A, Barrasso D, Pandey P, Wu H, Ramachandran R. Population balance model development, validation, and prediction of CQAs of a high-shear wet granulation process: towards QbD in drug product pharmaceutical manufacturing. J Pharm Innov 2014;9(1):53−64.

[13] Pal TK, Dan S, Dan N. Application of response surface methodology (RSM) in statistical optimization and pharmaceutical characterization of a matrix tablet formulation using metformin HCl as a model drug. Innoriginal: Int J Sci 2014;1(2).

[14] Shirazian S, Kuhs M, Darwish S, Croker D, Walker GM. Artificial neural network modelling of continuous wet granulation using a twin-screw extruder. Int J Pharm 2017;521(1):102−9.

[15] Tabora JE, Domagalski N. Multivariate analysis and statistics in pharmaceutical process research and development. Ann Rev Chem Biomol Eng 2017;8:403−26.

[16] Boukouvala F, Chaudhury A, Sen M, Zhou R, Mioduszewski L, Ierapetritou MG, Ramachandran R. Computer-aided flowsheet simulation of a pharmaceutical tablet manufacturing process incorporating wet granulation. J Pharm Innov 2013;8(1):11−27.

[17] Boukouvala F, Muzzio FJ, Ierapetritou MG. Methods and tools for design space identification in pharmaceutical development. Comp Qual DesignPharm Prod Dev Manuf 2017;95−123.

[18] Wang Z, Escotet-Espinoza MS, Ierapetritou M. Process analysis and optimization of continuous pharmaceutical manufacturing using flowsheet models. Comp Chem Eng 2017;107:77−91.

[19] Singh R, Muzzio FJ, Ierapetritou M, Ramachandran R. A combined feed-forward/feedback control system for a QbD-based continuous tablet manufacturing process. Processes 2015;3(2):339−56.

[20] Boukouvala F, Ierapetritou MG. Surrogate-based optimization of expensive flowsheet modeling for continuous pharmaceutical manufacturing. J Pharm Innov 2013;8(2):131−45.

[21] Biegler LT, Grossmann IE. Retrospective on optimization. Comp Chem Eng 2004;28(8):1169−92.

[22] Velásco-Mejía A, Vallejo-Becerra V, Chávez-Ramírez A, Torres-González J, Reyes-Vidal Y, Castañeda-Zaldivar F. Modeling and optimization of a pharmaceutical crystallization process by using neural networks and genetic algorithms. Powder Technol 2016;292:122−8.

[23] Monteagudo E, Langenheim M, Salerno C, Buontempo F, Bregni C, Carlucci A. Pharmaceutical optimization of lipid-based dosage forms for the improvement of taste-masking, chemical stability and solubilizing capacity of phenobarbital. Drug Dev Ind Pharm 2014;40(6):783−92.

[24] Chavez P-F, Lebrun P, Sacre P-Y, De Bleye C, Netchacovitch L, Cuypers S, Mantanus J, Motte H, Schubert M, Evrard B. Optimization of a pharmaceutical tablet formulation based on a design space approach and using vibrational spectroscopy as PAT tool. Int J Pharm 2015;486(1):13−20.

[25] Sheikholeslamzadeh E, Chen C-C, Rohani S. Optimal solvent screening for the crystallization of pharmaceutical compounds from multisolvent systems. Ind Eng Chem Res 2012;51(42):13792−802.

[26] Zhang C, Huang J. Optimization of process parameters for pharmaceutical wastewater treatment. Pol J Environ Stud 2015;24(1):391−5.

[27] Jolliffe HG, Gerogiorgis DI. Technoeconomic optimization of a conceptual flowsheet for continuous separation of an analgaesic active pharmaceutical ingredient (API). Ind Eng Chem Res 2017;56(15):4357−76.

[28] Abejón R, Garea A, Irabien A. Analysis and optimization of continuous organic solvent nanofiltration by membrane cascade for pharmaceutical separation. AIChE J 2014;60(3):931−48.

[29] Grossmann IE, Morari M. Operability, resiliency, and flexibility: process design objectives for a changing world. In: Westerberg AW, Chien HH, editors. Proc. 2nd international conference on foundations computer aided process design; 1983 [CACHE].

[30] Grossmann IE, Calfa BA, Garcia-Herreros P. Evolution of concepts and models for quantifying resiliency and flexibility of chemical processes. Comp Chem Eng 2014;70:22−34.

[31]   Rogers A, Ierapetritou M. Feasibility and flexibility analysis of black-box processes Part 1: surrogate-based feasibility analysis. Chem Eng Sci 2015;137:986—1004.

[32]   Rogers A, Ierapetritou M. Feasibility and flexibility analysis of black-box processes Part 2: surrogate-based flexibility analysis. Chem Eng Sci 2015;137:1005—13.

[33]   Adi VSK, Laxmidewi R. Design and operability analysis of membrane module based on volumetric flexibility. In: Computer aided chemical engineering, vol. 40. Elsevier; 2017. p. 1231—6.

[34]   Finnveden G, Hauschild MZ, Ekvall T, Guinée J, Heijungs R, Hellweg S, Koehler A, Pennington D, Suh S. Recent developments in life cycle assessment. J Environ Manag 2009;91(1):1—21.

[35]   Ott D, Kralisch D, Denčić I, Hessel V, Laribi Y, Perrichon PD, Berguerand C, Kiwi-Minsker L, Loeb P. Life cycle analysis within pharmaceutical process optimization and intensification: case study of active pharmaceutical ingredient production. Chem Sus Chem 2014;7(12):3521—33.

[36]   Ott D, Borukhova S, Hessel V. Life cycle assessment of multi-step rufinamide synthesis—from isolated reactions in batch to continuous microreactor networks. Green Chem 2016;18(4):1096—116.

[37]   Nha VT, Shin S, Jeong SH. Lexicographical dynamic goal programming approach to a robust design optimization within the pharmaceutical environment. Eur J Oper Res 2013;229(2):505—17.

[38]   Jeong SH, Kongsuwan P, Truong NKV, Shin S. Optimal tolerance design and optimization for a pharmaceutical quality characteristic. Math Prob Eng 2013:2013.

[39]   Yoshizaki R, Kano M, Tanabe S, Miyano T. Process parameter optimization based on LW-PLS in pharmaceutical granulation Process** this work was partially supported by Japan society for the promotion of science (JSPS), grant-in-aid for scientific research (C) 24560940. IFAC-Papers OnLine 2015;48(8):303—8.

[40]   Power G, Hou G, Kamaraju VK, Morris G, Zhao Y, Glennon B. Design and optimization of a multistage continuous cooling mixed suspension, mixed product removal crystallizer. Chem Eng Sci 2015;133:125—39.

[41]   Ridder BJ, Majumder A, Nagy ZK. Population balance model-based multiobjective optimization of a multisegment multiaddition (MSMA) continuous plug-flow antisolvent crystallizer. Ind Eng Chem Res 2014;53(11):4387—97.

[42]   Ardakani MK, Wulff SS. An overview of optimization formulations for multiresponse surface problems. Qual Reliab Eng Int 2013;29(1):3—16.

[43]   Deb K. Multi-objective optimization. In: Search methodologies. Springer; 2014. p. 403—49.

[44]   Brunet R, Guillén-Gosálbez G, Jiménez L. Combined simulation—optimization methodology to reduce the environmental impact of pharmaceutical processes: application to the production of Penicillin V. J Clean Prod 2014;76:55—63.

[45]   Derringer GC. A balancing act-optimizing a products properties. Qual Prog 1994;27(6):51—8.

[46]   Uttekar P, Chaudhari P. Formulation and evaluation of engineered pharmaceutical fine particles of Budesonide for dry powder inhalation (dpi) produced by amphiphilic crystallization technique: optimization of process parameters. Int J Pharm Sci Res 2013;4(12):4656.

[47] Sato H, Watanabe S, Takeda D, Yano S, Doki N, Yokota M, Shimizu K. Optimization of a crystallization process for orantinib active pharmaceutical ingredient by design of experiment to control residual solvent amount and particle size distribution. Org Proc Res Dev 2015;19(11):1655—61.

[48] Chakraborty P, Dey S, Parcha V, Bhattacharya SS, Ghosh A. Design expert supported mathematical optimization and predictability study of buccoadhesive pharmaceutical wafers of Loratadine. BioMed Res Int 2013:2013.

[49] Kermet-Said H, Moulai-Mostefa N. Optimization of turbidity and COD removal from pharmaceutical wastewater by electrocoagulation. Isotherm modeling and cost analysis. Pol J Environ Stud 2015;24(3).

[50] Li Z, Cho BR, Melloy BJ. Quality by design studies on multi-response pharmaceutical formulation modeling and optimization. J Pharm Innov 2013;8(1):28—44.

[51] Solomatine D, See LM, Abrahart R. Data-driven modelling: concepts, approaches and experiences. In: Practical hydroinformatics. Springer; 2009. p. 17—30.

[52] Razavi S, Tolson BA, Burn DH. Review of surrogate modeling in water resources. Water Res Res 2012;48(7).

[53] Wang GG, Shan S. Review of metamodeling techniques in support of engineering design optimization. J Mech Des 2007;129(4):370—80.

[54] Rogers AJ, Hashemi A, Ierapetritou MG. Modeling of particulate processes for the continuous manufacture of solid-based pharmaceutical dosage forms. Processes 2013;1(2):67—127.

[55] Jones DR. A taxonomy of global optimization methods based on response surfaces. J Glob Optim 2001;21(4):345—83.

[56] Boukouvala F, Muzzio FJ, Ierapetritou MG. Predictive modeling of pharmaceutical processes with missing and noisy data. AIChE J 2010;56(11):2860—72.

[57] Singh B, Kumar R, Ahuja N. Optimizing drug delivery systems using systematic" design of experiments." Part I: fundamental aspects. Crit Rev Therap Drug Carr Syst 2005;22(1).

[58] Elkhoudary MM, Abdel Salam RA, Hadad GM. Development and optimization of HPLC analysis of metronidazole, diloxanide, spiramycin and cliquinol in pharmaceutical dosage forms using experimental design. J Chromatograp Sci 2016;54(10):1701—12.

[59] Sahu PK, Swain S, Prasad GS, Panda J, Murthy Y. RP-HPLC method for determination of metaxalone using Box-Behnken experimental design. J Appl Biopharma Pharma 2015;2(2):40—9.

[60] Sharma D, Maheshwari D, Philip G, Rana R, Bhatia S, Singh M, Gabrani R, Sharma SK, Ali J, Sharma RK. Formulation and optimization of polymeric nanoparticles for intranasal delivery of lorazepam using Box-Behnken design: in vitro and in vivo evaluation. BioMed Res Int 2014;2014.

[61] Kalyani A, Naga Sireesha G, Aditya A, Girija Sankar G, Prabhakar T. Production optimization of rhamnolipid biosurfactant by streptomyces coelicoflavus (NBRC 15399T) using Plackett-Burman design. Eur J Biotechnol Biosci 2014;1(5):07—13.

[62] Agarabi CD, Schiel JE, Lute SC, Chavez BK, Boyne MT, Brorson KA, Khan MA, Read EK. Bioreactor process parameter screening utilizing a plackett—burman design for a model monoclonal antibody. J Pharma Sci 2015;104(6):1919—28.

[63] Sacks J, Welch WJ, Mitchell TJ, Wynn HP. Design and analysis of computer experiments. Stat Sci 1989:409—23.

[64] McKay MD, Beckman RJ, Conover WJ. Comparison of three methods for selecting values of input variables in the analysis of output from a computer code. Technometrics 1979;21(2):239—45.

[65] Huntington D, Lyrintzis C. Improvements to and limitations of Latin hypercube sampling. Probab Eng Mech 1998;13(4):245−53.

[66] Kalagnanam JR, Diwekar UM. An efficient sampling technique for off-line quality control. Technometrics 1997;39(3):308−19.

[67] Owen AB. Orthogonal arrays for computer experiments, integration and visualization. Statistica Sinica 1992:439−52.

[68] Fang K-T, Lin DK, Winker P, Zhang Y. Uniform design: theory and application. Technometrics 2000;42(3):237−48.

[69] Santner TJ, Williams BJ, Notz WI. The design and analysis of computer experiments. Springer Science & Business Media; 2013.

[70] Forrester AI, Keane AJ. Recent advances in surrogate-based optimization. Prog Aerosp Sci 2009;45(1):50−79.

[71] Box GE, Wilson KB. On the experimental attainment of optimum conditions. J Royal Stat Soc Series B (Methodological) 1951;13:1−45.

[72] Khuri AI, Mukhopadhyay S. Response surface methodology. Wiley Interdiscip Rev. Comput Stat 2010;2(2):128−49.

[73] Khuri AI, Cornell JA. Response surfaces: designs and analyses, vol. 152. CRC press; 1996.

[74] Boukouvala F, Dubey A, Vanarase A, Ramachandran R, Muzzio FJ, Ierapetritou M. Computational approaches for studying the granular dynamics of continuous blending processes, 2−population balance and data-based methods. Macromol Mat Eng 2012;297(1):9−19.

[75] Geladi P, Kowalski BR. Partial least-squares regression: a tutorial. Anal Chim Acta 1986;185:1−17.

[76] Wold S, Sjöström M, Eriksson L. PLS-regression: a basic tool of chemometrics. Chemom Intell Lab Syst 2001;58(2):109−30.

[77] Nelson PR, Taylor PA, MacGregor JF. Missing data methods in PCA and PLS: score calculations with incomplete observations. Chemom Intell Lab Syst 1996;35(1):45−65.

[78] Shao J. Linear model selection by cross-validation. J Am Stat Assoc 1993;88(422):486−94.

[79] Eriksson L, Johansson E, Kettaneh-Wold N, Wold S. Multi-and megavariate data analysis. Part I: Basic Princ Appl 2001;2:425.

[80] Rosipal R. Nonlinear partial least squares: an overview. Chemoinformatics and advanced machine learning perspectives: complex computational methods and collaborative techniques. 2010. p. 169−89.

[81] Tomba E, Facco P, Bezzo F, Barolo M. Latent variable modeling to assist the implementation of Quality-by-Design paradigms in pharmaceutical development and manufacturing: a review. Int J Pharm 2013;457(1):283−97.

[82] Agatonovic-Kustrin S, Beresford R. Basic concepts of artificial neural network (ANN) modeling and its application in pharmaceutical research. J Pharma Biomed Anal 2000;22(5):717−27.

[83] Simpson PK. Artificial neural systems: foundations, paradigms, applications, and implementations. Pergamon; 1990.

[84] Rumelhart DE, Hinton GE, Williams RJ. Learning internal representations by error propagation. In: California univ San Diego La jolla inst for cognitive science; 1985.

[85] Ripley BD. Pattern recognition and neural networks. Cambridge university press; 2007.

[86] Yu H, Fu J, Dang L, Cheong Y, Tan H, Wei H. Prediction of the particle size distribution parameters in a high shear granulation process using a key parameter definition combined artificial neural network model. Ind Eng Chem Res 2015;54(43):10825−34.

[87] Masters T. Practical neural network recipes in C++. Morgan Kaufmann; 1993.

[88] Basheer I, Hajmeer M. Artificial neural networks: fundamentals, computing, design, and application. J Microbiol Methods 2000;43(1):3–31.

[89] Orr MJ. Introduction to radial basis function networks. In: Technical report. Center for Cognitive Science, University of Edinburgh; 1996.

[90] Maher HM. Development and validation of a stability-indicating HPLC-dad method with ANN optimization for the determination of diflunisal and naproxen in pharmaceutical tablets. J Liquid Chroma Rel Technol 2014;37(5):634–52.

[91] Patel TB, Patel L, Patel TR, Suhagia B. Artificial neural network as tool for quality by design in formulation development of solid dispersion of fenofibrate. Bull Pharma Res 2015;5(1):20–7.

[92] Li Y, Abbaspour MR, Grootendorst PV, Rauth AM, Wu XY. Optimization of controlled release nanoparticle formulation of verapamil hydrochloride using artificial neural networks with genetic algorithm and response surface methodology. Eur J Pharma Biopharma 2015;94:170–9.

[93] Cressie N. The origins of kriging. Mathemat Geol 1990;22(3):239–52.

[94] Jones DR, Schonlau M, Welch WJ. Efficient global optimization of expensive black-box functions. J Glob Optim 1998;13(4):455–92.

[95] Rasmussen CE, Williams CK. Gaussian processes for machine learning, vol. 1. MIT press Cambridge; 2006.

[96] Huang D, Allen TT, Notz WI, Zeng N. Global optimization of stochastic black-box systems via sequential kriging meta-models. J Glob Optim 2006;34(3):441–66.

[97] Ankenman B, Nelson BL, Staum J. Stochastic kriging for simulation metamodeling. Oper Res 2010;58(2):371–82.

[98] Boukouvala F, Muzzio F, Ierapetritou MG. Dynamic data-driven modeling of pharmaceutical processes. Ind Eng Chem Res 2011;50(11):6743–54.

[99] Kleijnen JP, Sargent RG. A methodology for fitting and validating metamodels in simulation. Eur J Oper Res 2000;120(1):14–29.

[100] Meckesheimer M, Booker AJ, Barton RR, Simpson TW. Computationally inexpensive metamodel assessment strategies. AIAA J 2002;40(10):2053–60.

[101] Mitchell TJ, Morris MD. Bayesian design and analysis of computer experiments: two examples. Statistica Sinica 1992:359–79.

[102] Queipo NV, Haftka RT, Shyy W, Goel T, Vaidyanathan R, Tucker PK. Surrogate-based analysis and optimization. Prog Aerosp Sci 2005;41(1):1–28.

[103] Lin Y. An efficient robust concept exploration method and sequential exploratory experimental design. Georgia Institute of Technology; 2004.

[104] Shan S, Wang GG. Survey of modeling and optimization strategies to solve high-dimensional design problems with computationally-expensive black-box functions. Struct Multidiscip Optim 2010;41(2):219–41.

[105] Boukouvala F, Ierapetritou MG. Derivative-free optimization for expensive constrained problems using a novel expected improvement objective function. AIChE J 2014;60(7):2462–74.

[106] Wang Z, Escotet-Espinoza MS, Singh R, Ierapetritou M. Surrogate-based optimization for pharmaceutical manufacturing processes. In: Computer aided chemical engineering, vol. 40. Elsevier; 2017. p. 2797–802.

[107] Boggs PT, Tolle JW. Sequential quadratic programming. Acta Numer 1995;4:1–51.

[108] Schttfkowski K. More test examples for nonlinear programming codes. Lect Econ Math Syst 1987;282.

[109] Binder T, Blank L, Bock HG, Bulirsch R, Dahmen W, Diehl M, Kronseder T, Marquardt W, Schlöder JP, von Stryk O. Introduction to model based optimization of chemical processes on moving horizons. In: Online optimization of large scale systems. Springer; 2001. p. 295−339.

[110] Bonnans JF, Panier ER, Tits AL, Zhou JL. Avoiding the maratos effect by means of a nonmonotone line search. II. Inequality constrained problems—feasible iterates. SIAM J Num Anal 1992;29(4):1187−202.

[111] Bussieck MR, Meeraus A. General algebraic modeling system (GAMS). Appl Optim 2004;88:137−58.

[112] Fourer R, Gay DM, Kernighan BW. AMPL: a mathematical programming language. Citeseer; 1987.

[113] Sen M, Chaudhury A, Singh R, Ramachandran R. Two-dimensional population balance model development and validation of pharmaceutical crystallization processes. Am J Mod Chem Eng 2014;1:13−29.

[114] Acevedo D, Tandy Y, Nagy ZK. Multiobjective optimization of an unseeded batch cooling crystallizer for shape and size manipulation. Ind Eng Chem Res 2015;54(7):2156−66.

[115] Yang Y, Nagy ZK. Combined cooling and antisolvent crystallization in continuous mixed suspension, mixed product removal cascade crystallizers: steady-state and startup optimization. Ind Eng Chem Res 2015;54(21):5673−82.

[116] Gagnon F, Desbiens A, Poulin É, Lapointe-Garant P-P, Simard J-S. Nonlinear model predictive control of a batch fluidized bed dryer for pharmaceutical particles. Cont Eng Pract 2017;64:88−101.

[117] Wang J, Lakerveld R. Continuous membrane-assisted crystallization to increase the attainable product quality of pharmaceuticals and design space for operation. Ind Eng Chem Res 2017;56(19):5705−14.

[118] Emenike VN, Schenkendorf R, Krewer U. A systematic reactor design approach for the synthesis of active pharmaceutical ingredients. Eur J Pharma Biopharma 2018;126:75−88.

[119] Rios LM, Sahinidis NV. Derivative-free optimization: a review of algorithms and comparison of software implementations. J Glob Optim 2013;56(3):1247−93.

[120] Nelder JA, Mead R. A simplex method for function minimization. Comp J 1965;7(4):308−13.

[121] McKinnon KI. Convergence of the nelder–mead simplex method to a nonstationary point. SIAM J Optim 1998;9(1):148−58.

[122] Conn AR, Scheinberg K, Vicente LN. Introduction to derivative-free optimization. SIAM; 2009.

[123] Torczon V. On the convergence of pattern search algorithms. SIAM J Optim 1997;7(1):1−25.

[124] Hooke R, Jeeves TA. "Direct Search"Solution of numerical and statistical problems. JACM 1961;8(2):212−29.

[125] Audet C. A survey on direct search methods for blackbox optimization and their applications. In: Mathematics without boundaries. Springer; 2014. p. 31−56.

[126] Lewis RM, Torczon V. Pattern search algorithms for bound constrained minimization. SIAM J Optim 1999;9(4):1082−99.

[127] Lewis RM, Torczon V. Pattern search methods for linearly constrained minimization. SIAM J Optim 2000;10(3):917−41.

[128] Kolda TG, Lewis RM, Torczon V. Optimization by direct search: new perspectives on some classical and modern methods. SIAM Rev 2003;45(3):385−482.

[129] Grimard J, Dewasme L, Thiry J, Krier F, Evrard B, Wouwer AV. Modeling, sensitivity analysis and parameter identification of a twin screw extruder. IFAC-PapersOnLine 2016;49(7):1127−32.

[130] Besenhard MO, Chaudhury A, Vetter T, Ramachandran R, Khinast JG. Evaluation of parameter estimation methods for crystallization processes modeled via population balance equations. Chem Eng Res Des 2015;94:275−89.

[131] Paul J, Jensen S, Dukart A, Cornelissen G. Optimization of a preparative multimodal ion exchange step for purification of a potential malaria vaccine. J Chromatogr 2014;1366:38−44.

[132] Zuo J, Gao Y, Bou-Chacra N, Löbenberg R. Evaluation of the DDSolver software applications. BioMed Res Int 2014;2014.

[133] Xi J, Yuan JE, Si XA, Hasbany J. Numerical optimization of targeted delivery of charged nanoparticles to the ostiomeatal complex for treatment of rhinosinusitis. Int J Nanomed 2015;10:4847.

[134] Moudjari Y, Louaer W, Meniai A. Modeling of the solubility of Naproxen and Trimethoprim in different solvents at different temperature. In: MATEC web of conferences. EDP Sciences; 2013. p. 01057.

[135] Holland JH. Adaptation in natural and artificial systems. Ann Arbor: The University of Michigan Press; 1975.

[136] Whitley D. A genetic algorithm tutorial. Stat Comp 1994;4(2):65−85.

[137] Kumar M, Husian M, Upreti N, Gupta D. Genetic algorithm: review and application. Int J Inform Technol Knowl Manag 2010;2(2):451−4.

[138] Zaki MR, Varshosaz J, Fathi M. Preparation of agar nanospheres: comparison of response surface and artificial neural network modeling by a genetic algorithm approach. Carbohydr Polym 2015;122:314−20.

[139] Allmendinger R, Simaria AS, Turner R, Farid SS. Closed-loop optimization of chromatography column sizing strategies in biopharmaceutical manufacture. J Chem Technol Biotechnol 2014;89(10):1481−90.

[140] Rostamizadeh K, Rezaei S, Abdouss M, Sadighian S, Arish S. A hybrid modeling approach for optimization of PMAA−chitosan−PEG nanoparticles for oral insulin delivery. RSC Adv 2015;5(85):69152−60.

[141] Kalkhoran AHZ, Vahidi O, Naghib SM. A new mathematical approach to predict the actual drug release from hydrogels. Eur J Pharma Sci 2018;111:303−10.

[142] Wang Z, He Z, Zhang L, Zhang H, Zhang M, Wen X, Quan G, Huang X, Pan X, Wu C. Optimization of a doxycycline hydroxypropyl-β-cyclodextrin inclusion complex based on computational modeling. Acta Pharm Sinica B 2013;3(2):130−9.

[143] Huan X, Marzouk YM. Sequential Bayesian optimal experimental design via approximate dynamic programming. arXiv 2016:arXiv:1604.08320.

[144] Lam R, Willcox K, Wolpert DH. Bayesian optimization with a finite budget: an approximate dynamic programming approach. In: Advances in neural information processing systems; 2016. p. 883−91.

[145] Gutmann H-M. A radial basis function method for global optimization. J Glob Optim 2001;19(3):201−27.

[146] Regis RG, Shoemaker CA. Constrained global optimization of expensive black box functions using radial basis functions. J Glob Optim 2005;31(1):153−71.

[147] Regis RG. Constrained optimization by radial basis function interpolation for high-dimensional expensive black-box problems with infeasible initial points. Eng Optim 2014;46(2):218−43.

[148] Luna M, Martínez E. A Bayesian approach to run-to-run optimization of animal cell bioreactors using probabilistic tendency models. Ind Eng Chem Res 2014;53(44):17252−66.

[149] Mehrian M, Guyot Y, Papantoniou I, Olofsson S, Sonnaert M, Misener R, et al. Maximizing neotissue growth kinetics in a perfusion bioreactor: an in silico strategy using model reduction and Bayesian optimization. Biotechnol Bioeng 2018;115(3):617−29.

[150] Engisch WE, Muzzio FJ. Feedrate deviations caused by hopper refill of loss-in-weight feeders. Powder Technol 2015;283:389−400.

[151] Metta N, Ierapetritou M, Ramachandran R. A multiscale DEM-PBM approach for a continuous comilling process using a mechanistically developed breakage kernel. Chem Eng Sci 2018;178:211−21.

[152] Wang Z, Ierapetritou M. A novel surrogate-based optimization method for black-box simulation with heteroscedastic noise. Ind Eng Chem Res 2017;56(38):10720−32.

Chapter 14

Regulatory considerations for continuous solid oral dose pharmaceutical manufacturing

Douglas B. Hausner[1] and Christine M.V. Moore[2]

[1]*Thermo Fisher Scientific, Waltham, MA, United States;* [2]*Organon & Co., Jersey, NJ, United States*

1. Introduction

Over the past decade, there has been growing interest in continuous solid oral dose (SOD) pharmaceutical manufacturing. The interest has recently intensified with FDA's approval of the several products made by continuous manufacturing. Vertex's Orkambi was the first SOD product made by continuous manufacturing to be approved followed by Janssen's Prezista, Eli Lilly's Verzenio, and additional products manufactured by Pfizer and Vertex [1]. Many of these products have subsequently been approved by multiple regulatory authorities around the world, and several other products made by continuous manufacturing are anticipated to follow in the next few years.

While continuous manufacturing can provide a higher degree of control and manufacturing efficiency, at the end of the day, the advantages of continuous over batch manufacturing must be clear from a business perspective including a clear path to market. Regulatory considerations can factor heavily into the decision on whether an organization will adopt continuous manufacturing for their SOD pharmaceutical products. While the traditional approach has a known path to market, new technology has uncertainty and perceived risk. The more a new technology has been implemented, the clearer the pathway becomes.

Because of the highly regulated nature of pharmaceutical manufacturing, questions related to regulatory expectations and requirements inevitably arise when introducing new technologies. The good news is that regulatory expectations for a product made by continuous manufacturing are the same as those made by traditional batch manufacturing, namely, that the process reliably manufactures a quality product that is suitable for the patient. The challenging part is that demonstration of assurance of quality for continuous

How to Design and Implement Powder-to-Tablet Continuous Manufacturing Systems
https://doi.org/10.1016/B978-0-12-813479-5.00006-9

manufacturing can significantly differ than for traditional batch manufacturing. Part of the challenge stems from lack of familiarity within regulatory authorities and apprehension with new and unfamiliar approaches which can lead to requests for additional supporting information and the perception by industry that regulatory expectations are higher.

Regulators across the globe have largely encouraged continuous manufacturing technology, and while finalized industry guidance is still emerging, they have recognized that real or perceived hurdles may exist. Several regulatory agencies have positioned themselves so that new technologies, including continuous manufacturing processes, can be discussed much earlier than products using traditional technologies, well ahead of a submission. The US FDA established the Emerging Technologies Team (ETT), Europe's EMA tasked their process analytical technology (PAT) team to include continuous manufacturing within their remit, and Japan's PMDA established the Innovative Manufacturing Technology Working Group (IMT-WG). This "open arms" approach is intended to promote new technologies to assuage the uncertainty that adopters perceive.

At the time of this writing this chapter, the International Council on Harmonisation of Technical Requirements for Pharmaceuticals for Human Use (ICH) is in progress of developing ICH Q13 as a topic for continuous manufacturing of both small and large molecules [2]. Additional work is underway on industry standards through activities of ASTM [3] and the US Pharmacopeia [4]. In addition, draft guidelines have been published by US FDA [5] and Japan [6] and numerous presentations from regulators and several industry/academia white papers have been written on this topic [7,8] to form a strong basis for clarifying regulatory expectations on continuous pharmaceutical manufacturing.

2. Definitions

2.1 "Continuous manufacturing" versus "batch manufacturing"

Much confusion and contradiction exists on what is called continuous manufacturing in the pharmaceutical industry. The confusion in part is due to lack of clarity about whether the continuous system being discussed is a single unit operation or a manufacturing process comprised of several connected unit operations. To add to the confusion, the term continuous manufacturing is used by different communities to denote different things. Within the active pharmaceutical ingredient (API) or biotechnology communities, continuous manufacturing has traditionally been used to denote a single-unit operation, such as a flow reactor or perfusion cell culture system. For SOD forms, continuous manufacturing typically describes multiple connected and integrated unit operations, often including the entire manufacturing process from powder in to tablets out. The continuous manufacturing systems of subject in this chapter focus on integrated unit operations for manufacture of SODs, which provide special challenges from a process control and regulatory perspective.

A classic chemical engineering reference [9] defines a continuous reactor as one in which "reactants are introduced and products are withdrawn simultaneously in a continuous manner." This differs from a batch reactor, which "takes in all the reactants at the beginning and processes them according to a predetermined course of reaction during which no material is fed into or removed from the reactor." Other manufacturing operations where materials are added during process but not withdrawn are often called "semi-batch" or "semi-continuous," for example, a fed-batch fermentation or wet granulation. This thought process used for chemical reactors can be used to apply the term "continuous manufacturing" to any scenario where input material and product are simultaneously introduced and withdrawn. This definition is subject to the scale of scrutiny; the "box" can be drawn around a single unit operation or an entire process.

A traditional SOD manufacturing process is actually a series of unit operations that alone could be considered to be batch operations (e.g., bin blender), semi-batch (e.g., wet granulation), or continuous (e.g., tablet compression). In common terminology usage though, these operations are said to be run in a batch mode when all in-process materials are collected and production is halted at the end of each step before proceeding to the next step.

In contrast, continuous manufacturing for SODs connects and integrates all unit operations and essentially eliminates the "stop and go" operation. The process is continuous in that material flows in and out without stoppage. The majority of the individual unit operations in these systems are also "intrinsically" continuous but some exceptions can occur. In some cases, multiple "mini-batch" unit operations can be incorporated into the processing train. Examples include parallel mini-batch coaters or segmented driers where material is continually being added to one unit while the other unit(s) are processing or discharging material.

From a regulatory perspective, there are no specific requirements that come into play because a process is designated as fully or partly continuous. Different control strategies should be based upon the needs of the process to ensure product quality. In some cases, there is no need for modification of a standard control strategy approach, such as for a continuous flow reactor followed by a batch crystallizer. However, most operational designs for continuous manufacturing for SODs will likely lead to nontraditional control strategies to ensure quality, typically with a higher utilization of in-process measurements and real-time controls.

2.2 Regulatory definition of "batch"

Unfortunately, in the English language, the word "batch" has two meanings, which often leads to confusion related to continuous manufacturing. One meaning of "batch" is the mode of manufacturing as described above, whereas the second meaning is related to an amount of material produced. The regulatory definition of batch is solely based on the amount material produced, as defined in ICH Q7:

> **Batch (or lot):** *A specific quantity of material produced in a process or series of processes so that it is expected to be homogeneous within specified limits. In the case of continuous production, a batch may correspond to a defined fraction of the production. The batch size can be defined either by a fixed quantity or by the amount produced in a fixed time interval.*

A similar definition of "batch" or "lot" exists in the US FDA regulations in 21 CFR 210.3, which further specifies that a batch is made from a single manufacturing order during the same cycle of manufacture. **The fundamental regulatory concept of a batch is the same whether the product is produced via "batch processing" or "continuous processing."**

Defining a batch of product for a continuous manufacturing process has more options than in a batch manufacturing process, as there are a number of acceptable approaches. Some of the options for delineation of a batch include

- All material produced within a specific time interval
- A specific amount of material charged into the process, regardless of the time taken
- A specific amount of acceptable material discharged from the process, regardless of the time taken
- All of the material produced between two specific process events (for example, amount of material to fill a downstream operation, such as a batch crystallizer)
- Other scenarios as defined by the applicant (e.g., the quantity of product made from a single tote of drug substance, regardless of the tote weight)

Flexibility in batch size is an attractive feature for continuous manufacturing. Because changes in batch size are typically achieved by altering the runtime on the same equipment, little or no development work is needed to ensure successful operation. Theoretically, it could be possible for a continuous process to be operated with no predefined runtime or batch size. However, regulators have frequently stated their expectation publicly that the batch size be defined prior to the initiation of the run. Extension or reduction of the batch size once a run has started can be managed under the site's deviation procedures.

In many cases, continuous manufacturing units for SODs are operated such that only one batch is produced between the equipment's start-up and shutdown. Such runs usually last several hours or a few days. However, it is possible for a continuous rig to run for weeks or months before shutdown and produce multiple batches. The ability to trace materials is vital for this type of operation so that any problems later identified can be appropriately linked to the associated batches.

2.3 "State of control" and "steady state"

The terms "steady state" and "state of control" are both applicable to continuous manufacturing but have distinctly different meanings. ICH Q10 defines "state of control" as a condition in which a set of controls consistently provides assurance of continued process performance and product quality. Generally speaking, a "steady state" condition is one in which the process parameters and product output attributes are kept approximately constant, or the rate of change with respect to time of those variables is approximately equal to zero. **While continuous manufacturing will frequently operate in both a state of control and at (or near) steady state, achieving a state of control is paramount to assure product quality.**

While regulators have frequently articulated the need to achieve a state of control in continuous manufacturing, the expectations for achieving steady state is less clear. Logic supports that achieving a state of control should be the regulatory expectation for accepting quality product, independent of defining and achieving steady state operation. It is possible for a process to be operated at steady state with product of poor quality and thus not be in a desirable "state of control." For example, a feeder can be stable operating at a set point below its target value, thus causing a steady state to be reached in the system while producing product outside of specifications. As a counterexample, a process can achieve a dynamic state of control while not being at steady state, such as during the start-up phase of a continuous manufacturing line, where a highly automated sequence of events leads to high-quality product being produced early in a production run [10].

2.4 Scale-up

The traditional concept of "scale-up," using larger equipment to implement an equivalent process, is a potential but rarely used approach to increase the production volume in continuous manufacturing for SODs. More typical approaches are to increase the runtime, raise the throughput/flow rate, or "number up" by adding additional lines. Nevertheless, the term scale-up is commonly used to describe any increase in production volume for continuous manufacturing.

While it is possible to use larger-scale equipment to achieve scale-up of continuous manufacturing SOD processes, such an approach increases the scientific and regulatory challenges. When increasing production volume through larger equipment, development work typically would need to be done using equipment of multiple scales, and traditional validation of different batch sizes would be appropriate.

Because changes of product volume in continuous manufacturing are often very different than for batch, different regulatory approaches could apply for regulatory reporting and validation, as discussed in Section 4 of this document.

3. Regulatory considerations for designing and implementing a continuous manufacturing process for SODs

3.1 System dynamics and material traceability

One of the greatest differences between traditional batch and continuous manufacturing of SODs is the need to account for system dynamics. **Continuous manufacturing has time dependencies that are different than for batch manufacturing**. In a batch unit operation, variation in incoming materials over time is irrelevant because the transformation occurs after all materials are introduced and ends before the material exits the system. For example, in batch bin blending, it does not matter how fast excipients and API are added to the blender because the blending process only begins after all material transfer is complete. Assurance of quality is typically attained after the operation is complete (e.g., blend uniformity testing) and focuses on demonstrating spatial uniformity. In contrast, the rate of material addition is critical to blend composition for a continuous blender; disturbances in ingredient flow rates beyond what the system has been designed to handle can directly affect the quality of product exiting the system. An understanding of the time dependence of continuous systems is an essential component of assuring product quality. Thankfully, there is methodology for the assessment of time dependencies, and this is one of the advantages of continuous manufacturing. With continuous manufacturing, a disturbance of the type described above can be tracked through the system, its impact on quality can be predicted, and the portion of product affected can be segregated from the product collected. In batch manufacturing, a disturbance or manufacturing issue affecting the end product quality would typically necessitate that the entire batch be discarded.

The most common technique for characterizing the system dynamics of a continuous process is to measure and model the residence time distribution (RTD) [11,12]. The RTD describes the probability distribution of exit times for material entering the system. It also provides a characterization of the amount of mixing in the system, with the width of the RTD being proportional to the intensity of backmixing in the system. By measuring and modeling the RTD, material can be tracked through the system as a forward function of time.

RTDs can be highly beneficial in two manners. First, RTDs can provide traceability of raw material for multiple batches manufactured under the same campaign, which is a current Good Manufacturing Practice (CGMP) requirement. Second, RTDs can be used to model how disturbances travel through the system. Such models can be incorporated into the control strategy to isolate potentially nonconforming material. In this manner, a portion of the batch can be isolated and rejected if necessary without impacting the quality of the remaining material.

3.2 In-process monitoring and control strategy

The regulatory requirements for both traditional batch and continuous manufacturing for SODs are the same—to ensure that the process reliably manufactures a quality product suitable for the patient. The primary difference lies in the control strategies to achieve them. The pharmaceutical industry has had decades to understand the risks related to traditional batch SOD manufacturing and to develop standardized practices and control strategies demonstrating how those risks are managed. Control strategy development for continuous manufacturing of SODs is still in its infancy, though certain considerations have become clear, especially related to variability of the process with respect to time.

According to ICH Q10, a control strategy is a planned set of controls derived from current product and process understanding that ensures process performance and product quality. The controls can include parameters and attributes related to drug substance and drug product materials and components, facility and equipment operating conditions, in-process controls, finished product specifications, and the associated methods and frequency of monitoring and control [13].

One of the greatest differences between a traditional batch process for SODs and a continuous process is how blending is performed. Because of the potential impact of disturbances with time, most continuous manufacturing systems for SODs include a component of real-time monitoring for blend uniformity in the control strategy. The most common methods are spectroscopic (e.g., near-infrared [NIR], Raman) and/or RTD modeling.

In-process spectroscopy usually consists of one or more probes located throughout the processing train. NIR and Raman spectroscopy analyzers are most commonly used. The probes can be at different locations such as post-blending, prior to compression or encapsulation, or within a tablet feed frame. In order for control actions to occur, the spectroscopy results are interpreted and fed to a controller. The simplest design is to have a diversion point directly following the analyzer, but such diversion can cause system imbalances due to reduced flow downstream. RTD models are typically used to track potentially nonconforming material when diversion happens downstream of the analyzer. In that case, the material continues to be processed and is isolated at more convenient point later in the process, for example, after tablet compression.

As mentioned, RTD models can also be used to detect disturbances that could lead to nonconforming material directly from the feeders. In the RTD model approach, disturbances from the feeders are tracked through the system and an appropriate amount of material is diverted downstream at the diversion point. An underlying assumption of the RTD model approach is that the feeders are the sole source of inhomogeneity. This approach is only appropriate for compositions with a low potential to segregate or agglomerate during processing, as the resulting inhomogeneity from such sources would not be predicted by an RTD model.

Future approaches to assure blend uniformity or consistency of other critical quality attributes could use a "soft sensor" model that utilizes a combination of process data and software to describe the system predictively. Such modeling approaches include multivariate data analysis and multivariate statistical process control (MSPC) that incorporate a wide range of process parameters and material attributes to establish a range of normal operation. The central assumption is that operation within the normal range of operation, as defined by the model, will produce quality product. Operation outside of the normal range is marked as atypical and could indicate a loss of state of control and production of material of unacceptable quality. While the authors are unaware of any current examples of multivariate approaches as the primary approach to ensure product quality in commercial production, potential exists.

Measurement of blend uniformity either directly by a spectroscopic technique or indirectly by an RTD model can be considered an application of PAT. As defined in the FDA's Guidance for Industry PAT — A framework for innovative pharmaceutical development, manufacturing and quality assurance, "PAT is to be a system for designing, analyzing, and controlling manufacturing through timely measurements (i.e., during processing) of critical quality and performance attributes of raw and in-process materials and processes, with the goal of ensuring final product quality. It is important to note that the term analytical in PAT is viewed broadly to include chemical, physical, microbiological, mathematical, and risk analysis conducted in an integrated manner" [14].

While not chemically specific, nonspectroscopic process data and related models can often acquire and analyze information on a faster time scale, while simultaneously being more robust and more readily interpreted than spectroscopic data. Moreover, when properly aggregated with other data, nonspectroscopic information can provide a broader, more complete characterization of the state of the system.

Of course, appropriate controls need to be applied to other unit operations besides blending within the continuous manufacturing process train. Many of these unit operations are similar to traditional batch operations (e.g., tablet compression, encapsulators, roller compactors) and will have local-level controls similar to those used in batch manufacturing. RTD models or MSPC models can lie atop the individual unit operation controls to provide material tracking throughout the system and a measure of process consistency.

A control strategy for continuous manufacturing can optionally incorporate redundant controls, for example, measurement of blend uniformity at multiple locations. An advantage of redundancy is that the additional sensors can be beneficial for early detection and diagnosis of operational problems. However, redundant controls can also lead to an increased number of investigations due to inconsistent data, especially if one of the measurements is less robust. When incorporating redundant controls, it is important to have clear definition within the quality system on how to deal with inconsistent information and to pre-define the circumstances when the system can be run without the redundant controls.

Design of the overall control strategy should employ a holistic view to provide assurance of product quality. For many continuous manufacturing systems for SODs, the potential exists for short-term deviations to have an impact on product quality. Understanding the system dynamics, such as through RTD studies, will help determine what levels of disturbances will not dampen out by backmixing and require diversion. It will also inform the sampling frequency for in-process and/or end product testing. A thorough risk assessment that considers potential failure modes is essential to establish the suitability of the overall control strategy.

For the Chemistry, Manufacturing, and Controls (CMC) section of a reg-ulatory application, enough information should be provided for the assessor to understand how the control strategy assures product quality. The amount of information to be provided in the application is discussed in the ICH Q8/9/10 Points to Consider Document [15]. In general, the submission should provide scientific justification for the proposed control strategy and a discussion of the studies used to support its development, including rationale for the studies, description of the studies, and summary of the results and the conclusions.

3.3 Real-time release testing

Real-time release testing (RTRT) is not required for continuous manufacturing of SODs. However, the control strategies and high level of measurement associated with continuous manufacturing for SODs are often a natural fit for an RTRT approach. RTRT is defined in ICH Q8(R2) as

The ability to evaluate and ensure the quality of in-process and/or final product based on process data, which typically include a valid combination of measured material attributes and process controls. [16]

Major advantages of RTRT include shorter cycle times, a corresponding lower inventory, and the ability to detect problems in real time.

RTRT can be used for all or only some of the product attributes [15]. Many SOD continuous manufacturing processes include an in-process measurement of active ingredient concentration as part of the control strategy. In many cases, this information can directly support justification of an RTRT approach for assay and content uniformity. In addition, for closed systems, it may be possible to establish identity from in-process measurements. Other attributes, such as dissolution, are more challenging to assess in-process and usually rely on surrogate models incorporating process parameters and/or material attributes.

The RTRT strategy, like the overall control strategy, should be designed with a holistic view of the process and product needs. Elements contributing to product quality assurance can include in-process measurement of material attributes, monitoring of critical process parameters, and other process data such as soft sensors. While an RTRT approach can include elements of end product testing, in a well-designed process, the purpose of end product testing is confirmation that the process is working as intended, rather than as a primary assurance of product quality. In most cases, 100% testing of either in-process or end product is not achievable, not time- or cost-effective, and not value added from a statistical standpoint. Risk assessments should be conducted to ensure that the control strategy is effective in addressing potential failure modes. System dynamics should be considered when determining appropriate sampling frequencies and locations to ensure high probability of detection of faulty product [17].

In most cases, traditional offline tests of the in-line versions should be developed and validated as part of process development and implementation. These tests serve multiple purposes. They are often used as the reference test for developing and validating in-process methods, such as NIR spectroscopy. They also can serve as the test method for stability testing or nonstandard testing for investigations. An applicant may choose to include such tests in their application alternatives to an RTRT [15]. An approved alternative approach can allow production to continue when in-process monitoring equipment fails. Alternative control strategies to RTRT should be appropriately validated and also justified in the regulatory submission, including considerations of sample size.

The level of detail in the application for models should be commensurate with the importance of the model in assuring product quality and is described in the ICH Q8/9/10 Points to Consider Document [15]. A low impact model such as those used in development merits only a brief description of its use. Description of a medium impact model, such as a redundant in-process control, should include a description of the role of the model in the control strategy, model assumptions, model equations, and model results. Description of a high impact model that provides an indicator of product quality (e.g., NIR for tablet assay, surrogate model for dissolution) should include the information above plus details of the model validation and a high level discussion of model verification throughout the product life cycle. Most if not all models assuring product quality in an RTRT approach would be considered high impact models.

3.4 Stability data

The overall regulatory expectations for providing stability data are unchanged from traditional batch manufacturing, whether for a new continuous process or for manufacturing changes to an already approved process. These expectations for new products are outlined in ICH Q1A(R2), which states that formal stability studies should be provided on at least three primary batches of drug product with at least two of the batches being at pilot scale, where pilot scale is defined as a minimum of 1/10th full production scale or at least 100,000 dosage units, whichever is larger. Multiple lots of drug substance should be used whenever possible and bracketing and matrixing approaches may be utilized to reduce testing related to certain aspects (e.g., different strengths, batches, container closures), if justified. The manufacturing process for the primary batches is expected to simulate the production batches and provide the same quality as that intended for marketing [18].

Continuous manufacturing has some nuances that could allow for reduced data requirements and additional manufacturing flexibility related to manufacturing scale. As discussed in Section 2.4, scale-up in continuous manufacturing usually involves extending the length of time for a batch without a change of equipment size or type. For this scenario, it would be very rare to see a change in stability characteristics based on the length of the run. **Therefore, the minimum 1/10th limitation of "pilot batches" should be irrelevant for continuous manufacturing**. Similarly, it would be atypical to have different stability characteristics for changes in process parameters, such as flow rates. While scientifically supportable, circumventing the 1/10th scale expectation from ICH Q1A(R2) is largely untested.

In general, the approach used to determine representative stability samples should be justified through a risk assessment. The risk assessment should evaluate the potential for the stability characteristics to change at different conditions (e.g., batch size/runtime, flow rate). An additional consideration is the time within the manufacturing batch for obtaining the stabilities of samples (e.g., the start, middle, end of batch). A similar risk assessment approach can be used to determine appropriate samples for annual stability testing.

3.5 Process validation

The question that inevitably arises related to process validation for continuous manufacturing of SODs is, how many process performance qualification (PPQ) batches are needed and at what scale? This level of detail is not currently available in regulatory guidance nor is it anticipated to be in the future. In general, the manufacturer needs to determine the number of batches necessary to provide justification that the process is capable of consistently producing acceptable quality product at commercial manufacturing conditions. The process validation activities confirm that the control strategy is adequate for the process design and quality of the product. PPQ activities must be complete before releasing product to the market.

While FDA guidance does not specifically state a minimum number of PPQ batches, other regulatory authorities such as the EMA have a general expectation of a minimum of three production scale batches, although some may allow fewer if appropriately justified and supported by pilot scale batches. **While theoretically all PPQ batches could be from a single continuous manufacturing run, this approach likely would not sufficiently demonstrate consistency of start-ups and shutdowns.** In general, it is recommended to run at least one PPQ batch at the maximum anticipated runtime [7]. This approach can help to reveal issues associated with material buildup or heat generation over an extended period of time.

Most considerations for process validation for continuous manufacturing are the same as traditional batch process validation such as utility verification, cleaning considerations, consideration of raw material variability, and applying bracketing approaches. Some special considerations for continuous manufacturing of SODs relate to qualification of system pauses and restarts and successful demonstration of diversion of potentially nonconforming material; these aspects are typically verified during equipment qualification activities.

Following PPQ, both European Union (EU) and US agencies have an expectation for an extended period of monitoring to support the validated status. In the EU, this phase is called "ongoing process verification," while in the United States it is termed "continued process verification." The goal of this third stage of process validation is to provide assurance that the process remains in a state of control within the validated state during commercial manufacture. Enhanced monitoring and sampling are recommended until sufficient data are generated to support the frequency of routine sampling.

Continuous process verification (CPV) is an alternative approach to traditional process validation in which manufacturing process performance is continuously monitored and evaluated [16]. In essence, CPV combines Stage 2 (PPQ) and Stage 3 (continued/ongoing process validation) activities such that every batch is analogous to a PPQ batch. Many continuous manufacturing designs are well poised for a CPV approach because of their high level of integrated measurements. Indeed, the EU process validation guideline states that "*continuous process verification would be considered the most appropriate method for validating continuous processes.*" Even when utilizing a continuous process validation approach, the manufacturer is expected to predefine the criteria at which the process is considered to be under control and the basis to release product to the market [19].

In essence, the requirements for inclusion of process validation and life cycle management information in the regulatory dossier are the same for continuous manufacturing as for traditional batch manufacturing. However, while continuous manufacturing is in its early stages, EMA could consider it "new technology" and require inclusion of production scale validation data in the dossier, unless otherwise justified [19].

3.6 CGMP considerations

Most of the CGMP considerations for continuous manufacturing are identical to those for batch manufacturing. When first implementing continuous manufacturing at a facility, the existing procedures and documents should be reviewed for consistency with the future operations. Minor issues may arise such as how the word "batch" is used within manufacturing procedures and quality systems documentation. More significant gaps are likely to be found related to the ability to divert potentially nonconforming material and perform a partial lot rejection within the quality system.

Not all diversions will necessarily result in a deviation. For example, diversions during normal start-up and shutdown are not deviations. Start-up and shutdown procedures are typically included in the site's quality system, often in operating procedures. Clear criteria should be established for when to collect or divert material during start-up, shutdown, and disturbances. Diverted material that is potentially nonconforming could be determined to be acceptable if, following an appropriate investigation, it is determined that its quality was not compromised. Examples of such scenarios could include erroneous readings from a fouled sensor or data interruption. Care needs to be taken to separate and track different diversion events if reintroduction of diverted material into the rest of the product is possible. It may also be necessary to streamline practices and procedures to be able to respond to deviations in near real time.

The ability to perform partial lot rejections in continuous manufacturing is substantially different than in batch manufacturing where, because of the process dynamics of the system, rejection of part of the batch implicates the quality of the rest of the batch. The material traceability in a continuous manufacturing system provides a high level of confidence that potentially nonconforming material can be detected and separated from the material of known acceptability. In the end, the quality of the product is enhanced by careful monitoring and diversion of material of marginal or poor quality. Regulators have publicly expressed that a continuous manufacturing process should have a defined minimum yield to provide assurance that the diversion and rejection events are not excessive and that the process as a whole remains in a state of control.

Additional CGMP considerations will be needed for sites that have not previously utilized models, for example, for PAT or RTRT. These procedures should cover issues such as model maintenance, how to deal with out of specification or out of trend results, and when to perform model recalibration.

The final product disposition decision at the end of production is identical for batch processing and for continuous manufacturing; it involves review of all the manufacturing documents, investigations, and the in-process/release data to confirm that the manufacturing process was in a state of control and, as such, high-quality product has been produced which is acceptable for release. All unexpected diversion events and other deviations should be resolved prior to release of batch.

4. Manufacturing changes for continuous processes

4.1 Conversion of already approved batch process to a continuous process

As continuous manufacturing for SODs is introduced into the pharmaceutical industry, some of the first applications of this new technology will be conversions of already approved batch processes to continuous processes. By starting with an already approved product, the manufacturer has the opportunity to gain experience with the new equipment and to navigate an uncertain regulatory landscape without the tight timelines associated with a new product introduction.

Having both a batch process and a continuous process in the same application allows maximal flexibility to respond to changes in market demand and agility for unforeseen circumstances with equipment or manufacturing sites. Right now, there are very few continuous manufacturing equipment rigs existing that can serve in a backup function. As adoption grows, additional equipment is expected to come online both within pharmaceutical companies and within contract manufacturing organizations. In addition, as the technology matures, more information will become available related to the interchangeability of equipment and control systems from different vendors. Finally, over that same time period, regulatory familiarity will increase, likely providing additional clarity on certain topics.

The approach to have multiple processes in the same dossier is supported by the ICH Q8/9/10 Points to Consider document [15], which discusses that different control strategies for the same product are possible, including for different sites and/or different technologies. However, the pathway is not totally clear. The ability to include multiple processes/control strategies within the same dossier is untested in the emerging regulatory regions that have not adopted ICH guidelines. In addition, a guideline on Manufacturing of Finished Dosage Form from EMA states that "alternative manufacturing processes, which used different principles and may or may not lead to differences in the in-process controls and/or finished product quality are not acceptable" [20]. The position in this guideline could be a major impediment for converting some batch processes to continuous processes (e.g., a wet granulation batch process to a direct compression continuous process).

Patient considerations are essential when looking to have product from two different manufacturing processes/control strategies in the same application. The dosage forms made from the different processes need to be indistinguishable to patients, pharmacists, and other health care givers. The visual appearance between the two products should be identical and the product's performance (e.g., dissolution) should be the same. From a regulatory perspective, the same label is needed for the products. A change in formulation between batch and continuous processes for SODs should not add or remove any ingredient so as not to change the qualitative list of ingredients.

Flexibility for slight formulation changes between batch and continuous in many cases could be important for robust operation and consistent product quality. Examples include slight changes in lubricant (e.g., magnesium stearate) or film coating thickness. Although the concept of "formulation design space" is not mentioned in ICH Q8(R2) or other regulatory guidance, adopting such an approach may help increase the robustness of the process and decrease product variability. Of course, appropriate in vitro and/or in vivo studies would need to be performed to ensure that the bioavailability of the differing formulations remains the same. Future advances in material understanding may support feed forward control where the composition of the formulation is slightly modified for each batch depending on raw material properties. So far, regulators in some of the major regulatory regions (i.e., United States, EU, and Japan) seem receptive to the concept of flexible formulations, but its acceptability in other parts of the world remains unknown.

Changes related to the equipment used in continuous manufacturing merit the same considerations as other equipment changes. The changes would be evaluated through approaches such as chemical equivalency, comparative batch data, impurity profiles, drug release profiles, and stability data. It is anticipated that simply linking the same unit operations together without changing the operational principle of the equipment should not constitute a major change, for example, coupling of a roller compactor to a tablet press. However, changes from a batch bin blender to a typical continuous blender (e.g., tubular paddle blender) would constitute a change in the operating equipment principle and subsequently merit the appropriate regulatory considerations.

Quality standards remain the same for a product whether manufactured by batch or continuous manufacturing. However, the specifications for the two processes could vary. For example, a continuous manufacturing process could demonstrate content uniformity through a Large N sampling test criteria [21] of core tablets, whereas the corresponding batch process uses standard uniformity of dosage units determination [22] on coated tablets. At this time, there is no clear regulatory standard on how to communicate different specifications for multiple processes/control strategies. Some regulators may prefer multiple specification tables; others may prefer a single table with multiple columns and/or footnotes. Discussions with regulators are recommended prior to the submission.

The considerations provided above for conversion of an already approved batch process to a continuous process also hold for a batch to continuous conversion in the development phase. It is expected to be common, at least during early adoption, for early clinical trial material to be made by a batch process and later phase material to be made by continuous processing. To minimize bridging studies, it is best to use the anticipated commercial process when manufacturing pivotal clinical trial batches that are used to support safety and efficacy studies. Formal stability studies should be conducted using product manufactured at the commercial manufacturing site using the intended commercial process or processes.

4.2 Changes to an approved continuous manufacturing process site

In general, the laws and regulations related to changes to an approved application are the same regardless of the mode of manufacturing. Most aspects of changes for continuous systems remain the same as for traditional system; for example, movement within a design space does not need regulatory reporting. However, much of the more detailed guidance from regulatory agencies was developed with batch manufacturing in mind. For example, the FDA Guidance on Scale-Up and Post-Approval Changes for Immediate Release Solid Oral Dosage Forms (SUPAC IR) provides that a change in scale up to $10\times$ can be reported as a Level 1 change via Annual Report if the equipment is of the same design and operating principle [23]. The $10\times$ constraint does not make sense for increasing the production volume of a continuous manufacturing system where the increased production volume is achieved simply by extending the runtime, especially for products and processes that do not have time dependencies (e.g., temperature instabilities). Future regulatory certainty should bolster adoption of continuous manufacturing technologies.

5. Discussions with regulators

Several regulatory agencies have developed specialized groups to support innovative technologies, including continuous manufacturing, as described below.

The US FDA established the ETT in 2014 and finalized a guidance related to its procedures [24] in 2017. The ETT includes representation from all FDA pharmaceutical quality functions. The team functions as a centralized point of contact for the applicant on the new technology and a team member serves as the lead or co-lead of the review team. In addition to meetings with the US FDA, a program exists for a Pre-Operational Review (also known as a Pre-Operational Visit), which provides guidance to industry for construction of facilities prior to commercial production [25].

The EMA directs questions related to continuous manufacturing to its PAT team, which was initially set up in 2003. The PAT team acts as a forum for dialogue between the working parties for small molecules, biologics, and GMP and as appropriate provides specialist input into dossier assessment and scientific advice.

In 2016, the Japan Health Authority PMDA established the Innovative Manufacturing Technology Working Group (IMT-WG). The purpose of this group is to establish a common perspective on new technologies that may lead to new regulatory frameworks. The Japanese regulators have partnered with industry and academics to recently release a Points to Consider Document on continuous manufacturing [6].

In general, it is recommended that an applicant arrange discussions with the regulatory agency early in the development process. Typically, enough information needs to be known to prepare a briefing packet for the regulators explaining the general approach at a high level and providing regulatory questions related to implementation. The applicant should provide a clear proposal for their path forward and ask specific questions related to their regulatory concerns.

Through open and early conversations with regulators, manufacturers can help plan development of their equipment, processes, and procedures and lower their regulatory risks.

References

[1] Lalloo AK. Regulatory progress in global advancement of continuous manufacturing for pharmaceuticals. Pharm Eng 2019;39(3):29−31.

[2] ICH Draft Guideline Q13. Continuous Manufacturing of Drug Substances and Drug Products 2021.

[3] ASTM E2968. Standard guide for application of continuous processing in the pharmaceutical industry: ASTM.

[4] http://qualitymatters.usp.org/exploring-continuous-manufacturing-technology-and-applications-pharmaceutical-industry.

[5] Draft FDA Guideline. Quality Considerations for Continuous Manufacturing. Feb 2019.

[6] NIHS Points to Consider Regarding Continuous Manufacturing http://www.nihs.go.jp/drug/section3/AMED_CM_PtC.pdf.

[7] Regulatory perspectives on continuous pharmaceutical manufacturing: moving from theory to practice- https://iscmp2016.mit.edu/regulatory-white-paper.

[8] USP Accelerating adoption of pharmaceutical continuous manufacturing.

[9] Perry Robert H, Green Don. Perry's chemical engineers' handbook. 6th ed. McGraw-Hill; 1984.

[10] Almaya A, et al. Control strategies for drug product continuous direct compression—state of control, product collection strategies, and startup/shutdown operations for the production of clinical trial materials and commercial products. J Pharm Sci 2017;106(4):930−43.

[11] Gao Y, Muzzio FJ, Ierapetritou MG. A review of the Residence Time Distribution (RTD) applications in solid unit operations. Powder Technol. 2012;228:416−23.

[12] Weinekötter R, Gericke H. Mixing of solids. Springer; 2000.

[13] ICH Q10 pharmaceutical quality system. 2009. p. 22.

[14] USDHHS, FDA, CDER, CVM, ORA. Guidance for industry. PAT- A framework for Innovative pharmaceutical development, manufacturing, and quality assurance. 2004. p. 19.

[15] ICH Quality implementation working group points to consider (R2) ICH-endorsed guide for ICH Q8/Q9/Q10 implementation.

[16] ICH. Q8(R2) Pharmaceutical development. 2009. p. 29.

[17] Lee SL, et al. Modernizing pharmaceutical manufacturing: from batch to continuous production. J Pharm Innov 2015;10(3):191−9.

[18] ICH Stability testing of new drug substances and products Q1A(R2). 4 ed. ICH; 2003. p. 24.

[19] EMA Guideline on process validation for finished products - information and data to be provided in regulatory submissions. Nov 2016.

[20] European Medicines Agency Guideline on manufacture of the finished dosage form.

[21] European Pharmacopeia 2.9.47 Demonstration of uniformity of dosage units using large sample sizes.

[22] USP <905> Uniformity of Dosage Units.

[23] FDA SUPAC-IR Immediate-Release Solid Oral Dosage Forms: Scale-Up and Post-Approval Changes: Chemistry, Manufacturing and Controls in Vitro Dissolution Testing and in Vivo. Bioequivalence Documentation.

[24] FDA. CDER advancement of emerging technology applications to modernize the pharmaceutical manufacturing base guidance for industry. September 2017.

[25] FDA Field Management Directive 135 Pre-Operational Reviews of Manufacturing Facilities.

Chapter 15

Continuous manufacturing process development case study

Eric Sánchez Rolón, Mauricio Futran and William Randolph

Janssen Supply Chain, The Janssen Pharmaceutical Companies of Johnson and Johnson, Raritan, NJ, United States

1. Introduction

The Food and Drug Administration (FDA) recently approved several new products and supplement changes of existing products on conversions from batch to continuous manufacturing (CM) process technologies. These approvals included the CM mode of operation and real-time release using process analytical technology (PAT) data collected through near-infrared (NIR) spectroscopy in substitution of the traditional wet chemistry analytical testing [1].

This milestone was possible through strong collaboration between industry and academic researchers. In our experience, the product conversion project described in this chapter required a great deal of internal collaboration between our company teams in Puerto Rico and Raritan, NJ, supply chain organizations, including regulatory, quality, operations, engineering, and other technical operations.

The interaction with academia in this project was a win/win opportunity for our company. It enhanced our knowledge base and helped us identify and develop students with great professional potential who interacted with our employees and resulted in jobs within the industry for some of the student participants.

The objective of this chapter is to describe our collaboration approach. It defines the different stages of technology implementation from the feasibility studies to the technology transfer and validation in the Janssen supply chain facility in Gurabo, PR. The chapter sections include engineering design, analytical development, process development, and technology transfer, including project management for implementation of the technology in the commercial facility up to the successful health authorities' supplemental new drug application submission process.

How to Design and Implement Powder-to-Tablet Continuous Manufacturing Systems
https://doi.org/10.1016/B978-0-12-813479-5.00016-1
319

2. Criteria for product selection

We considered four criteria in selecting our first batch-to-CM conversion product: product lifecycle stance, financial benefits, current product design, and product robustness and knowledge.

2.1 Product lifecycle stance

At the initiation of the project, the selected product was in the growth phase of its pharmaceutical product lifecycle.

2.2 Financial benefits

Projected benefits were determined based on % internal rate of return (% IRR) and net present value (NPV) calculations. Both financial indicators were positive for the project initiation.

2.3 Current product design

The product formulation consisted of a dry blend of five well-known and well-characterized ingredients. The formulation's simplicity and its material composition were favorable for processability as well as for implementation of real-time release testing (RTRt) capabilities.

2.4 Product robustness and knowledge

Prezista development was part of the 2007−08 FDA program on quality by design (QbD) registration process [2]. During QbD filing, we gained valuable process knowledge that served as a foundation for this product's transition into a continuous manufacturing process with real-time release testing. The QbD submission included a thorough design of experiments (DoEs) foundation that served as a baseline for the material characteristics, including active pharmaceutical ingredient particle size distribution optimization for dissolution performance and downstream CM process information. This also included leveraging risk analysis elements. In addition, a risk analysis focusing on the specifics of the CM process and the integration of PAT provided assurance of the process robustness of Prezista moving into CM and PAT technologies.

3. General approach and description of the process development, Process Analytical Technology development, and feasibility studies with academic partners

Throughout process development, supported by our academic partners, we focused on assessing the impact of each critical process parameter (CPP) on the product's critical quality attributes (CQAs). This also allowed us to identify sources of variability that could impact product quality. Material characterization focused on the physical and spectral properties of the blend of materials and subsequent manufacturing unit operations, compression, and coating. This provided significant insight into the ways these properties affect the performance of down-stream process.

The integrated nature of a CM process requires a detailed characterization of processing conditions with respect to the critical intermediate material attributes and finished product. These material attributes are a fingerprint of the product's susceptibilities to material/equipment interactions, and indicative of the stress conditions that the material suffers throughout the different manufacturing unit operations. CM process conditions considered the material information collected from development studies relative to these attributes and translated it into CPPs. These CPPs in interaction with the defined CQAs served as the basis for the CM process design space (and subsequent control space) that ensured equivalence between the product manufactured using continuous and batch processing modes. This design space allowed for maximum product quality prediction and less variability.

The development of the parametric design space of the unit operations involved in the CM process, namely feeding (materials loading and transfer), blending, compression and tablet coating, required focusing on the resolution of the flow and stress fields within the equipment, and assessing the resulting changes to material quality attributes. The vast number of feasible combinations of CM process conditions, operating parameters, and intermediate material attributes required a multivariable and multilevel experimental design approach.

Our academic partner's labs served as facilities for initial process development and feasibility studies along with PAT development activities.

This chapter section defines the approach followed for the development and feasibility studies on the CM line and the work performed to develop the necessary PAT methods.

3.1 Continuous manufacturing process development with academic partners

The following section provides a discussion of the approach followed for process development:

3.1.1 Raw material characterization and feeder characterization

The evaluation of material properties, like powder rheology or electric properties, ensures material feedability through twin screws and that the feeding operation does not affect material functionality. Fig. 15.1 shows a composite of the results for some of the relevant rheology tests performed. The raw material (RM) rheology data provided an excellent basis for evaluating the effect of

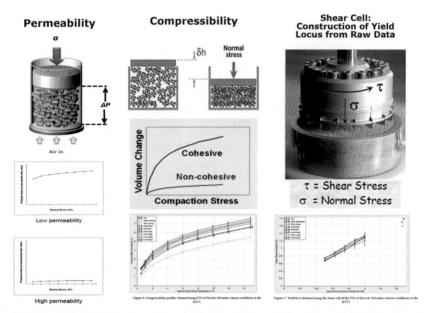

FIGURE 15.1 FT4 Rheometer test used for the characterization of materials rheology and analysis of unit operations effect on powder rheology.

powder properties on feeding performance. It enabled the selection of optimal setup conditions to achieve accurate material dispensing. Fig. 15.2 shows an example of the material rheology characterization pre- and post- the twin screw feeding operation. All RM achieved optimal feeding conditions with negligible effect over material rheological properties as ingredients traveled through their feeders.

3.1.2 Loss-in-weight feeder performance characterization for raw material

A DoE approach identified the optimal feeder setup that resulted in the best feeding performance for each material. The feeder dispensing rate determination used an independent, automated register of the feeder feed rate (catch scale). Feeding equipment conditions included the type of screw (auger vs. concave), the screw cavity volume (coarse vs. fine), the feeder outlet screen configuration, and the feeder speed settings. Fig. 15.3 shows a composite picture of the evaluated feeder elements. The DoE's approach was adequate for determining the twin screw feeder elements used for each RM. The selection criteria for the screw design combination were the smallest RSD% of the feed rate test ranges (Fig. 15.3D).

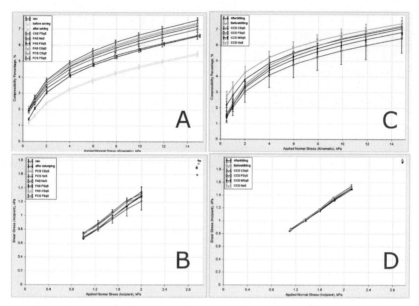

FIGURE 15.2 Powder rheology test showing relevance of material changes due to passage through a feeder. For major excipients, figures (A–B) show the effect of screw feeder setup conditions on compressibility (A) and Shear Yield Loci (B). For active pharmaceutical ingredient (C–D), the same tests show decreased effects.

3.1.3 Blending system performance characterization

Another DoE provided the evidence to define the blender paddles configuration, geometry, and orientation. The tubular blender provides a paddle shaft that needs configuration optimization. The blender study included geometric properties of the paddle elements, the combinations of paddle types, direction of the flow (forward or reverse), and their distribution within the shaft. In addition, the studies included an evaluation of the blender's rotational speed. An Acetaminophen (APAP) surrogate formula provided a test material with properties similar to the product formula. Low and high APAP concentrations in the formula provided the test range for the evaluation of the blender's setup conditions. Fig. 15.4 shows a composite picture of the blender's study elements. The experimental design (Fig. 15.4C) provided an adequate evaluation of the effect that paddle configuration had on blending performance and enabled us to identify an effective paddle configuration. The study showed that a configuration where the large, angled paddle geometry installed in all paddle positions and distributed with a third of paddles placed in a forward position, a third in a reverse position and a third in a forward position provided the best

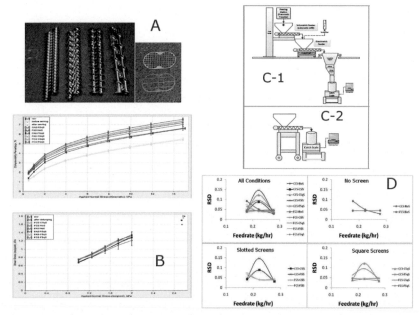

FIGURE 15.3 Empirical test approach followed in the determination of gravimetric screw feeder setup conditions. The approach starts with the selection of a combination of screws and screen to be tested from four distinct twin screws and two options of outlet screens (A). The material properties were analyzed to define the effect of the conditions on the material rheology (B). The flow rate variability during gravimetric dispensing is evaluated using calibrated scale system that provides information of the material dispensed independently from the gravimetric feeder control loop on a automated feedding layout (C-1) or manual feeding layout (C-2). All conditions are evaluated with respect to the RSD of the material dispensing rate (D).

distribution of API at the two concentration levels tested. Fig. 15.4D shows the results obtained in the blending study. Fig. 15.4A illustrates the experimental setup, while Fig. 15.4B shows the different blade shapes evaluated in the study.

3.1.4 Blending transfer and process analytical technology integration

Blend sensing using NIR spectrometry was part of the academic lab's evaluation. The fundamental aspects of sensing interface design and integration to the CM line evaluation occurred first at the academic partner's laboratory. The evaluation included sensing frequency interval, blend presentation to the spectrometer, and geometry of the sensing interface with respect to sensing accuracy and relevance to the control strategy. Spectral model evaluation included the impact of process/sensing critical variables on their predictability. Fig. 15.5 shows a composite picture of the sensing elements evaluated during this study. The study showed that the best mode of presentation of the blend to

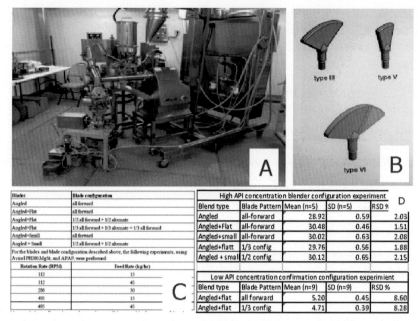

FIGURE 15.4 The experimental approach followed for the optimization of the blender (A) included the analysis of the diverse blender paddle configuration (B) in a structured design of experiment approach (C). The results obtained in the study showed that a configuration that distributes the paddles orientation with 1/3 forward, 1/3 reverse, and 1/3 forward provided the smallest blend concentration variability.

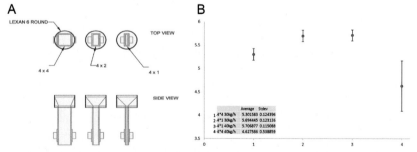

FIGURE 15.5 The sensing interface studies included the evaluation of two main geometric configurations, a rectangular configuration where powder flow was confined in a cubic tube (A) and a diamond configuration (B) where powder slides through the inner walls of the sensing interface from the blender powder stream. The accuracy of the blend predictions was optimized for the one-inch gap rectangular configuration as shown in the center of pane (A). The diamond configuration was not viable due to material flow issues on side surfaces. Results of the cube configuration showed that the one-inch path provided higher accuracy and precision of predicted concentration values (C and D).

the spectrometer consisted of vertical transfer of the blend through a tubular passage of 1 inch at target mass flow rate. Fig. 15.5A and B show the different interface geometries that were explored. A subset of the results is shown in Fig. 15.5C and D.

3.1.5 Compression design of experiment, core tablet process analytical technology development, and integrated line runs

This was the final stage of CM development and feasibility studies at our academic partner's lab. This study provided the means for the integration of all the CM process knowledge built throughout the previous evaluations into the final process design and feasibility studies. The DoEs approach provided the basis for the evaluation of the blending unit operation (blender speed and throughput), with elements of the compressibility studies and final product quality attributes. Fig. 15.6 provides a summary of the DoE variables included in the study. The results obtained from the experimental design and integrated line testing confirmed the equivalence of the CM process to the legacy batch process. A comparison of the product's physical and chemical properties between both processes provided the basis of confirmation for the design space. Fig. 15.7 provides a summary of the obtained results. The basis of comparison included a tablet compressibility analysis and a dissolution profile comparability evaluation in the regulatory and physiological media. Results of the compressibility analysis are shown in Fig. 15.7A, and dissolution results are shown in Fig. 15.7B and C. The results obtained confirmed the equivalence of the product manufactured at the academic partner's CM line and the batch product manufactured at our facilities (Fig. 15.7D).

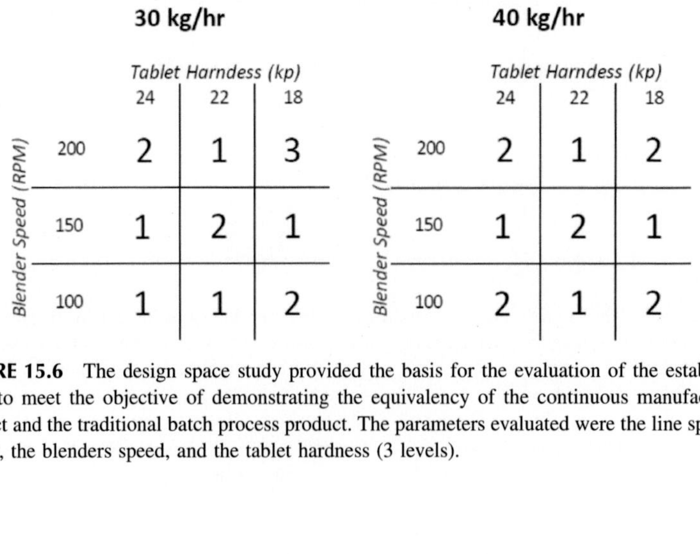

30 kg/hr

Tablet Harndess (kp)

Blender Speed (RPM)	24	22	18
200	2	1	3
150	1	2	1
100	1	1	2

40 kg/hr

Tablet Harndess (kp)

Blender Speed (RPM)	24	22	18
200	2	1	2
150	1	2	1
100	2	1	2

FIGURE 15.6 The design space study provided the basis for the evaluation of the established CPPs to meet the objective of demonstrating the equivalency of the continuous manufacturing product and the traditional batch process product. The parameters evaluated were the line speed (2 levels), the blenders speed, and the tablet hardness (3 levels).

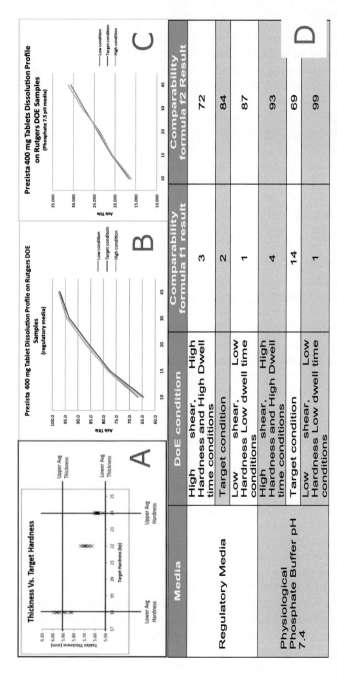

FIGURE 15.7 The design space study results showed that the physical properties of the tablets obtained for the continuous manufacturing line product correlated to the space of variability defined by the current batch process specification (A). The dissolution of tablets in either the regulatory media or physiological media showed small variability within the design space CPP's ranges evaluated (B and C). The dissolution profiles were also compared to a batch process reference batch with confirmed SUPAC guideline model independent comparability results meeting the comparability requirement.

3.2 Process analytical technology and near-infrared spectrometry feasibility studies

This section discusses the strategy for PAT method development and transfer from our academic partner's laboratory. The analytical academic partner team provided the resources and knowledge needed for the evaluation of spectroscopic technologies for the analysis of in-process materials and the finished product manufactured during the process development exercise.

Even though this project was based on previous work [3−11], the PAT transfer required additional research activities to adapt methods to the commercial pharmaceutical manufacturing facility.

Furthermore, as the project evolved, significant process knowledge and understanding resulted in the optimization of the spectrophotometric chemometric models. The optimization enabled these models to be used in substitution of wet chemistry analytical test methods in a full RTRt implementation.

Activities done by the academic partner's analytical team were as follows:

3.2.1 Predevelopment activities (feasibility studies)

The team performed spectral analysis of RM and intermediate materials with different spectral systems (PAT analyzers at different resolution levels and wavelengths, Raman spectroscopy). During this stage, the analytical team evaluated the spectral properties of the RM, intermediate, and core tablet product to propose the best spectroscopy option to develop the predictive models. Fig. 15.8 shows a composite of the information collected during this stage. In conclusion, the spectroscopic properties of the RM, particularly API and formulation properties, revealed that NIR spectroscopy was an adequate source for chemometric evaluation of the intermediate (blend) and finished products (core tablet) to determine their identity and API concentration predictions.

3.2.2 Familiarization with near-infrared process analytical technology analyzers (near-infrared process analytical technology tools)

During the feasibility stage, the academic partner's analytical team lab had the same spectrometer as the one used during the CM line implementation and helped define spectral collection parameters and procedures in Janssen's facility. For blend sensing, we used the Bruker Matrix diffuse reflectance spectrometer, and for tablet measurement, we used the Bruker MPA transmission spectrometer. Fig. 15.9 provides additional information on the NIR spectrometers and their application in our control strategy design. Fig. 15.9A and B show the instrument and the sensing element. Fig. 15.9C shows the transmission NIR instrument used to determine the API composition of the tablets, which used in conjunction with the tablet auto-tester (Fig. 15.9D) formed the basis of the control strategy.

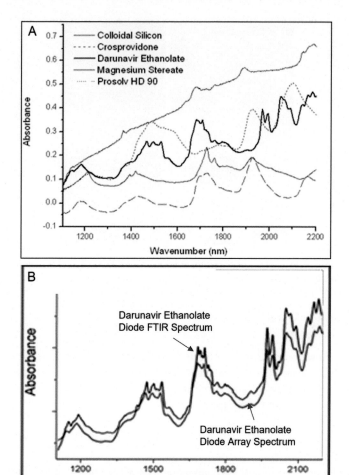

FIGURE 15.8 Process analytical technology development included the analysis of raw materials spectral properties (A), the analysis of different options of NIR spectroscopy (B), and Raman spectroscopy. In addition, the determination of concentration sensitive FT-NIR bands (D) were also examined during the feasibility stage.

3.2.3 Development of gravimetric-based chemometric model in formulation design of experiment

The lab used the *diffuse reflectance* NIR mode to evaluate blends and *transmittance* for tablet evaluation. The calibration samples set (CSS) were prepared by mixing excipients, API, and lubricant. Prior to the preparation of these calibration samples, a total of five placebos were prepared. The statistical DoEs software "Modde Software" provided a D-Optimal DoE approach used to develop the multivariate analysis partial least squares (PLS) model used for

FIGURE 15.9 Near-infrared spectrometer equipment was evaluated by the analytical academic collaborator lab team, including the spectrometer with the sensing/emission head and fiber optics system (A and B) and the bench tablet spectrometer (C) which provided along tablet API concentration prediction which in combination with the automated tablet testing system provided the in process controls strategy for the line (D).

API concentration prediction. The model validation followed a leave-one-out cross-validation technique. Method transfer to the CM line required the refinement of the model and spectral collection parameters, details of which are discussed later in this chapter.

3.2.4 Evaluation of the developed calibration model under dynamic sampling

The academic collaborator PAT lab team used a conveyor belt to simulate the blend movement from the CM mixer. The analysis of the NIR calibration model developed by the academic analytical team included a challenge of the predictions at several linear velocities. The conveyor belt permitted the acquisition of spectra of a single representative unit dose while in movement. The experimental conditions were as follows:

The test used a box 11.25 inches long, 3 inches wide, and 2 inches deep to place the powder.
The box full of blend material moved over a conveyor belt aligned with the Bruker FT-NIR Matrix Instrument Q412 light emission head. The spectra acquisition simulated the blend movement at the continuous line.
The conveyor belt moved at a linear velocity of about 9 mm per second, a speed similar to the powder movement at the CM line.

Fig. 15.10 shows the experimental setup for this study. Fig. 15.10A shows the number of scans that can be acquired at different linear velocities of the

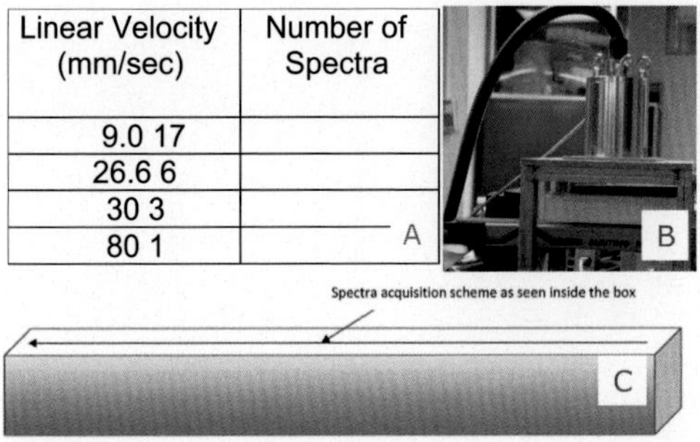

Linear Velocity (mm/sec)	Number of Spectra
9.0 17	
26.6 6	
30 3	
80 1	

FIGURE 15.10 The experimental setup for the evaluation of the dynamic sensing performance of the bench-based active pharmaceutical ingredient concentration prediction model demonstrated the influence of sample movement on the predictability of the concentration and the need to include this sample presentation aspect in the model calibration sample set. The experiment consisted of the spectra collection at different linear velocities (A). The spectrometer emission head (B) was placed on top of a conveyor belt where a box (C) containing an already homogeneous blend pass through the path of the emission head beam allowing spectral collection at different linear speeds.

powder. Fig. 15.10B is a picture of the spectral signal of the instrument while Fig. 15.10C is a rending of the conveyor belt. Table 15.1 and Fig. 15.11 show the score plots of the PLS model prediction scores of the still (Fig. 15.11D) versus dynamic samples (Fig. 15.11A−C). The PLS model under-predicts the concentration of the active ingredient when the powder is moving compared to static sampling. The latter predicts API concentration values much closer to the actual concentration. The score plots show that the scores obtained from most of the dynamic samples fall outside of the principal components analysis (PCA) model Mahalanobis distance 95% confidence limit ellipse. On the other hand, the static samples showed scores within the score plot ellipse. This was a very important contribution of this experiment, as it provided the first insight on the importance of chemometric model development in a dynamic environment.

3.2.5 Development of core tablet calibration model to predict active pharmaceutical ingredient concentration in the Bruker MPA

The academic collaborator's PAT lab team provided a preliminary model for the prediction of API concentration in core tablets as part of the feasibility stage objectives. For the CSS sample preparation, model development, and cross-validation, the team followed a similar approach to the blends process (D-Optimal design with cross-validation leave-one-out approach). In the case of the tablets, a comparison of models based on transmission or diffuse reflectance provided information on the model's accuracy and precision of predictions. A NIR PLS calibration model was prepared using high pressure

TABLE 15.1 The quantitative analysis of the partial least squares model based on static samples confirmed the bias imposed by the moving sampling scheme (top table portion) as compared to the static sample presentation (bottom table portion). Although all of these measurements were taken over the same sample in repeated and independent events, the moving sample predictions showed to be under predicted in concentration compared to the theoretical concentration. On the other hand, the samples tested under static conditions showed a prediction closer to the theoretical concentration.

		Dynamic samples predictions (2 factors)			
Lot. No.	# Samples (n)	Concentration	Ave. Predicted	Ave. Residual	Standard deviation
A	153	52%	48.02	3.98	1.70

		Static samples predictions (2 factors)			
Lot. No.	# Sample (n)	Concentration	Ave. Predicted	Ave. Residual	Standard deviation
A	33	52%	52.32	−0.32	1.15
A	42		50.25	1.75	2.27

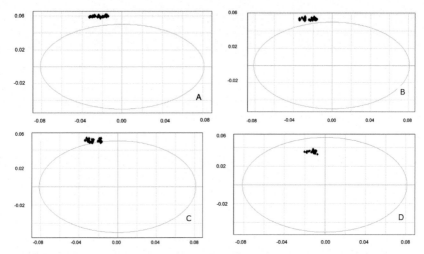

FIGURE 15.11 The results are presented here as principal components analysis (PCA) score plots on the main two components of variation. The PCA model built with static samples predicted different samples concentration levels of the moving conveyor belt presented blend samples outside the model Mahalanobis distance ellipse (A through C). On the other hand, when the samples of the same blend are collected static the samples scores fall inside the model ellipse (D).

liquid chromatography as the primary reference method for the PLS model regression. The conclusion was to give preference to the transmission model over the diffuse reflectance model due to the richness of information from the tablet spectra obtained using the transmission mode of spectral acquisition.

The series of activities discussed above forms the aggregated contribution of the academic partner's lab to the feasibility and early development stages of the CM process and PAT development for our product. The academic partner's lab achieved all objectives set by Janssen and the accumulated learnings of this stage provided an excellent foundation to design the process development activities needed at the Janssen Gurabo facility.

The following section discusses the final development stage of the selected product conversion to the CM technology with PAT and RTRt capabilities.

3.3 Final process and process analytical technology development activities at Janssen

The lessons from the process development activities performed at our academic partner's lab on pharmaceutical dosage direct compressible tablet CM technology were the basis for the design of our facility. The line built at Janssen's Gurabo facility closely resembled the prototype line built at our academic partner's lab. The design of the CM line built by our academic partner was the product of many years of research by the partner [12]. Janssen representatives as the manufacturing partner also contributed elements to the design, particularly in relation to engineering controls for explosion risk mitigation and safety aspects of the line. It also ensured the building, and the facility were ready for the introduction of our product into the academic laboratory environment.

4. Janssen's continuous manufacturing line design was based on academic partner's continuous manufacturing line

Upon completion of the CM feasibility program, we achieved a robust foundation for the design of the commercial manufacturing line installed at Janssen's facility at Gurabo, PR. Fig. 15.12 compares the two lines; the left pane (Fig. 15.12A) shows a photo of the academic lab line while the right pane (Fig. 15.12B) shows a design architectural drawing of the Gurabo line.

There are three main differences between the lines. The first difference is in the feeding operation: Janssen's line had a more complex layout, with multiple material feeders, additional material transfer capabilities, and refill stations not included in the academic lab design.[1] The second difference is in the tablet

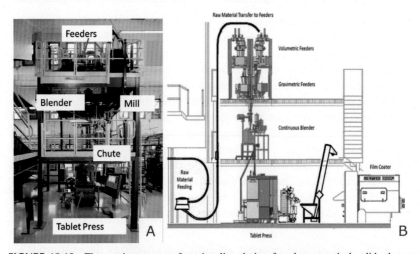

FIGURE 15.12 The continuous manufacturing line design for pharmaceutical solids dosage forms in direct compressible process modality build at the academic collaborator's lab (A) provided the basis of design prototype for the continuous manufacturing line build at Janssen (B). In general terms, the feeding, mill, blender, blend transfer chute, and tablet press are vertically installed in a gravity feed design in both lines. Additional features are included in Janssen's line (noncolored elements in schematic (B)) including an upstream material transfer vacuum system and a downstream tablet transfer and film coating unit operation.

---

1. Further discussed in the Feeding and Blending Operation section.

transfer and film coating operation that is absent in the academic lab line design. The CM process at the Janssen line included tablet transfer and coating operations to increase the benefits of integrating these operations in the continuous process. The third difference was the integration of the line using state-of-the-art automation technology. The academic lab line was mechanically integrated when we performed our feasibility studies. The demands of a commercial manufacturing environment required the incorporation of a distributed control system (DCS) in the design of Janssen's line. A Siemens PCS seven DCS with the Siemens Integration PAT solution provided the basis for full automation of our commercial CM line.

The differences between the prototype and the commercial lines, although minimal, required additional evaluation from a design test perspective as well as a structured qualification. The activity also required disciplined project management along with a significant amount of time and effort by a dedicated team of professionals working to achieve our project goal.

5. A project that led to the first batch-to-cm process transformation approved by the US Food and Drug Administration

The project leading to the design, installation, commissioning, and technology transfer of Janssen's first product conversion from batch to CM manufacturing required sound project management tools to achieve our goal. The RTRt component of the project eliminated the time-consuming laboratory tests of the sampled product and permitted the analysis of the product concurrently during its production.

The benefits related to operational efficiencies intrinsic to a CM process, combined with advantages from RTRt, spearheaded the business case. It resulted in a favorable projection of financial metrics (NPV and % IRR), including the now proven cost of goods reduction. Table 15.2 provides a summary of how we built the project's business case.

The use of the flawless project execution (FPX) methodology was a key for the success of this project. FPX is a set of tools that promotes sound project management practices, commonly used by Janssen project managers. The FPX methodology provides guidelines for the execution of complex projects with a certain level of assurance for success (Fig. 15.13). FPX is based on four good project management principles (Fig. 15.13A−D), which were applied as follows:

5.1 Team structure and governance

The CM line project team was a multidisciplinary team, formed in a matrix approach, where team members may or may not have sub-teams working with them. Team leadership was well defined, ensuring that the team stayed on

TABLE 15.2 Continuous manufacturing benefits compared to batch manufacturing process.

Process element	Batch to Continuous Net benefit
Analytical test cycle time	33% reduction
Manufacturing area footprint	50% reduction
Product accumulated waste	50% reduction
Capital investment	40% reduction
Manufacturing cycle time	75% reduction
Product release time	30% reduction
Cost of goods sold	50% reduction
Process sample evaluation	93% increase
Process knowledge and in process control points	50% increase

The project business case included several elements impacting cost and quality elements. Among the benefits we identified operational efficiencies (decrease in analytical and manufacturing cycle time), capital optimization (manufacturing footprint minimization and capital investment reduction), waste reduction (yield and cost of goods sold), and quality improvements (process appraisal and process knowledge improvements).

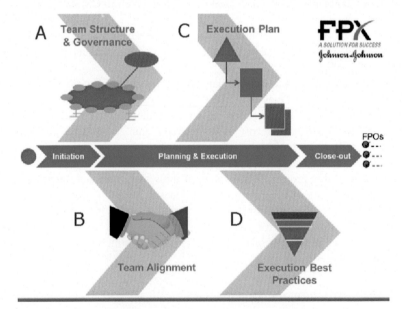

FIGURE 15.13 The continuous manufacturing line project was managed using Janssen's flawless project execution program, which provided a sound project management foundation to the project. The program principles include the formation of a governance structure that support all levels of project execution (A), it promotes a collaborative environment that starts by a principle of alignment and consensus of project execution (B) around a plan (C) that is performed under sound project managements tools and practices (D).

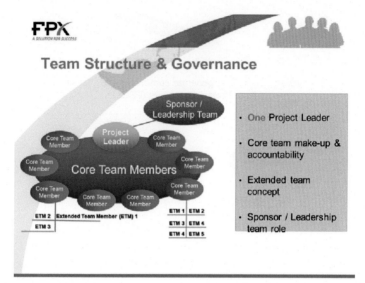

FIGURE 15.14 The project team governance structure is defined early at the project initiation. It is composed of multifunctional members that act on their own or as representatives of execution sub-teams. In the case of the process development activities, the key deliverables were reached by a sub-team of engineers and scientists.

track to achieve the final project objectives (FPOs). The team also had a sponsor that facilitated team interactions between business units. The sponsor is usually a senior leader in the organization (Fig. 15.14).

5.2 Team alignment

Upon project initiation, the team created a series of FPOs based on project requirements. The team democratically agreed on these objectives and used them as guidance for the project's execution and completion (Fig. 15.15).

5.3 Execution plan

After agreeing on the project FPO's, each team member analyzed the project activities that their sub-team or function needed to accomplish. These activities form the key deliverables (Fig. 15.16).

5.4 Execution best practices

Each team member received training in a series of best practices in project management to guide them toward success in execution of the project.

The next sections describe key process development deliverables and how they contributed to the overall FPO—successful approval of the Prezista CM process (Fig. 15.17).

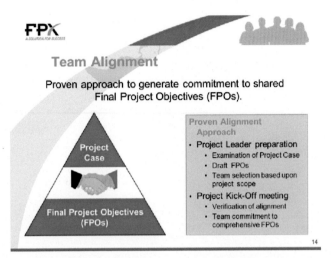

FIGURE 15.15 This picture shows the principle of team alignment that follows the team formation process. Upon completion of the kick-off meeting, the alignment process concludes with an agreed set of final project objectives. This is a key step toward a harmonious completion of the project objectives.

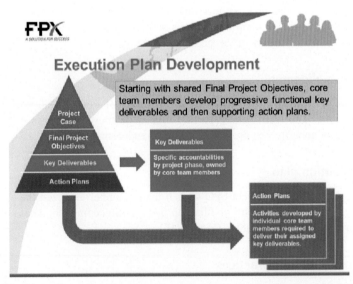

FIGURE 15.16 The project execution plan defines each team function key deliverables that forms the series of actionable elements that need to be performed to complete the final project objectives. Action plans are then created to support each team member or sub-team groups to complete the project.

FIGURE 15.17 Each team member is trained in a series of project management tools that are intended to facilitate the project execution.

6. Project development team plan and key deliverables

The development sub-team started its endeavor in the project by preparing a process and a PAT development plan. This was a starting document, subject to revision as project needs evolved, which described the key deliverables associated with development activities for the process and the analytical spectrophotometric procedures. Fig. 15.18 shows a diagram of the stages followed for the development and implementation of the CM process. The approach started with the predevelopment activities performed at the academic collaborator's lab, which prepared the stage for the early development

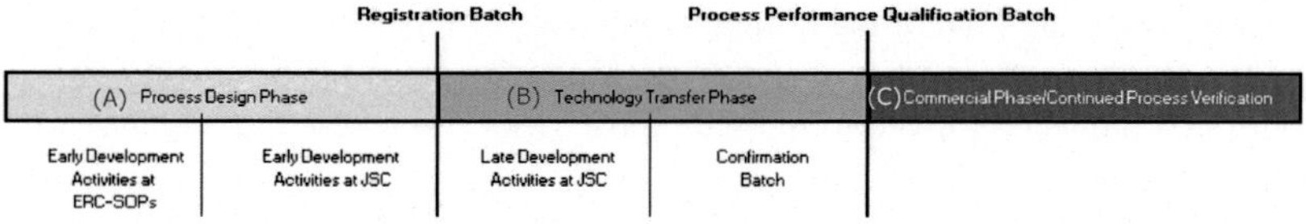

FIGURE 15.18 The process development plan provides a thorough description of the project in general. It includes the definition of key deliverables for each significant technology project phase up to commercialization of product. The phases are (A) Process Development, which consisted of two separate sub-phases, the academic collaboration stage, and the early development activities performed at Janssen, this stage ended with the Registration batch completion. (B) Technology Transfer Phase, which included the late development activities and a confirmation batch that consolidates the process knowledge base into the final process design, this stage ends with the filing of the supplemental dossier to the regulatory agencies and set the stage for the process performance qualification (PPQ) phase. Finally, (C) the Process commercial phase includes the continued process verification requirement, a set of prospective product/process evaluation activities that further evaluate the consistency of the continuous manufacturing process design.

activities performed at Janssen's CM line. These two activities bridged the knowledge base needed to ensure the manufacturability of the registration batch, prior to completing subsequent development activities. The registration batch was an important project FPO and a significant milestone for the project. This batch provided the basis for comparability against the batch process; a key element of the regulatory submission and a key deliverable that ensured the success of the project. It was also a time-sensitive element as registration batch tablets provided the material needed for the accelerated and long-term stability studies for submission purposes.

After the registration batch, we performed the in-depth process evaluation that constituted the technology transfer stage. This stage encompassed late development activities, such as determining the residence time distribution (RTD) and design space studies. It resulted in the determination of the target manufacturing conditions, including automation elements, in preparation for the confirmation batch. The confirmation batch demonstrated an adequate integration of the process control strategy elements within the DoE space. The process performance qualification (PPQ) stage was the last stage of process design and validation, before transitioning to continued process verification.

The key deliverables described in the development plan were the following:

6.1 Criticality analysis

The criticality analysis document, based on the International Conference on Harmonization (ICH) Q9 guideline, defines the risk elements of the process. This document evaluates each stage in the process to define their failure modes. It is a living document that requires continued revision during the life cycle of the process. The criticality analysis provides the basis for the DoE justification and execution, and it requires updates at major stages such as DoE completion and PPQ. It is also fundamental for the health authority's filing process. The main goal of the criticality analysis is to provide a comprehensive failure mode analysis in a matrix that illustrates process risks along with control elements employed to mitigate those risks and assess the effectiveness of the control elements.

6.2 Continuous process development

This project component includes the action items needed to develop the Prezista 600 mg tablets' direct compressible process in the CM line. The action items included the following:

Material characterization and feeder performance study.
Engineering run, the first evaluation of the mechanically integrated line.
Residence time distribution studies.

Design space study.
Full automation and design space center batch manufacturing study.

6.3 Process analytical technology methods development

Feasibility stage (conducted in academic collaboration and already discussed).

Integration of PAT tools in the CM process.
Blend calibration model development: ID and API concentration determination.
Tablet calibration model development: ID and API concentration determination.

7. The control strategy and the evaluation of failure modes as an integral part of process development activities

Following ICH Q8(R2), the project development sub-team discussed the criticality and risk base analysis of the variables associated with the batch-to-CM process transition. This is because our approach to introduce continuous manufacturing to our facility was the transformation of Prezista 600 mg tablets, a preexisting batch-manufactured product that was initially part of the 2007−2008 FDA QbD program. This made it easier to assess the criticality and risk analysis, allowing us to update the batch process control strategy with relevant changes based on CM process considerations.

The criticality analysis drove the development plan and provided an excellent scenario to understand and confirm process risk. The highly automated nature of the line, combined with large amounts of online analytical data obtained from the process, provided new elements to consider in the risk analysis. This allowed improvement of detectability and control of process circumstances that may result in out-of-control conditions. Fig. 15.19 shows a fish bone analysis of the CM process indicating the relationship of process elements considered in our criticality analysis.

In the next sections, we will provide a description of each key deliverable of the project plan and how they supported reaching our FPO of approval of the first batch-to-CM product conversion by the FDA.

8. Feeder performance study and material transfer study

The feeder performance study provided the data to establish the accuracy of the feeders' dispensing flow rates. Feeder accuracy is based on the ability of the feeder to consistently dispense each material at the desired throughput with minimum or no variation outside of the established tolerances. Deviations

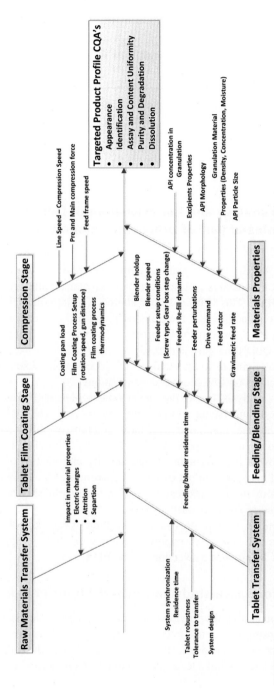

FIGURE 15.19 The criticality analysis provides the basis for the evaluation of the control points and defining the needs for experimentation and design improvements needed to warrant adequate process performance and the definition of the process design space. A fish bone diagram is an excellent tool to establish the causal relationship between materials, systems, and unit operations failure modes and the targeted product profile critical quality attributes.

from this expectation were subject to engineering evaluations and modifications to improve the system's accuracy and precision. The feeder environmental stability evaluation included the following tests:

8.1 Sources of variability on feeder performance

The CM line installed at Janssen's facility at Gurabo, PR, was a first-of-a-kind design experience for the project team. As such, it required careful examination of the design elements to ensure minimal disruption of operations, particularly the gravimetric feeder's performance. The gravimetric feeding and blending operation are the heart of the direct compression CM process, and as such, the feeder system design deserves a great deal of attention. The commissioning activities evaluated several elements of the line's design. The feeder design evaluation stage provided the basis for refining specific elements of the design that did not show up prior to this stage. Fig. 15.20 shows the six

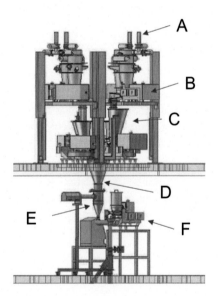

FIGURE 15.20 As part of the line commissioning evaluation, the project team performed a series of analysis of the feeder/blender system design characteristics to determine sources of perturbation of the gravimetric feeders under normal process operation. The following sources of perturbation were evaluated: The transfer of powders from the vacuum transfer system (A), the transfer of powders from the volumetric feeder to the gravimetric feeder (B), the control scheme used by the K-tron controls system for each material (C), the transfer of powders from the gravimetric feeder to the materials collection hopper (D), evaluation of the combined transfer of all raw materials from the feeders, through the transfer hopper into the blender for accuracy of the full material transfer (E), and the evaluation of the fifth feeder next to the blender under actual processing conditions (F).

design aspects evaluated to minimize sources of feeder perturbation. The performance elements evaluated were the following:

Effect of materials' vacuum transfer (Fig. 15.20A).
Transfer of materials during refill operation (Fig. 15.20B).
Feeder control procedure selection (Fig. 15.20C).
Transfer of powders from the gravimetric feeder to the material collection hopper (Fig. 15.20D).
The combined transfer of all RM from the feeders, through the transfer hopper into the blender (Fig. 15.20E).
The evaluation of the fifth feeder next to the blender under actual processing conditions (Fig. 15.20F).

8.2 Gravimetric feeder maximum hopper capacity

We performed a study to determine the maximum hopper capacity for the gravimetric feeder for each material's designated feeder in the continuous line. This study defined the maximum fill level. Table 15.3 summarizes the findings of this study section.

TABLE 15.3 The feeder maximum use level was calculated based on an empirical test based on the equipment geometry and material density information.

Material description	Feeder ID	Hopper volume (dm^3)	Material density (kg/dm^3)	Calculated hopper capacity (kg)[1]	Confirmed hopper capacity (kg)
Raw material 1 (filler)	F1	25	0.49	11.0	11.0
Raw material 2 (disintegrant)	F2	25	0.33	7.4	7.4
Raw material 3 (API)	F3	25	0.61	13.7	13.7
Raw material 4 (lubricant)	F5	15	0.14	1.9	1.9

8.3 Gravimetric feeder and refill accuracy tests per raw material

The feeder vendor provided a catch scale system used to determine the feeder's accuracy. The catch scale provides a gravimetric signal to associated software that organizes and provides a means to collect and analyze the data. Upon optimization of the material transfer setup, the feeder accuracy test performance provided an adequate estimate of the gravimetric feeder's accuracy for material dispensing. In addition, the test evaluated how two different levels of refill operation, 60% and 80%, affected material dispensing accuracy. Fig. 15.21 shows the feeder's variability at different test conditions. Fig. 15.21A (refill at 60%) and Fig. 15.21B (refill at 80%) show the results obtained at 40 kg/h, while Fig. 15.21C (refill at 60%) and Fig. 15.21D (refill at 80%) show results at a line throughput of 60 kg/h.

8.4 Feeder refill dynamics in gravimetric feeder accuracy

The study included the evaluation of refill dynamics for each RM. The test evaluated in detail the effect of the hopper level on feeding accuracy at the initiation of the refill. The endpoint of the study provided an optimum level of fill for the hopper which minimized feed rate variability. Fig. 15.22 shows a composite figure of the progression of improvement in feeder accuracy based on different levels of refill in the gravimetric feeder hopper [13].

9. Engineering run and first evaluation of the mechanically integrated line

Upon completion of the feeder's evaluation studies, the next step was to perform an integrated test of the line to verify its performance with actual product and the performance of in-process controls.

The engineering study had the following goals:

Evaluate the feeding, blending, and compression process integration factors that contribute to the adequate performance of these processing steps under normal operating conditions at the expected target processing conditions.

Establish an initial RTD model for the line. The RTD model study included the regression of the RTD model parameters based on step concentration change deconvoluted for feeding, blending, and compression operation as a preparatory step for the establishment of the control strategy. Fig. 15.23 shows the RTD plot and the characteristic time determination. The RTD model regression used the Taylor dispersion equation shown in Table 15.4. Collection of calibration samples for the PAT methods development for blend and tablets.

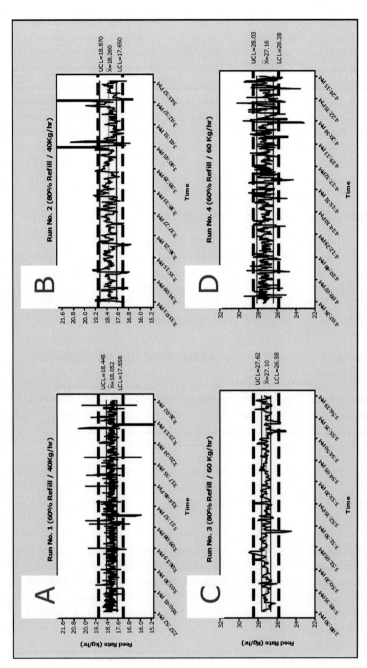

FIGURE 15.21 During feeder studies, we determine the feeder accuracy under various dispensing speeds (throughputs) and refill dynamic settings. In this figure, we show the feeder study perform on one of Prezista excipients (filler), the test was conducted at two throughput rates, 40 kg/h (A and B) and 60 kg/h (C and D). The test was replicated on each throughput to test the effect of refills on feeding variability, at 60% (A and C) versus 80% level refills (B and D) Variability observed was acceptable for each condition tested, nevertheless, a tendency was observed to have decreased feeding variability when a more frequent refill was used on the high throughput while a tighter variability was observed when a less frequent refill is used for lower throughput.

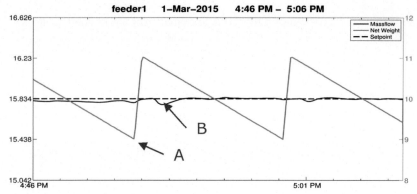

FIGURE 15.22   The refill event was closely observed with respect to the feeder variability attenuated by the blender variation. A Fourier transformed equation was used to transform the feeder data to establish clearly how the feeder refills event perturb the feed rate under the established optimum refill level. The data shown in this figure show that after the refill event (A) and within a delay equivalent to the residence time in the blender, there is a small but noticeable perturbation (B) that is much smaller than the allowable range of variation for this material.

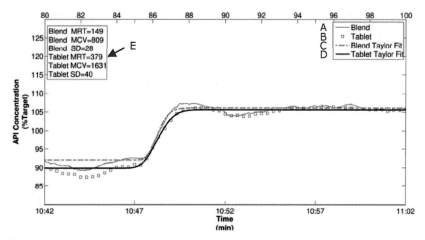

FIGURE 15.23   The residence time distribution data (A and C) and the Taylor dispersion model fit (B and D) shows nicely fit curves on either blend and tablets data. Blend data were collected at 4 s interval and the tablets were collected at 5 s interval. The model distribution statistics are shown for blends and tablets data (E).

10. Residence time distribution studies and evaluation of process factors' effect on the Residence Time Distribution model

The determination of the RTD of a continuous line is specific to the line design and construction elements, and depends on the process parameters used during

TABLE 15.4 Equations used for the determination of the Residence Time Distribution using the Taylor dispersion model (A). The fit data from equation A was used to determine the Mean Residence Time (B), Mean Center Variance (C), and Standard Deviation (D).

	Description	Formula
A	Taylor dispersion model	$C(\varepsilon, \theta) = \frac{C_0 Pe^{1/2}}{(4\pi\theta)^{1/2}} e^{-Pe(\varepsilon-\theta)^2/4\theta}$ where; C_0 is the initial concentration of the pulse, Pe is the peclet number (ratio of convection to diffusion). ε is the relative location to the end of the mixer, and $\theta = \frac{t-t_0}{\tau}$, where t, is time, to is a time delay and τ is the mean Residence time.
B	Mean residence time distribution (MRT)	$\tau = \int_0^\infty tE(t)dt$
C	Mean centered variance (MCV)	$\sigma_\tau^2 = \int_0^\infty (t-\tau)^2 \cdot E(t)dt$
D	Standard deviation (SD)	$\sigma_\tau = \sqrt{\int_0^\infty (t-\tau)^2 \cdot E(t)dt}$

manufacturing and the materials' properties. This section describes the approach followed to determine the RTD of the CM line at Janssen's Gurabo, PR facility.

10.1 Approach to residence time distribution at the project line

We evaluated the line's RTD at three points of interest that are essential for the control strategy. Fig. 15.24 illustrates these points of evaluation. The method followed to determine the residence time was based on a concentration step change approach. The concentration gradient formed by performing a coordinated change in concentration from the API and the major excipient filler material provided the means to establish the concentration gradient formation in the NIR sensing interface. The samples collected at the rotary valve and the tablets collected at the tablet press outlet provided the information to determine the following RTD deconvolutions:

Residence time between the feeder material collection hopper and the blender outlet NIR sensing interface.

Residence time between the NIR sensing interface and the outlet of the rotary valve.

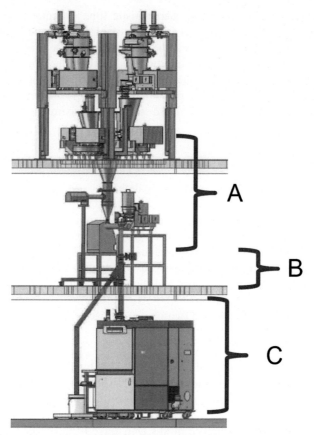

FIGURE 15.24   The residence time distribution studies included the following elements of the line for the individual unit operation and integrated line residence time distribution (RTD) modeling. We determined the RTD model from the Gravimetric feeder to the blend sensing interface (A), the RTD model between the blend sensing interface to the Rotary valve outlet (B), and the RTD model between the Rotary valve outlet to the tablet press outlet.

Residence time between the outlet of the rotary valve and the tablet press outlet.

10.2 Experimental design to determine how process variables affect residence time distribution

A factorial experimental design described the effect of three main variables on the residence time. Factors were selected based on the feasibility studies performed at the academic collaborator's labs. The evaluated factors were: the line throughput, blender speed, and feed frame speed as a factor confounded with the throughput related to the tablet RTD.

The RTD study provided the characteristic time extremes based on the operational interval defined by the process variables. The experimental variable combination that maximizes the time distribution initiation, the 1 percentile, and the maximized distribution finalization time, the 99th percentile, defined timers in the automated control system that initiates product rejection (minimum time at 0.01st percentile—T0R), and the timer for product acceptance upon dissipation of the disturbance (maximum time at 99th percentile—T99R).

10.3 Design space study: objective, approach, and results

The design space study's objective was to define the criticality of the selected process parameters using an empirical data collection approach. A statistically based DoE provided the specific factor effect relationship between the presumptive CPPs and the CQAs. The DoE provided the information needed to ensure the control strategy risk assumptions and process parameters criticality. The final goal of the DoE data analysis was to establish statistical significance and correlation between the CQAs and the process parameters, and validate their design and control spaces.

The DoE strategy followed a two-phase approach:

10.3.1 Phase A

This stage facilitated the execution of the final DoE by minimizing the number of variables considered. This was a screening design intended to determine the important compression variables. Fig. 15.25 shows a diagrammatic representation of Phase A DoE.

The potential CPPs examined in this DoE stage included the following:

10.3.1.1 Pre-design of experiment compression study

This study determined the level of compression force. This provided the basis for the Phase A DoE operation conditions needed to achieve the tablet's desired physical properties.

10.3.1.2 Mill speed

This is the speed of the Quadro Comil's impeller located before the blending operation. This factor was a potential CPP due to its shearing effect on materials passing through the mill on their way to the blender. Even though the materials passing through the mill do not flood the mill but pass through in an aerated state (starved regime), the shearing action of the mill may alter the properties of the powder mixture.

10.3.1.3 Throughput

This is the line's total mass flowrate. It can potentially affect the variability in the total material shear experienced at different throughputs.

4 Factor Factorial DoE (2⁴) with 4 center points

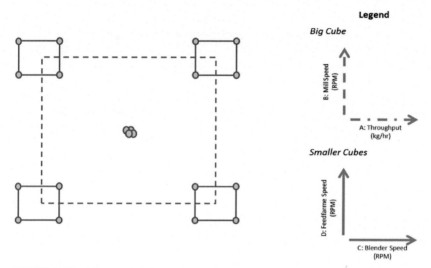

FIGURE 15.25 The stage 1 design of experiment consisted of two level four factor half factorial level IV design with four center point repetitions. The experiment was intended to determine the important factors related to the blend compressibility as a preliminary evaluation of factors related to compression stage. The results of this study were included in the design space study.

10.3.1.4 Blender speed

This is the rotational speed of the paddle/impeller in the blender. We included this factor in the study recognizing its potential contribution to process/material interaction effects on the tablet's physical properties.

10.3.1.5 Feed frame speed

This is the rotational speed of the tablet press feed frame. In combination with the line speed, the feed frame's RPM setting might have an impact on the consistency of the tablet's weight, as well as a significant effect on the shear that the material experiences.

Fig. 15.26 shows the compression parameters' statistical analysis indicating the statistically relevant parameters selected for the Phase B DoE stage. Our findings indicated that the ratio between the tablet press speed and the feed frame speed was statistically relevant to tablet physical properties. Upon completion of this first stage study, we learned that the feed frame speed and the throughput had a significant effect on tablet weight control. Therefore, the inclusion of these two parameters in the Phase B DoE optimized the parameters' effect on the tablet's physical properties.

Estimated Effects and Coefficients for Phase A Part 2 Filtered Weight (coded units)

Term	Effect	Coef	SE Coef	T	P
Constant		1230.25	3.203	384.11	0.000
Through Put	-27.08	-13.54	3.581	-3.78	0.019
Mill Speed	5.83	2.92	3.581	0.81	0.461
Blender Speed	2.30	1.15	3.581	0.32	0.765
Feed Frame Speed	47.66	23.83	3.581	6.66	0.003
Through Put*Mill Speed	1.97	0.98	3.581	0.27	0.797
Through Put*Blender Speed	-3.08	-1.54	3.581	-0.43	0.690
Through Put*Feed Frame Speed	20.73	10.37	3.581	2.90	0.044
Mill Speed*Blender Speed	5.30	2.65	3.581	0.74	0.500
Mill Speed*Feed Frame Speed	-0.69	-0.35	3.581	-0.10	0.928
Blender Speed*Feed Frame Speed	6.80	3.40	3.581	0.95	0.396
Through Put*Mill Speed*Blender Speed	-0.35	-0.17	3.581	-0.05	0.964
Through Put*Mill Speed* Feed Frame Speed	-4.08	-2.04	3.581	-0.57	0.599
Through Put*Blender Speed* Feed Frame Speed	3.00	1.50	3.581	0.42	0.697
Mill Speed*Blender Speed* Feed Frame Speed	3.23	1.61	3.581	0.45	0.675
Through Put*Mill Speed* Blender Speed*Feed Frame Speed	0.32	0.16	3.581	0.04	0.967

S = 14.3235   PRESS = 907793
R-Sq = 94.61%   R-Sq(pred) = 0.00%   R-Sq(adj) = 74.38%

FIGURE 15.26 The stage 1 design of experiment results indicated that for the compression in-process testing (tablet weight control) the statistically significant factors were throughput and feed frame speed and their interaction. The study consisted of two level four factor half factorial type IV design with four center point repetitions. The experiment was intended to determine the important factors related to the blend compressibility as a preliminary evaluation of factors related to compression stage.

10.3.2 Design of experiment phase B

Determining the overall design space for Prezista 600 mg tablet continuous manufacturing process:

With the information collected in DoE Phase A, the number of factors for the overall line design space became more manageable for a fractional factorial DoE. Our Phase B study included the line throughput and the feed frame speed as confounded factors (lowest and largest ratio of both factors). The factors included in phase B DoE were the following:

Stage 1 ratio: maximum and minimum ratio of levels used in the phase A study.
Tablet weight: in-process test equivalent to a range of 3% of target tablet weight variation.
Tablet Hardness: based on the maximum hardness attainable for each formulation base on API concentration.[2]

---

2. The API concentration range used in the experiment produces significantly different compressibility, resulting in different hardness maxima depending on the API concentrations.

API concentration: the formula was set to deliver concentration combinations of API between 90% and 110%.

Magnesium stearate concentration: to study the range of concentration established in a previous formulation DoE, producing a variation of a range of 81%–122% of the target.

The half factorial design for these five variables resulted in 32 experimental conditions. However, we had a total number of 38 runs with the addition of two center points and four experiments representing extreme conditions—all variables low and high in duplicate.

Fig. 15.27 shows a diagrammatic representation of the DoE. Tables 15.5 and 15.6 show the statistical analysis of the experimental data for the Hardness test and the dissolution test, respectively. Fig. 15.28 shows the raw dissolution test results, while Fig. 15.29 shows the tablet weight normalized dissolution test results.

The results obtained in the design space study revealed that all presumptive CQA's meet design requirements under the experimental variable ranges. All factors included in the DoE showed marginal or full statistical significance for the hardness test results. Regression analysis using Phase B DoE data provided evidence of the design space's adequacy, supporting the hardness specification

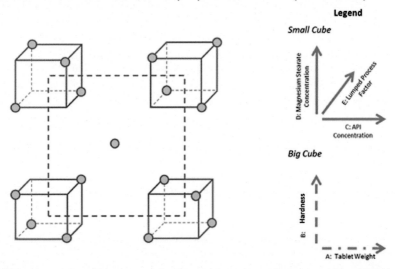

FIGURE 15.27 The phase B design of experiment was intended to define the design space of the continuous manufacturing process that includes formulation variables and process parameters. The design was a half factorial of five factors in two levels, resolution level V. A total of 16 factorial runs and the center point were carried out. All experimental conditions were replicated with center points quadruplicated.

TABLE 15.5 The Phase B design of experiment hardness factor effect showed that most of the factors were statistically significant, with only the magnesium stearate concentration being marginally important. The active pharmaceutical ingredient concentration level was the most significant factor along with the hardness level from an experimental factor perspective, and the tablet weight experimental level.

Type 3 tests of fixed effects				
Effect	Num DF	Den DF	F value	Pr > F
API	1	29	85.19	<0.0001
Mg Stearate	1	29	3.55	0.070
Hard Level	1	29	652.72	<0.0001
TabletWt	1	29	10.91	0.003
API*HardLev	1	29	52.45	<0.0001
PhRatioA*HardLev	1	29	6.65	0.015
Curvature	1	29	1.41	0. 245

TABLE 15.6 The Phase B design of experiment (DoE) Dissolution critical quality attribute factor effect showed that active pharmaceutical ingredient concentration and tablet weight were statistically significant for the 30 min dissolution results. The dissolution profile results obtained showed also that the 30 min dissolution minimum value obtained in the DoE support that dissolution results will always meet test requirements throughout the design space.

Raw values type 3 tests of fixed effects				
Effect	Num DF	Den DF	F value	Pr > F
Api	1	30	937.30	<0.0001
Twt	1	30	59.18	<0.0001
api2	1	30	0.43	0.5147

Normalized values type 3 tests of fixed effects				
Effect	Num DF	Den DF	F value	Pr > F
Api	1	30	4.33	0.0461
Twt	1	30	0.89	0.3541
api2	1	30	0.30	0.5879

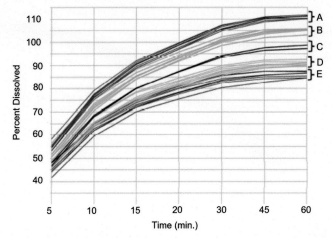

FIGURE 15.28   The stage 2 design of experiment dissolution critical quality attribute analysis included the observation that except for the active pharmaceutical ingredient concentration and the tablet weight level, no other experimental variable had a statistically significant effect upon the dissolution results. Dissolution profiles meet the minimum dissolution requirement at the test Q time of 30 min (dissolution values above 80%). At the same time, the dissolution response observed as a function of the content uniformity variables, tablet weight and tablet concentration, an adequate predictor of dissolution on this product. The following sample concentration and tablet weight combinations were subject to the dissolution test (n = 6), 110 % API / 1280 mg tablet (A), 110 % API / 1220 mg tablet (B), 100 % API / 1250 mg tablet (C), 90 % API / 1280 mg tablet (D), and 90 % API / 1220 mg tablet (E).

range. The dissolution CQA showed statistical significance on two CPP factors, tablet weight, and API concentration. From a control strategy perspective, the API concentration depends on the control of the API and the major excipient feeders. The tablet press control includes a weight feedback system that controls the force set point to ensure tablet weight control. This tablet press feature addresses the tablet weight CPP.

The DoE data analysis provided the framework to evaluate the effect of presumptive CPPs on IPCs and CQAs. From an IPC perspective, the DoE data provided the basis for parametric modeling that allowed evaluation of the design space's effect on tablet hardness, tablet thickness, compression force, and other tablet properties like disintegration and friability.[3]

To determine the regression model to define tablet hardness, thickness, and compression force limits, we followed an approach based on a best subset of regression models. The best subset model was the model with the highest R^2 and the smallest number of statistically significant independent variables.

---

3. Although Tablet weight is an official IPC for the purpose of this study, it was taken as a DoE independent variable. Upon evaluation of the DoE data, hardness was taken as a dependent variable, compression force was taken as a dependent variable due to the strong influence of API concentration on the compressibility response to force.

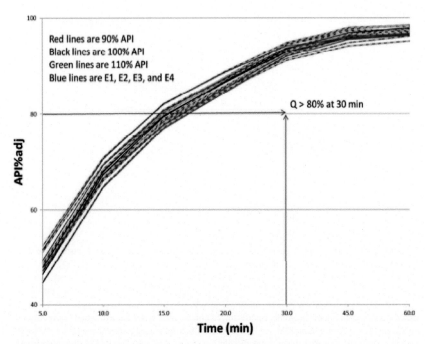

FIGURE 15.29   The dissolution data were subjected to a normalization against the tablet weight. These data transformation clearly showed that the dissolution profiles are not affected by any other experimental variable included in the phase B design of experiment. This information further supports the criticality of the tablet weight and active pharmaceutical ingredient concentration as the sole main factors affecting dissolution in this product within the design space evaluated.

The identified compression presumptive process factors evaluated included the following variables:

API concentration
Mg stearate concentration
Feed frame speed
Tablet weight
Main compression edge to edge thickness

A regression model based on API concentration, tablet weight, and main compression edge to edge thickness showed the greatest predictability for tablet hardness, thickness, and compression force simultaneously. Table 15.7 shows the regression model best subset analysis results.

The results obtained in the regression analysis showed the design space study's presumptive CQAs meet the CQA's design requirements under the experimental variable ranges. Two factors showed a statistically significant effect for the dissolution CQAs; the tablet weight and the API concentration, which depend on feeding operation control. From a practical perspective, the

TABLE 15.7 The Phase B design of experiment Hardness factor effect showed that most of the factors were statistically significant, with only the magnesium stearate concentration being marginally important. The active pharmaceutical ingredient concentration level was the most significant factor along with the hardness level from an experimental factor perspective, and the tablet weight experimental level.

	Vars	R²	S	API concentration	MG-ST concentration	Feed frame	Tablet Weight	Main compression edge-edge
Regression	1	16.1	5.3	X				X
	1	91.9	1.6				X	X
	2	82.0	2.5	X				X
	2	96.5	1.1	X			X	—
Hardness	3	92.8	16			X	X	X
	3	97.1	10	X	X		X	X
	4	96.8	1.0	X		X	X	X
	4	97.4	0.9	X	X	X	X	X
	1	95.8	0 06					X
	1	23.8	0.26				X	
	2	98.4	0.04				X	X
	2	95.9	0 06			X		X

Continued

TABLE 15.7 The Phase B design of experiment Hardness factor effect showed that most of the factors were statistically significant, with only the magnesium stearate concentration being marginally important. The active pharmaceutical ingredient concentration level was the most significant factor along with the hardness level from an experimental factor perspective, and the tablet weight experimental level.—cont'd

Regression	Vars	R^2	S	API concentration	MG-ST concentration	Feed frame	Tablet Weight	Main compression edge-edge
Thickness	3	98.5	0.04			X	X	X
	3	98.4	0.04	X			X	X
	4	98.5	0.04	X		X	X	X
	4	98.5	0.04		X	X	X	X
	5	98.5	0.04	X	X	X	X	X
	1	86.6	2.2					X
	1	2.2	6.0	X				
	2	95.0	1.4					X
Main compression force	2	86.7	2.2			X	X	X
	3	95.2	1.3	X			X	X
	3	95.0	1.4			X	X	X
	4	95.3	13	X		X	X	—X
	4	95.3	1.3	X	X		X	X
	5	95.4	1.3	X	X	X	X	X

API concentration translates to the criticality for the control of the feeding operation for the API and the major excipient. The tablet weight control is still another CPP for which the tablet press control strategy provides weight feedback control and force-based control.

11. Observations from fully automated confirmation batch

The results of the DoE allowed us to confirm the establishment of the CPPs and their target values and range. The next step was to confirm that the CM line can manufacture acceptable product reliably and operate continuously for an extended run time. In addition, the confirmation batch provided the basis to verify the automated control system integration.

The automated control systems were based on a Siemens PCS seven Line Control and SCADA system (LC&S) with IEEE-S88 recipe-based operation. It provided full integration of the line including unit operations and PAT tools. The LC&S commands the formula recipe, the gravimetric feeder controls, the unit operation recipe, the de-lumper operation, the blender, the blending interface for NIR spectroscopic measurements, the tablet press, the in-process controls for tablet measurements, the tablet quarantine hopper, the tablet transfer system, and the film coating operation.

The confirmation batch consisted of a continued line operation for a total of 14 coating pans equivalent to a 14-h run time at the target throughput of 40 kg/h. It also included planned interruptions to demonstrate the line holding time and to test that the line consistently returns to a state of control upon resuming, after an interruption.

The process performance on this confirmation batch provided additional evidence of the line's stability and consistency on a long manufacturing run, meeting all control expectations with regards to the CPPs, in-process controls, and CQA limits. Figs. 15.30−15.32 show the consistency of results of the API and major excipients feeders (CPPs), the in-process NIR testing, tablet physical properties, and tablet assay, CU, and dissolution results.

12. Validation and continued process verification stages

Our approach to validation was based on the current FDA guideline for industry on process validation: general principles and practices, of January 2011 [14]. Fig. 15.33 shows a diagram that summarizes the stage-based guideline principles. As demonstrated in the previous sections of this chapter, the design aspects of the line fully addressed guideline stage (1). The commissioning and qualification activities addressed guideline stage. (2). The PPQ stage strategy provided a series of processing challenges under the fully automated operation mode. Table 15.8 shows a summary of the batches in the PPQ stage.

The selected PPQ approach provided a series of incremental run times of four batches manufactured at this stage. The PPQ stage started with one batch

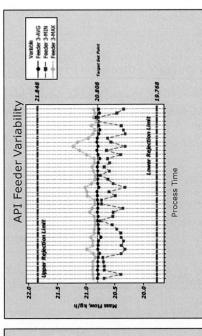

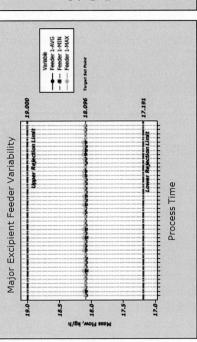

FIGURE 15.30 The declared critical process parameter, feeder mass flow (kg/h) was monitored for each major component, filler and active pharmaceutical ingredient, the observed variability confirms the allowable variations limits established in the design of experiment.

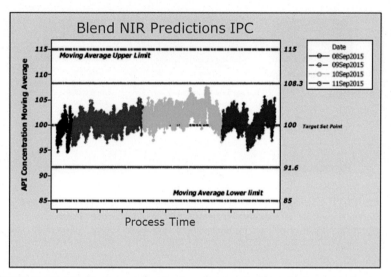

FIGURE 15.31 The blend near-infrared active pharmaceutical ingredient concentration in process control showed an adequate control and performance throughout the confirmation batch run, the observed predictions were within the control limits defined in our preliminary statistical evaluation of the development data.

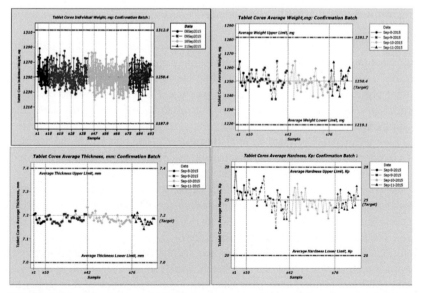

FIGURE 15.32 The tablet compressibility in process control showed adequate control during the confirmation lot, all tests, including individual tablet weight, average tablet weight, thickness, and hardness met the product specification requirements.

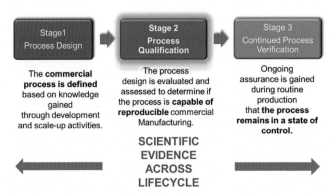

FIGURE 15.33 The approach followed for validation meets the current Food and Drug Administration guideline for industry on process validation: principles and practices. Stage 1 elements were already discussed, and the discussion of stage 2 and 3 is the subject of this chapter section.

of 8-h run time, followed by one 16-h run time, and ending with two 32-h run time batches. These last two batches defined the maximum run time for Prezista, based on product demand monthly requirements. Complete line cleanings intercalated between the PPQ runs serve as the cleaning validation challenge. In addition, this study included an evaluation of the effect of process interruptions, which occurred either purposely, since the manufacturing area worked over two shifts, or unintendedly, as process interruptions occurred due to fluctuation of CPPs.

The PPQ event also included cleaning validation activities, which included full cleaning validation of the line and a dirty equipment holding time evaluation, which defined the amount of time the line can be "dirty" or with residual product before cleaning activity occurs. This challenged the cleaning procedures under an incremental degree of difficulty for the operation and the equipment cleaning procedures.

Line efficiencies or yield were also evaluated. The first batch had a series of material losses related to the automation system's performance that needed further actions and modifications. This first batch suffered losses on efficiency[4] of up to 50% during the run time. None of the losses were related to product quality, but to logic defects and failures of the automation execution that caused rejection of good product. An analysis of the rejection identified the need for several modifications to the control system logic. The short-term solutions implemented before starting the second batch achieved an increase in process efficiency toward the last batch, which showed an absolute increase yield of 37% compared to the first batch for a total efficiency of 87% (refer to

---

4. Efficiency is defined as the percent of good product manufactured in the line against the incoming amount of materials for production.

TABLE 15.8 The Phase B design of experiment Hardness factor effect effect showed that most of the factors were statistically significant with only the magnesium stearate concentration being marginally important. The active pharmaceutical ingredient concentration level was the most significant factor along with the hardness level from an experimental factor perspective, and the tablet weight experimental level.

Process performance qualification batch	Dates	Theo. Runtime	Actual runtime	Control state time	Line start-ups	Longest interruption	Actual yield in tablets	Process efficiency
1	02-25-16 to 02-26-16	8 h	6.5 h	5.65 h	4 (feeding) 5 (blend)	9.0 h	130,597	50%
2	04-07-16 to 04-09-16	16 h	15.9 h	13.8 h	14 (feeding) 11 (blend)	13.0 h	355,035	67%
3	04-25-16 to 04-28-16	32 h	32.0 h	28.3 h	14 (feeding) 26 (blend)	9.7 h	858,782	82%
4	05-09-16 to 05-11-16	32 h	31.9 h	29.7 h	11 (feeding) 15 (blend)	11.4 h	910,788	87%

Table 15.8 for additional information on the PPQ batches' performance). Several other solutions implemented at a later stage, improved line efficiency up to approximately 98%.

The PPQ stage resulted in adequate results on all process and cleaning requirements. Fig. 15.34A and B show consistent results for the API and major excipients feeders (CPPs). Table 15.9 shows the in-process NIR testing results. Tables 15.10A and B summarize the tablet physical testing results. Fig. 15.35 shows the tablet in-process release NIR test results for assay while Fig. 15.36 shows the tablet content uniformity results. Table 15.11 shows the single point dissolution Q30 min results.

The four PPQ batches set the foundation for the long-term analysis of the Prezista CM process, enabling the transition from the PPQ stage into stage 3 of the continuous process verification (CPV) guideline. Up until the preparation of this chapter, three CPV reports have been issued over 2017 on six batches produced during the first and third quarters of 2017. The CM commercial batch sampling and testing level is the same as the PPQ batches. This is a great advantage of the CM process over batch manufacturing, as the latter is subject to a reduced testing regime on commercial batches when compared to the validation PPQ batches.

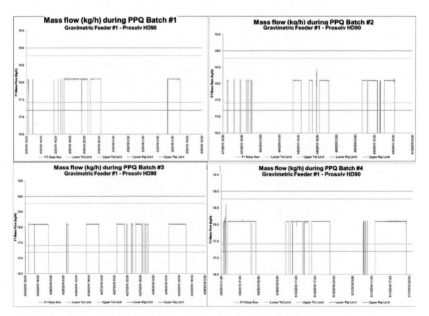

FIGURE 15.34A The major excipient feeder control demonstration per each process performance qualification batch. Major excipient feed rate was declared as a critical process parameter for the selected product continuous manufacturing process, during the manufacture of each batch this material feed rate showed consistent control.

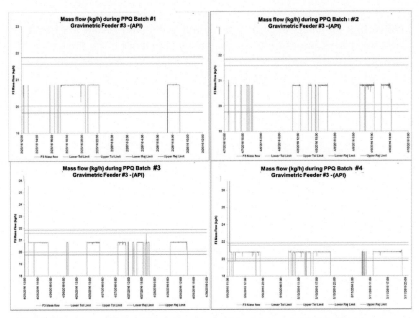

FIGURE 15.34B The active pharmaceutical ingredient feeder control demonstration per each process performance qualification batch. The API feed rate was declared as a critical process parameter for the selected product continuous manufacturing process; during the manufacture of each batch, this material feed rate showed consistent control.

TABLE 15.9 Summary of process analytical technology results per batch, instances where results observed are higher than limits (*, **, +, #) occur in rare occasion and trigger reject of blend through the diverter valve, as expected for the control strategy.

PPQ batch no.	Statistic parameter	Mean	Minimum	Maximum
1	Principal components analysis (PCA) DModX	0.0007	0.0003	0.0014
	partial least squares (PLS) T^2Range	0.4	0.0	3.5
	Blend active pharmaceutical ingredient (API) moving block average (%LC) (sample size $n = 3$)	99.9	94.6	104.5
	Blend API moving block range (%LC) (sample size $n = 3$)	N/A	0.0	9.6
	PCA DModX	0.0007	0.0002	0.0014
	PLS T^2Range	0.5	0.0	3.2

Continued

TABLE 15.9 Summary of process analytical technology results per batch, instances where results observed are higher than limits (*, **, +, #) occur in rare occasion and trigger reject of blend through the diverter valve, as expected for the control strategy.—cont'd

PPQ batch no.	Statistic parameter	Mean	Minimum	Maximum
2	Blend API moving block average (%LC) (sample size $n = 3$)	101.6	96.2	106.5
	Blend API moving block range (%LC) (sample size $n = 3$)	N/A	0.0	9.5
	PCA DModX	0.0007	0.0002	0.0019
	PLS T^2 Range	0.3	0.0	3.2
3	Blend API moving block average (%LC) (sample size $n = 3$)	101.2	94.3	108.4*
	Blend API moving block range (%LC) (sample size $n = 3$)	N/A	0.0	12.1**
	PCA DModX	0.0007	0.0002	0.0018
	PLS T^2 Range	0.3	0.0	3.4
4	Blend API moving block average (%LC) (sample size $n = 3$)	102.0	96.5	110.8[†]
	Blend API moving block range (%LC) (sample size $n = 3$)	N/A	0.0	14.4[††]

Acceptance criteria:

Blend PCA DModX: Less than 0.0026 (for NIR03A-CML101)

Blend PLS T2 range: Less than 6.2 (for NIR03A-CML101)

Blend API concentration moving block average: 91.6 to 108.3 %LC

Blend API concentration moving block range: ≤ 11.82 %LC

In general, the batches produced during the CPV report period showed consistent results when compared to the PPQ batches as indicated by the three levels of information compared before, namely, the CPP variability, in-process controls, and CQA variability. Table 15.12 shows the consistency of the CPP on API and major excipient feed rate. Table 15.13 and Table 15.14 show the consistency of blend NIR concentration and tablet physical properties, both in-process controls. Fig. 15.37 shows plots of the average NIR concentration and

TABLE 15.10A Summary of tablet physical properties, weight, thickness and hardness, results show the consistency of the tablet press operation.

PPQ batch no.	Statistic parameter	Individual weight (mg)	Average weight (mg)	Average thickness (mm)	Average hardness (kp)
1	Mean	1252.0	1252.0	7.2	24
	Minimum	1226.9	1233.4	7.2	23
	Maximum	1280.8	1277.5	7.3	25
2	Mean	1251.4	1251.4	7.2	25
	Minimum	1200.7	1220.9	7.1	22
	Maximum	1295.0	1273.2	7.3	27
3	Mean	1249.6	1249.6	7.2	25
	Minimum	1230.2	1241.3	7.1	21
	Maximum	1276.0	1260.6	7.2	27
4	Mean	1250.9	1250.9	7.2	25
	Minimum	1230.8	1239.2	7.1	24
	Maximum	1237.7	1262.8	7.2	26

Acceptance criteria:

Individual weight (mg): 1250.4 (1187.9–1312.9)

Avg. Weight (mg) ($n = 10$): 1250.4 (1219.1–1281.7)

Avg. Thickness (mm) ($n = 10$): 7.2 (7.0–7.4)

Avg. Hardness (kp) ($n = 10$): 25 (20–28)

TABLE 15.10B Summary of tablet physical properties, disintegration, and friability. The test results show the consistency of the tablet press operation.

PPQ batch no.	Sampling point	Disintegration (min: sec)	Friability (%)
1	Beginning	00:19	0.1
	Middle	00:17	0.2
	End	00:18	0.2
2	Beginning	00:27	0.2
	Middle	00:27	0.3
	End	00:49	0.1
3	Beginning	00:22	0.1
	Middle	00:25	0.1
	End	00:26	0.1
4	Beginning	00:26	0.1
	Middle	00:25	0.3
	End	00:27	0.1

Acceptance criteria:

Disintegration ($n = 10$): NMT 5 min

Friability ($n = 6$): NMT 0 5%

NIR content uniformity results for the CPV batches. Table 15.15A and B shows the dissolution single point results for the beginning, middle, and end of each CPV batch.

13. Prezista 600 mg continuous manufacturing supplemental new drug application approved: this is just the beginning

Upon successful completion of Prezista 600 mg CM process filing and approval in the United States in April 2016, we confirmed the expectations of a successful regulatory approval process in other countries, including US, EU, Switzerland, Canada, New Zealand, Taiwan, Serbia, Israel, Mexico, Ukraine, Algeria, Singapore, and Brazil. The regulatory team is currently pursuing approvals in Colombia, Turkey, Paraguay, Hong Kong, Japan, Uruguay, and Argentina.

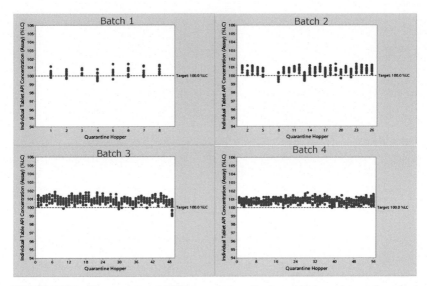

FIGURE 15.35 The near-infrared (NIR) assay test results showed high consistency throughout the four process performance qualification batches. Results shown correspond to the individual tablet NIR concentration test per quarantine hopper.

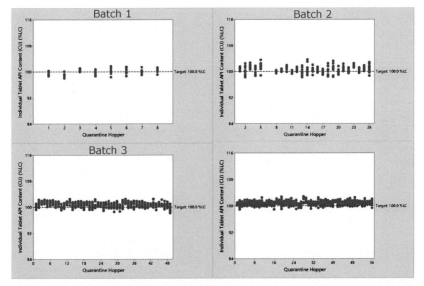

FIGURE 15.36 The near-infrared (NIR) tablet content uniformity test results showed high consistency throughout the four process performance qualification batches. Results shown correspond to the individual tablet NIR content test per quarantine hopper.

TABLE 15.11 Summary of process performance qualification (PPQ) stage tablet dissolution single point results. Three samples were collected per PPQ batch (beginning, middle, and end) all batches met robustly the dissolution test criteria.

PPQ batch	Beginning	Dissolved drug (%)	Middle	Dissolved drug (%)	End	Dissolved drug (%)
1	Tablet 1	93.523	Tablet 1	94.654	Tablet 1	93.618
	Tablet 2	94.795	Tablet 2	92.340	Tablet 2	87.530
	Tablet 3	94.159	Tablet 3	93.394	Tablet 3	95.941
	Tablet 4	93.996	Tablet 4	91.936	Tablet 4	93.256
	Tablet 5	92.581	Tablet 5	93.824	Tablet 5	91.018
	Tablet 6	93.087	Tablet 6	92.705	Tablet 6	94.233
	Mean	94	**Mean**	93	**Mean**	93
	Min	93	**Min**	92	**Min**	88
	Max	95	**Max**	95	**Max**	96
2	Tablet 1	96.941	Tablet 1	96.297	Tablet 1	96.251
	Tablet 2	94.641	Tablet 2	94.568	Tablet 2	96.378
	Tablet 3	94.557	Tablet 3	93.189	Tablet 3	94.445
	Tablet 4	94.445	Tablet 4	94.753	Tablet 4	97.476
	Tablet 5	95.875	Tablet 5	93.741	Tablet 5	95.440
	Tablet 6	95.262	Tablet 6	92.942	Tablet 6	93.468
	Mean	95	**Mean**	94	**Mean**	96
	Min	94	**Min**	93	**Min**	94
	Max	97	**Max**	96	**Max**	98

3	Tablet 1	96.481	96.694	90.264
	Tablet 2	95.656	95.899	97.151
	Tablet 3	97.109	96.907	97.033
	Tablet 4	96.005	96.468	96.804
	Tablet 5	95.900	94.966	98.972
	Tablet 6	96.982	95.667	96.864
	Mean	96	96	96
	Min	96	95	90
	Max	97	97	99
4	Tablet 1	95.471	95.203	94.465
	Tablet 2	96.102	94.738	93.883
	Tablet 3	94.63	98.044	96.442
	Tablet 4	95.818	94.735	96.158
	Tablet 5	98.305	98.276	91.913
	Tablet 6	95.359	99.019	94.015
	Mean	95	97	95
	Min	98	95	92
	Max	96	99	96

TABLE 15.12 Summary of the continuous process verification (CPV) stage on the control strategy declared critical process parameter, the feed rate for the major excipient (A) and the active pharmaceutical ingredient (B). Feed rates are shown for the six batches included in the continuous process verification report feed rate variability between and within batches was extremely low RSD, within a range of 0.0%—1.3% RSD.

A	Continuous process verification 3a Period	Batch	Mass flow mean (kg/h)	Mass flow RSD (%)	Overall mass flow mean (kg/h)	Overall mass flow RSD (%)
	Q1	5	18.095	0.3	18.095	0.1
		6	18.096	0.1		
		7	18.095	0.0		
	Q3	8	18.095	0.0		
		9	18.091	0.1		
		10	18.095	0.0		

B	Continuous process verification 3a Period	Batch	Mass flow mean (kg/h)	Mass flow RSD (%)	Overall mass flow mean (kg/h)	Overall mass flow RSD (%)
	Q1	5	20.803	0.3	20.804	0.3
		6	20.814	1.3		
		7	20.808	0.0		
	Q3	8	20.805	0.1		
		9	20.801	0.3		
		10	20.807	0.1		

TABLE 15.13 Summary of the continuous process verification stage on the control strategy for the in process control moving blend active pharmaceutical ingredient concentration moving block average and range. Results obtained showed comparable and consistent values to those observed in process performance qualification, the mean value ranged between 100.3% and 101.7% while the observed range values maximum ranges between 8.2% and 12.4%. The moving block average limits are 91.S and 108.3, while the range limit is 11.82. During manufacturing of batches 8, 9, and 10, a total of four instances were observed with moving block range values above the limit. These values were observed during startup events and material was rejected during this stabilization period as designed in the control system.

| Continuous process verification 3a | | | | | | |
Period	Batch	Statistic parameter	Mean (% LC)	Min (% LC)	Max (% LC)
A	5	Blend active pharmaceutical ingredient (API) moving block Average (n = 3)	101.1	96.3	106.6
Q1		Blend API moving block range (n = 3)	N/A	0.0	9.5
	6	Blend API moving block Average(n = 3)	100.3	92.7	107.0
		Blend API moving block range (n = 3)	N/A	0.0	9.0
	7	Blend API moving block Average(n = 3)	100.6	95.8	105.0
		Blend API moving block range (n = 3)	N/A	0.0	8.2
B	8	Blend API moving block Average(n = 3)	101.3	92.2	108.2
		Blend API moving block range (n = 3)	N/A	0.0	12.2
Q3	9	Blend API moving block Average(n = 3)	101.7	93.5	107.4
		Blend API moving block range (n = 3)	N/A	0.0	11.8
	10	Blend API moving block Average(n = 3)	101.5	95.5	107.8
		Blend API moving block range (n = 3)	N/A	0.0	12.4

TABLE 15.14 Summary of the continuous process verification stage on the control strategy for the in process control tablet physical properties. Results obtained showed comparable and consistent values to those observed in process performance qualification, all results met the individual and average test requirements.

CPV 3a Period	Batch	Statistic parameter	Average weight (mg)	Average thickness (mm)	Average hardness (kp)
Q1	5	Mean	1250.2	7.2	25
		Minimum	1229.4	7.1	22
		Maximum	1268.0	7.3	28
	6	Mean	1252.1	7.2	25
		Minimum	1238.1	7.2	23
		Maximum	1266.9	7.3	26
	7	Mean	1250 4	7.2	25
		Minimum	1238.9	7.2	23
		Maximum	1261.6	7.3	26
	8	Mean	1249.7	7.1	25
		Minimum	1229.3	7.1	23
		Maximum	1268.1	7.2	27

Q3	9	Mean	1251.4	7.2	25
		Minimum	1237.5	7.1	23
		Maximum	1261.9	7.2	27
	10	Mean	1250.7	7.2	25
		Minimum	1237.4	7.1	22
		Maximum	1265.8	7.2	27

Acceptance criteria:

Individual weight (mg): 1250.4 (1187 9–1312.9) - Avg. Thickness (mm) ($n = 10$): 7.2 (7.0–7.4) Avg weight (mg) ($n = 10$): 1250 4 (1219 1–1281 7) - Avg hardness (kp) ($n = 10$): 25 (20–28)

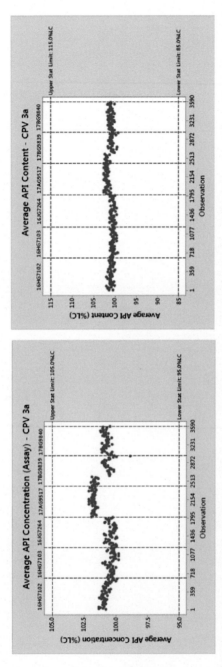

FIGURE 15.37 The near-infrared (NIR) tablet concentration and content uniformity test results showed high consistency throughout the six continuous process verification batches. Results shown correspond to the average tablet NIR concentration and content uniformity per quarantine hopper.

TABLE 15.15A Summary of the first quarter CPV stage batches for the dissolution critical quality attribute. The dissolution results showed similarity to the process performance qualification batches dissolution performance and demonstrate the consistency of the dissolution test performance through each continuous manufacturing batch manufactured during 2017 met the individual and average test requirements.

CPV 3a Period	Batch	Beginning	Dissolved drug (%)	Middle	Dissolved drug (%)	End	Dissolved drug (%)
Q1		Tablet 1	92	Tablet 1	91	Tablet 1	97
		Tablet 2	90	Tablet 2	91	Tablet 2	97
		Tablet 3	93	Tablet 3	94	Tablet 3	98
		Tablet 4	93	Tablet 4	91	Tablet 4	97
	5	Tablet 5	94	Tablet 5	91	Tablet 5	94
		Tablet 6	92	Tablet 6	92	Tablet 0	98
		Mean	92	Mean	92	Mean	97
		Min	90	Mm	91	Mm	94
		Max	94	Max	94	Max	98
		RSD	1.2%	RSD	1.0%	RSD	1.5%
		Tablet 1	95	Tablet 1	94	Tablet 1	96
		Tablet 2	94	Tablet 2	94	Tablet 2	96
		Tablet 3	95	Tablet 3	95	Tablet 3	96
		Tablet 4	94	Tablet 4	95	Tablet 4	96

Continued

TABLE 15.15A Summary of the first quarter CPV stage batches for the dissolution critical quality attribute. The dissolution results showed similarity to the process performance qualification batches dissolution performance and demonstrate the consistency of the dissolution test performance through each continuous manufacturing batch manufactured during 2017 met the individual and average test requirements.—cont'd

CPV 3a Period	Batch	Beginning	Dissolved drug (%)	Middle	Dissolved drug (%)	End	Dissolved drug (%)
	6	Tablet 5	93	Tablet 5	96	Tablet 5	98
		Tablet 6	95	Tablet 6	95	Tablet 6	92
		Mean	94	Mean	95	Mean	95
		Min	93	Min	94	Mm	92
		Max	95	Max	96	Max	98
		RSD	0.8%	RSD	0.6%	RSD	2.0%
	7	Tablet 1	95	Tablet 1	96	Tablet 1	97
		Tablet 2	94	Tablet 2	95	Tablet 2	96
		Tablet 3	94	Tablet 3	96	Tablet 3	97
		Tablet 4	96	Tablet 4	95	Tablet 4	97
		Tablet 5	96	Tablet 5	97	Tablet 5	97
		Tablet 6	93	Tablet 6	95	Tablet 6	96
		Mean	95	Mean	96	Mean	97
		Min	93	Min	95	Min	96
		Max	96	Max	97	Max	97
		RSD	12%	RSD	0.7%	RSD	0.5%

TABLE 15.15B Summary of the third quarter continuous process verification stage batches for the dissolution critical quality attribute. The dissolution results showed similarity to the process performance qualification batches dissolution performance and demonstrate the consistency of the dissolution test performance through each continuous manufacturing batch manufactured during 2017 met the individual and average test requirements.

CPV 3a Period	Batch	Beginning	Dissolved drug (%)	Middle	Dissolved drug (%)	End	Dissolved drug (%)
	8	Tablet 1	92	Tablet 1	93	Tablet 1	94
		Tablet 2	92	Tablet 2	90	Tablet 2	90
		Tablet 3	92	Tablet 3	94	Tablet 3	91
		Tablet 4	93	Tablet 4	92	Tablet 4	92
		Tablet 5	92	Tablet 5	92	tablet 5	92
		Tablet 6	92	Tablet 6	92	Tablet 6	91
		Mean	92	Mean	92	Mean	92
		Min	92	Min	90	Min	90
		Max	93	Max	94	Max	94
		RSD	0.6	RSD	1.4	RSD	1.2
	9	Tablet 1	92	Tablet 1	95	Tablet 1	96
		Tablet 2	97	Tablet 2	94	Tablet 2	97
		Tablet 3	97	Tablet 3	95	Tablet 3	98
		Tablet 4	96	Tablet 4	95	Tablet 4	99
		Tablet 5	96	Tablet 5	94	Tablet 5	99

Continued

TABLE 15.15B Summary of the third quarter continuous process verification stage batches for the dissolution critical quality attribute. The dissolution results showed similarity to the process performance qualification batches dissolution performance and demonstrate the consistency of the dissolution test performance through each continuous manufacturing batch manufactured during 2017 met the individual and average test requirements.—cont'd

CPV 3a Period	Batch	Beginning	Dissolved drug (%)	Middle	Dissolved drug (%)	End	Dissolved drug (%)
Q3		Tablet 6	95	Tablet 6	95	Tablet 6	95
		Mean	96	Mean	95	Mean	97
		Min	95	Min	94	Min	95
		Max	97	Max	95	Max	99
		RSD	0.9	RSD	0.7	RSD	1.5
		Tablet 1	95	Tablet 1	96	Tablet 1	96
		Tablet 2	94	Tablet 2	96	Tablet 2	96
		Tablet 3	94	Tablet 3	96	Tablet 3	96
	10	Tablet 4	97	Tablet 4	95	Tablet 4	96
		Tablet 5	95	Tablet 5	96	Tablet 5	96
		Tablet 6	98	Tablet 6	92	Tablet 6	96
		Mean	96	Mean	95	Mean	96
		Min	94	Min	92	Min	96
		Max	96	Max	96	Max	98
		RSD	1.7	RSD	1.5	RSD	1.0

Janssen has built additional CM process capabilities at our pharmaceutical process development site in Belgium and our operations facility in Latina, Italy. Janssen has committed to continue developing products in CM moving forward as the primary means for pharmaceutical process development of solids dosage forms.

References

[1] FDA approves tablet production on Janssen continuous manufacturing line.

[2] Van Arnum P. A FDA perspective on quality by design. Pharm Tech 2007;3(12). https://www.pharmtech.com/view/fda-perspective-quality-design.

[3] Vanarase AU, Alcalà M, Jerez Rozo JI, Muzzio FJ, Romañach RJ. Real-time monitoring of drug concentration in a continuous powder mixing process using NIR spectroscopy. Chem Eng Sci Nov. 2010;65(21):5728−33. https://doi.org/10.1016/j.ces.2010.01.036.

[4] Barajas MJ, et al. Near-Infrared spectroscopic method for real-time monitoring of pharmaceutical powders during voiding. Appl Spectrosc May 2007;61(5):490−6. https://doi.org/10.1366/000370207780807713.

[5] Beach LB, Ropero J, Mujumdar A, Alcalà M, Romañach RJ, Davé RN. Near-Infrared spectroscopy for the in-line characterization of powder voiding part II: quantification of enhanced flow properties of surface modified active pharmaceutical ingredients. J Pharm Innov 2010. https://doi.org/10.1007/s12247-010-9075-1.

[6] Alcalà M, Ropero J, Vázquez R, Romañach RJ. Deconvolution of chemical and physical information from intact tablets NIR spectra: two- and three-way multivariate calibration strategies for drug quantitation. J Pharmacol Sci Aug. 2009;98(8):2747−58. https://doi.org/10.1002/jps.21634.

[7] Alcalà M, León J, Ropero J, Blanco M, Romañach RJ. Analysis of low content drug tablets by transmission near infrared spectroscopy: selection of calibration ranges according to multivariate detection and quantitation limits of PLS models. J Pharmacol Sci Dec. 2008;97(12):5318−27. https://doi.org/10.1002/jps.21373.

[8] Ropero J, Beach L, Alcalà M, Rentas R, Davé RN, Romañach RJ. Near-infrared spectroscopy for the in-line characterization of powder voiding Part I: development of the methodology. J Pharm Innov Dec. 2009;4(4):187−97. https://doi.org/10.1007/s12247-009-9069-z.

[9] Meza CP, Santos MA, Romañach RJ. Quantitation of drug content in a low dosage formulation by transmission near infrared spectroscopy. AAPS PharmSciTech Mar. 2006;7(1):E206−14. https://doi.org/10.1208/pt.070129.

[10] Boukouvala F, Muzzio FJ, Ierapetritou MG. Design space of pharmaceutical processes using data-driven-based methods. J Pharm Innov Oct. 2010;5(3):119−37. https://doi.org/10.1007/s12247-010-9086-y.

[11] Næs T, Isaksson T. Selection of samples for calibration in near-infrared spectroscopy. Part I: general principles illustrated by example. Appl Spectrosc Feb. 1989;43(2):328−35. https://doi.org/10.1366/0003702894203129.

[12] Ierapetritou M, Muzzio F, Reklaitis G. Perspectives on the continuous manufacturing of powder-based pharmaceutical processes. AIChE J 2016;62(6):1846−62. https://doi.org/10.1002/aic.15210.

[13] Engisch WE, Muzzio FJ. Loss-in-weight feeding trials case study: pharmaceutical formulation. J Pharm Innov Mar. 2015;10(1):56−75. https://doi.org/10.1007/s12247-014-9206-1.

[14] Process validation: general principles and practices. p. 22.

Chapter 16

Orkambi: a continuous manufacturing approach to process development at Vertex

Stephanie Krogmeier, Justin Pritchard, Eleni Dokou, Sue Miles, Gregory Connelly, Joseph Medendorp, Michael Bourland and Kelly Swinney
Vertex Pharmaceuticals, Boston, MA, United States

1. Introduction

Cystic fibrosis (CF) is a rare, life-shortening genetic disease that currently affects approximately 75,000 people in North America, Europe, and Australia. CF is caused by mutations in the cystic fibrosis transmembrane conductance regulator (CFTR) gene, which encodes the CFTR chloride ion channel. These defects cause CF patients to produce abnormally thick sticky mucus that obstructs the airways, intestine, and ducts of the pancreas.

At Vertex, the approach to treating CF became an innovation model for precision medicine and manufacturing. Specifically, the first two Breakthrough Therapy Designations from the FDA were granted to ivacaftor monotherapy and to the combination of lumacaftor and ivacaftor for the treatment of CF. Following this, the importance of high-quality, robust manufacturing capabilities able to keep pace with the clinical development became evident and the Vertex initiative for continuous manufacturing (CM) was launched. Therefore, Orkambi, the fixed dose combination of lumacaftor drug substance (DS) and ivacaftor spray dried dispersion, was developed under both a CM and Quality by Design (QbD) paradigm. While the simultaneous development of both CM and Orkambi was challenging, it was ultimately a critical component of the success, providing the motivation and clarity needed to bring this innovative and important technology to the pharmaceutical industry. **Overall, the use of the CM led to early finalization of the formulation and process at commercial scale and a data-rich QbD commercial design space.**

How to Design and Implement Powder-to-Tablet Continuous Manufacturing Systems
https://doi.org/10.1016/B978-0-12-813479-5.00003-3
383

2. Continuous manufacturing equipment and process development

Two CM systems are employed by Vertex to manufacture Orkambi. The Continuous Tableting Line (CTL-25) is a continuous system from the intra-granular (IG) blend to core tablets, while the Development and Launch Rig (DLR) at Vertex is a continuous system from individual components to film coated tablets. The manufacturing process is illustrated in Fig. 16.1 with the boxes indicating the boundaries of the continuous process equipment used at both the sites.

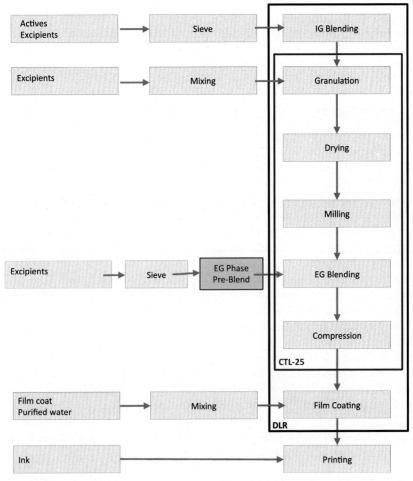

FIGURE 16.1 Manufacturing process flow diagram. *DLR*, Development and Launch Rig; *EG*, extragranular; *IG*, intragranular.

The granulation, drying, milling, compression, and printing operations are the same on both systems, whereas the materials handling and feeding, IG blending, extragranular (EG) blending, and film coating processes as well as process analytical technology (PAT) capabilities are different. Specifically, on the CTL-25, the materials are blended using traditional bin blending for IG blending and a ribbon blender for EG blending, while the Vertex DLR uses continuous IG and EG blending. The composition and line rate are defined in the batch recipe, which sets the mass flow rate of the individual components. The ratio of the set point mass flow rates of the four feeders in the IG feeder bank is kept constant to maintain composition. The desired quantity of each component is fed into an in-line blender for both IG and EG components. The components are added to the blender simultaneously by LIW feeders through a central header, as shown in Fig. 16.2. Experiments have shown that blend uniformity is achieved independent of the paddle speed of the blender and manufacturing line rate. Finally, for film coating, the CTL-25 uses a conventional perforated pan film coating process, while Vertex DLR uses a semi-continuous film coating system, also with a perforated pan.

Regarding similarities, the CTL-25 and DLR equipment employ a combination of fully continuous unit operations and semi-continuous processing steps. Twin screw wet granulation is an example of a truly continuous process, where the IG blend and the binder solution are continuously dosed into the granulator and granules continuously exit the granulator. The fluid bed dryer used at both sites consists of drying cells that operate in parallel. Wet granules are charged into a drying cell for a defined time. The granules in each filled cell are dried until the desired water content is reached. At that time, the contents of that cell are evacuated leaving the cell ready for the next filling cycle.

The final blend is transferred by gravity to the rotary tablet press for compression, and the resulting tablets are mechanically transferred to a

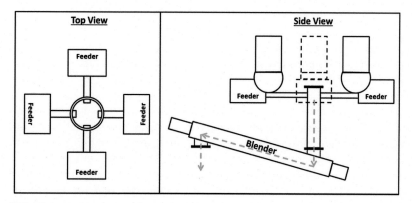

FIGURE 16.2 Illustration of continuous blender used on the Development and Launch Rig.

deduster/metal checker followed by the tablet relaxation unit (TRU). The same tablet press equipment is utilized in both process lines. On the DLR, the TRU conveys the tablets to the tablet coaters (perforated pan coating system) for application of the film coat. The continuous coater, incorporated in the DLR, allows film coating within a shorter time than traditional pan film coating. To accommodate the desired range of line rates of the DLR, two identical continuous film coaters are incorporated in the system, running in parallel. The continuous tablet coater is designed to load, coat, polish, and discharge small pan loads of tablets back-to-back as part of the continuous system. A detailed schematic of the DLR in the wet granulation configuration is given in Section 4, Fig. 16.4.

Vertex's approach to process development and QbD for CM includes process understanding at the unit operation level as well as integrated runs utilizing the entire line. Key equipment is available either as standalone (e.g., second press similar to that on the DLR) or can be operated in standalone mode (e.g., coaters) in order to evaluate and optimize the process for that unit operation. For Orkambi, QbD studies were executed at the unit operation level and models were built connecting process parameters and material attributes to product critical quality attributes. Confirmatory runs were then executed at the full line to confirm the design spaces and test the control strategy.

3. Continuous manufacturing and cGMPs

In establishing a GMP facility to implement CM, Vertex's approach was to evaluate all of its existing Manufacturing and Quality Systems. This evaluation was required, as the US Code of Federal Regulations, European Commission Directives, and various regulatory guidance documents did not directly address CM GMP operations. In addition, Vertex wanted to ensure CM GMP operations were performed within a state of control as expected by the International Council for Harmonisation (ICH) Guidelines. Vertex assessed each Manufacturing and Quality System to determine if modifications or supplemental procedures were required.

The DLR and automated control system allow for effective monitoring and control of process performance and product quality. The control system allows for complete automated monitoring and control of the Orkambi process control strategy with system warnings within in-process control (IPC) limits and normal operating ranges (NORs) for critical process parameters. The warnings provide operator awareness of manufacturing conditions allowing adjustments within defined ranges, as needed. In addition, system alarms are set for critical process parameters outside the design space limits (DSLs) and IPCs outside of specifications. Once outside a critical DSL or IPC specification, material is identified as nonconforming. An in-depth discussion on the implementation of Orkambi control strategy is provided in Section 4. Material traceability through the CM rig is essential to ensure conforming material is processed forward and nonconforming material is segregated from the process. The DLR is designed to move material on a plug flow basis, and a product key (PK) is the material ID for the smallest unit mass that is tracked by the process control system and can be segregated from the process. The PKs within batch processing are traceable across unit operations and are well characterized. Where there is potential for intermixing of PKs through processing, the extent of intermixing was evaluated using residence time distribution (RTD) studies. The outcome of these studies was incorporated into the strategy for segregation of nonconforming material for the Orkambi process. PKs can be deemed nonconforming by failure to meet IPC specifications or failure to be manufactured within a critical DSL. Once a PK is deemed nonconforming, the segregation strategy from the RTD studies is applied; the nonconforming PK and any potentially affected adjacent PKs are segregated from the batch at the next segregation point. As these Production and Material Controls are necessary, the CM control system provides validated PK Tracking and Line Extraction Reports to supplement the executed batch records.

The DLR and control system coupled with the PAT IPC analysis allows for effective monitoring and control of process performance and product quality on the PK level; naturally, the question of batch definition and batch size needs to be addressed. Vertex is aligned with regulatory definitions of batch which refer to the quantity and quality of material, regardless of the mode of manufacture. The Orkambi process design space was established using a range of manufacturing line rates. As a result, a batch size range was submitted to the regulatory marketing applications. The Orkambi process was validated using a batch size within the filed batch range.

As aforementioned, in addition to evaluating the Manufacturing Systems, Vertex assessed the existing Quality Systems. Most of the Quality Systems (Change Control, Deviations, CAPAs, Risk Management, etc.) did not require supplemental procedures or modifications, except for the procedure for investigating out of specifications. The FDAs Guidance for Industry, investigating the out-of-specification (OOS) test results for pharmaceutical production, specifically states it is not intended to address PAT. Vertex's OOS procedure ensures compliance with the FDA Guidance and includes the requirement of the Phase I laboratory investigation as prescribed in the guidance. Vertex supplemented the procedure with an internal process for Phase I laboratory investigations for PAT IPC and real-time release testing (RTRT) OOS results.

The implementation and approval of RTRT also required Vertex to supplement the QP release process. In alignment with the European Medicines Agency (EMA) Guideline on RTRT, Vertex does not perform importation product testing for Orkambi (except for identification) when the batch is released by RTRT.

Implementation of CM into GMP operations resulted in modifications to elements within the Product and Material Controls within the Manufacturing

Systems, as well as adjustments to Quality System procedures. **Although the US Code of Federal Regulations, European Commission Directives, and various regulatory guidance documents do not specifically address CM, general principles of the GMPs can be successfully applied**.

4. Control strategy implementation at Vertex

The CM process automated control strategy consists of four levels of control. These four control strategy building blocks are depicted in Fig. 16.3. (1) The lowest level of control relies on automated feedback loops used to drive each unit operation toward its target set point. (2) Monitoring of design space parameters throughout manufacturing is performed to ensure product is manufactured within the design space. (3) IPCs are in place at various points throughout the manufacturing process. Material manufactured within the design space and conforming to IPC acceptance criteria is included in the batch for release consideration. (4) Finally, the manufactured batch is evaluated to assess conformance to the finished product specifications.

Automated feedback loops are used to drive each unit operations toward the target set points. For example, loss-in-weight feeder is controlled through an automated feedback loop that is continuously driving toward set point (100% target), actively ensuring that each feeder delivers to target. Critical process parameters, as well as process parameters with defined DSLs, are monitored in real time. Excursions from the NOR and/or the DSLs are noted by the system and the operator is alerted. Material manufactured outside the DSLs is segregated and investigated.

For IPC, the DLR is equipped with both spectroscopic and non-spectroscopic PAT elements as shown in Fig. 16.4. Acceptance criteria are defined for each IPC, which are monitored in real time. Failure of IPC acceptance criteria results in a control system alarm and automated segregation of the associated material from the process followed by an investigation.

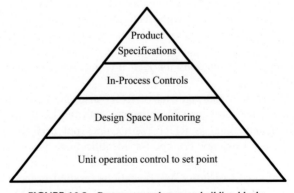

FIGURE 16.3 Process control strategy building blocks.

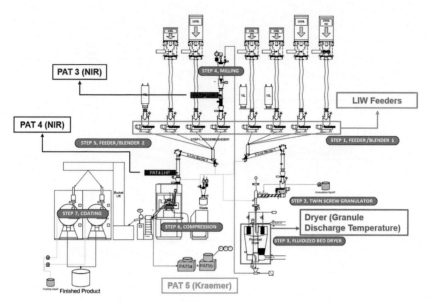

FIGURE 16.4 Development and Launch Rig process analytical technology (PAT) locations used for in-process control.

Operator action limits, where appropriate, are set within the IPC acceptance criteria to proactively constrain the within-batch variability to within the IPC limits. IPC results exceeding the operator action limits trigger a control system alarm, and subsequent operator actions, defined in the batch record, are taken to bring the process back to target. A sampling plan is defined and justified for each IPC.

For RTRT, spectroscopic and nonspectroscopic PAT measurements are also implemented as shown in Fig. 16.5.

A batch release test result is reported based on repetitive measurements throughout processing. Material manufactured within the DSLs and conforming to IPC acceptance criteria is included in the batch and considered for release.

4.1 Assay

The average lumacaftor content and average ivacaftor content in final blend, determined by near-infrared (NIR) spectroscopy (PAT 4) for the batch, along with the batch average core tablet weight (PAT 5), are subsequently used to calculate the batch assay for lumacaftor and ivacaftor. The capability of the RTRT method to appropriately characterize batch assay was assessed across the design space. A direct comparison of the results obtained for the RTRT assay method using a composite sample tested by the regulatory (HPLC) method and mean results obtained from stratified sampling of core tablets are

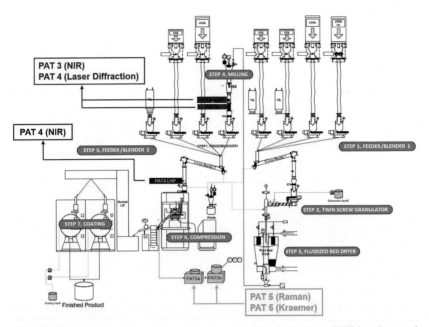

FIGURE 16.5 Development and Launch Rig process analytical technology (PAT) locations used for real-time release testing.

shown in Fig. 16.6 for the design space confirmation runs. The larger difference observed between the regulatory method and the RTRT method is explainable by the difference in sample size, as the RTRT method extensively samples the batch over the time course of the manufacture. Comparable results are obtained by the three methods.

4.2 Content uniformity

NIR spectroscopy of final blend (PAT 4) and core tablet weight (PAT 5) are used to determine the lumacaftor and ivacaftor batch content uniformity. The RTRT CU method is based on USP <905> and describes the measurement inputs for the calculation of acceptance value (AV). The standard deviation of the batch is calculated by first calculating the combined variance of DS content in final blend (PAT 4, NIR) and tablet weight (PAT 5, Kraemer tablet tester) and then converting this value into standard deviation (%LC).

The capability of the RTRT method to appropriately characterize the content uniformity of the batch by calculating AV was assessed for the DLR design space confirmation runs as shown in Table 16.1. A direct comparison of the results obtained for the RTRT assay method using a composite sample tested by the regulatory (HPLC) method and results obtained from stratified sampling of core tablets are shown in Fig. 16.6 for the design space confirmation runs. Comparable results are obtained by the three methods.

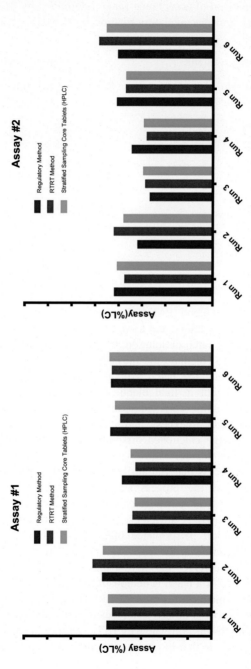

FIGURE 16.6 Comparison of regulatory and real-time release testing (RTRT) assay method results with stratified sampling results for the Development and Launch Rig design space confirmation runs.

TABLE 16.1 Comparison of acceptance value determined by RTRT method stratified sampling, and the regulatory method for the DLR design space confirmation runs.

	Drug substance I			Drug substance II		
Run	RTRT method	Stratified sampling	Regulatory method	RTRT method	Stratified sampling	Regulatory method
1	3.3	4.9	5.8	3.2	4.5	4.8
2	7.5	6.9	6.1	2.8	5.2	6.2
3	5.5	5.4	5.3	8.1	8.6	8.0
4	7.8	5.5	7.3	8.0	7.7	8.3
5	3.6	4.6	4.1	4.8	3.7	2.7
6	4.1	6.2	4.4	7.2	7.0	4.4

4.3 Dissolution

NIR spectroscopy of final blend (PAT 4), laser diffraction of granules (PAT 3), and core tablet weight, thickness, and hardness (PAT 5) are used to assess the lumacaftor and ivacaftor batch dissolution. Chemometric partial least squares models are employed to calculate the mean batch dissolution rate of lumacaftor and ivacaftor using the batch average lumacaftor content, ivacaftor content, and water content in final blend as model inputs along with the average batch granule volume weighted particle size distribution and the batch average core tablet weight, thickness, and hardness. Using a modified Noyes Whitney equation and the calculated dissolution rate, the batch dissolution profile of each DS is calculated.

The measured in-process material attributes were selected as model inputs based on knowledge of the process and factors influencing dissolution performance at the time of batch release. The relationship between the measured in-process materials attributes, process parameters, raw material attributes, and dissolution performance that were studied during method development is described using the Ishikawa diagram presented in Fig. 16.7.

The capability of the RTRT dissolution models to properly characterize dissolution performance of a batch is demonstrated by the results obtained for the design space confirmation runs on the DLR. The dissolution models were developed across the process design space wherein dissolution is acceptable. The RTRT dissolution results correlate with HPLC dissolution across the range of dissolution performance that has been seen throughout the process design space (Figs. 16.8 and 16.9).

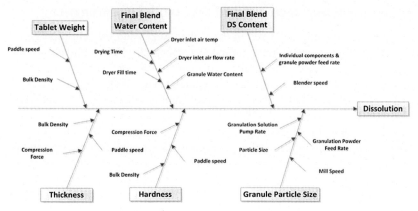

FIGURE 16.7 Real-time release testing dissolution model Ishikawa diagram. *DS*, drug substance.

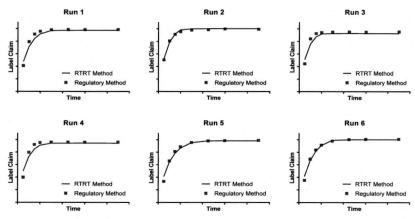

FIGURE 16.8 Comparison of real-time release testing (RTRT) method and regulatory method results for the design space confirmation runs for ivacaftor.

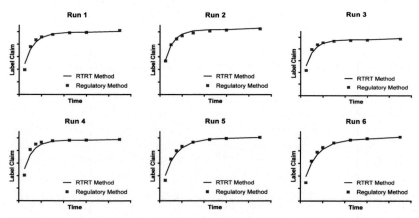

FIGURE 16.9 Comparison of real-time release testing (RTRT) method and regulatory method results for the design space confirmation runs for lumacaftor.

5. Life cycle management and PAT model maintenance

Orkambi was initially approved in the United States and the European Union in 2015 and received multiple subsequent approvals. Vertex utilized a QbD approach for process development, supported by the current regulatory framework and guidance documents, including QbD ICH Q8 (R2), Quality Risk Management (Q9), Pharmaceutical Quality System (Q10), and PAT guidance documents from the FDA and EMA. Vertex leveraged various health authority interactions, including face-to-face meetings, Pre-Operational Visits (Field Management Directive 135), and consultative advice as an opportunity, to share progress and seek guidance. The submission was written using the Common Technical Document (CTD) format, with multiple leafs created to facilitate the review of the CM and RTRT sections such as 3.2.P.2.3 Introduction to CM and 3.2.P.5.2 Introduction to Analytical Methods.

Following the approval of the initial application, the product has been maintained using the existing regulatory framework for post-approval changes of each individual country. As an example, Vertex previously submitted a post-approval change in multiple regions to the testing strategy on the CTL-25. The initial control strategy included an at-line NIR IPC measurement at the compression stage to measure active pharmaceutical ingredient (API) concentration (% target) in final blend. The control strategy was updated to replace the NIR IPC method with two fully automated gravimetric IPCs, which collectively provided better control of the tablet API concentration and greatly reduced operational complexity. This change effectively moved the control point upstream from the compression step to the blending step, and material segregation became fully automated and required no manual intervention. Secondly, the gravimetric IPCs provided a tighter control than the NIR IPC and reduced variability across the batch. The change was submitted and approved in the aforementioned countries as a post-approval change after the initial application.

In CM systems that use PAT in the control strategy for IPC or RTRT, another example of a typical post-approval submission Vertex manages is the ongoing model maintenance of PAT methods. Many of the most common updates can be managed under the quality systems. However, some of these updates can require regulatory filings and prior approval, which are managed according to regional guidance and have different review and approval timelines. This can be challenging when a post-approval submission is required, especially if there are sequential updates that require prior approval before implementation. In global pharmaceutical operations, qualification of second sources for excipients and APIs, changes in material attributes, changes to the manufacturing process, equipment changes, or even spectrometer preventive maintenance schedules all have the ability to impact spectroscopy models and may necessitate PAT model updates. Beyond the technical challenges, in order to make this process as scalable and manageable as possible from the regulatory standpoint, Vertex is exploring mechanisms such as post-approval change management protocols to facilitate the life cycle management of Orkambi; however, potential challenges remain, as not all countries have mechanisms such as change management protocols. Due to these post-approval challenges, compounded with the complexity of factors capable of disrupting PAT model functionality, it was advantageous to devise the model maintenance and filing strategy with life cycle management requirements in mind. With CM and a PAT-based control strategy, the Orkambi submission did provide some examples for the design of PAT method submissions with increased long-term flexibility to operate. For example, in place of the fixed numerical limit for model diagnostics criteria, the confidence limit used to calculate model diagnostics criteria can be presented, which does not change over time as models are updated.

Orkambi PAT models are assessed routinely throughout their life cycles (e.g., annual parallel testing). The PAT model evaluation consists of a formal and direct comparison between the PAT methods and their respective reference methods. Parallel testing includes the evaluation of each IPC method and each of the RTRT methods which employs a chemometric model. For each method, the number of samples used for evaluation and the acceptance criteria are predefined in sampling and parallel testing protocols. Failure to meet these acceptance criteria requires model updates.

Beyond the commitment to perform annual parallel testing, some other drivers for model assessments have included the results of long-term commercial trending activities, known operation changes (e.g., process changes, material changes, nonroutine PAT instrument changes, etc.), and results of short-term or long-term model diagnostics evaluations. Through extensive development efforts prior to commercialization and sample sets comprised of the expected range of process parameters and material attributes, the initial PAT models still predict accurately and have remained sensitive to the intended analytical properties of interest throughout commercial production.

6. Conclusion and lessons learned

With the implementation of CM for Orkambi, the promised benefits of CM were realized. The Orkambi process was developed and finalized on commercial-scale equipment early in development with the commercial equipment and CM process used to supply pivotal clinical trials. By leveraging the small-scale nature of the CM equipment and the data-rich environment, QbD drug product process development was conducted over a short period of time with limited DS demand and in parallel with DS process scale-up. The ability to remove the dependency of drug product process development on DS process scale-up enabled the development timeline to be reduced by more than a year.

In addition, multiple interactions including various face-to-face meetings and transparency with health authorities facilitated a successful inspection and approval. Agency confidence in the proposed process and advanced control strategy was achieved by demonstrating a high level of process understanding on commercial equipment using process models and QbD data-rich experimentation results and conclusions.

As Vertex continues the life cycle management of Orkambi, the development of additional assets is underway. Therefore, the opportunity for a multiproduct rig has been realized and the benefits of CM are being leveraged in both development and commercial.

Chapter 17

Outlook—what comes next in continuous manufacturing (and in advanced pharmaceutical manufacturing)

Fernando J. Muzzio

Engineering Research Center for Structured Organic Particulate Systems (C-SOPS), Department of Chemical and Biochemical Engineering, Rutgers, The State University of New Jersey, Piscataway, NJ, United States

A favorite quote of ours, attributed after some paraphrasing both to Yogi Berra and to Niels Bohrs (and we would side with Niels Bohrs on this one), is that

Prediction is very difficult, especially about the future.

That said, we will nonetheless share some thoughts regarding where we might be going as a community of practice in implementation of advanced pharmaceutical manufacturing methods.

To provide context, this book took 3 years to write, and in the last 10 months, COVID-19 changed almost every idea regarding the role of pharmaceutical manufacturing on the health security of the US population. The fragility of complex international supply chains was made evident, and the risk to the US population due to disruptions in availability not just of medicines but also of mundane supplies such as masks and hand sanitizer became undeniable. In light of the pandemic, the need to develop and approve medicines faster became a matter of common sense. Moreover, the sharp contrast between the industry's ability to develop multiple vaccines in just a few months (an immense accomplishment by any standard) and an industry that is unable to manufacture such a vaccine in sufficient amounts for many months, failing to prevent hundreds of thousands of avoidable deaths, has served to propel the need to reinvent process development and product manufacturing to the forefront of our collective attention.

Thus, given the history of the last few months, we expect that the implementation of continuous manufacturing methods will accelerate until it represents a large fraction of all processes. The pharmaceutical industry's

How to Design and Implement Powder-to-Tablet Continuous Manufacturing Systems
https://doi.org/10.1016/B978-0-12-813479-5.00007-0

transition to manufacturing a large fraction of its output using knowledge-intensive continuous technologies will take place not only in the United States and in Europe but also in all advanced economies and in many emerging economies with large populations that require access to low-cost medications. Designing and implementing the technology platforms to enable this transformation will likely create tens of billions of dollars in economic activity. Importantly, growth in continuous pharmaceutical manufacturing would trigger significant activity in many other industries up and down the supply chain, including (1) raw materials suppliers, which are already indicating an interest in developing ingredients in grades optimized for continuous processing; (2) equipment and instrumentation companies, many of which are actively commercializing specialized manufacturing equipment, sensors, and process analyzers; (3) companies that commercialize closed-loop control systems; and many others. Moreover, effective implementation of continuous manufacturing for pharmaceuticals will also provide an opportunity for enhancing other industries that rely of powder processing, including cosmetics, food, consumer products, dietary supplements, etc. These industries utilize manufacturing processes that are very similar to those used by the pharmaceutical industry and in many cases the technology is directly portable.

We expect that for solid dose products, applications will quickly migrate into over-the-counter and generic products, where the reduced cost of manufacturing will have the largest impact on profitability. We further expect that other industries that use very similar manufacturing processes, such as supplements and cosmetics, will adopt the same or similar methods as currently developed for pharmaceutical products. Our prediction is that within the next 10−15 years, continuous manufacturing methods will become very common to the point where the products manufactured by continuous approaches will rival, in annual sales and in numbers of product units, those manufactured by batch methods. Adoption will first be driven by a desire for lower cost, faster development timelines, and higher quality. Eventually, however, as the advantages of continuous manufacturing become increasingly evident, use of continuous manufacturing methods and other advanced manufacturing technologies will become an implicit (or perhaps, explicit) expectation by regulators.

At that point, implementation will quickly reach full maturity for solid dose products and will provide renewed impetus for similar developments in other areas of pharmaceutical manufacturing. Efforts are already under way to develop continuous processes for active pharmaceutical ingredient (API) synthesis and isolation, including both organic synthesis and biological synthesis. For synthetic APIs, implementation of continuous systems is quickly reaching maturity. The main area of effort at the present time is continuous crystallization, which still requires some effort to become a mature process, but where progress is being made rapidly. Slightly behind in its degree of implementation is continuous bioprocessing, but this area is attracting huge interest, and it also benefits from the widespread perception that biological products are likely to experience faster growth in the next few years. We believe that the technical problems preventing full implementation of integrated continuous manufacturing in both of these areas are surmountable, in particular as they would also benefit from a trail already blazed by solid dose products.

How will the technology evolve for solid dose products? It is now generally accepted that direct compression is the least expensive, simplest, and therefore preferred approach to solid dose continuous manufacturing. Our team adopted this view back in 2006, based simply on firsthand knowledge that the level of complexity of integrated manufacturing processes increases very fast with the number of integrated unit operations. One of the main obstacles of implementing direct compression in batch processes is fear of blend segregation. However, work by us [1] and by others [2,3] has demonstrated that a properly designed continuous line is much less vulnerable to segregation than its batch counterpart. Currently, the main limitations to implementation of continuous direct compression are (1) for very potent compounds, where the API would typically be micronized, be very cohesive, have a high tendency to agglomerate, and be very difficult to feed accurately and to monitor via PAT, and (2) at the opposite extreme, for very high-dose products, where the API is the majority component of the dosage form, if the API is poorly compressible. For both of these systems, companies pursue various forms of roller compaction or wet granulation. In both of these situations, we anticipate that API preprocessing methods, able to generate API material that flows well, does not agglomerate, and is highly compactible, will be developed to broaden the applicability of "direct compression" methods, now modified slightly to incorporate API materials that have experienced preconditioning.

In a nutshell, preconditioning is all about managing API properties, and this also includes the need to ensure that APIs flow well enough to allow accurate feeding. However, the need to characterize, understand, and manage material properties goes well beyond pure APIs and extends to excipients, blends, granulated intermediates, etc. Currently, the industry experiences a technical gap in this area—the material properties of interest have not been univocally identified, and the methods for measuring them have not been standardized, and as a result, the industry currently lacks access to a material property database that would enable digital design, meaningful ingredient specification, and many other benefits. The need to fill this gap is evident to practitioners, so we expect that this situation will evolve rapidly in the next few years.

Modeling is central to the predictive implementation of continuous systems, and the community of practice is actively engaged in creating model libraries and flowsheet systems that enable rapid and reliable modeling of integrated systems for a variety of applications. This process will continue, and models will become more mature and more robust. The main opportunity for growth in this area is the ability to incorporate material property measurements into the models. Another opportunity is to create models where blend properties can be predicted as a function of blend composition and shear history. More recently, the FDA has provided significant funding for the development of models useful for quality control. All of these topics are generating substantial interest and we expect substantial progress.

Another question currently unanswered is to what extent closed-loop quality control, real-time quality assurance, and real-time release testing will become standard. In our opinion, they should, and they will. In its full expression, advanced manufacturing will enable companies to monitor their processes closely and to assure the quality of every product unit released to the market. The technology to enable this desirable outcome already exists for many products and can be developed for many more. The economic arguments against implementation of real-time quality assurance methods will lose potency as more companies demonstrate that this is feasible and also as the cost of implementation decreases (or the cost of nonimplementation increases).

Importantly, we believe that we are at the beginning of the evolution of the technology. In years to come, we expect that continuous manufacturing systems will become smaller, modular, portable, highly efficient, and highly flexible. We also expect that other types of advanced manufacturing methods will emerge as major options. Already there is one approved product that is manufactured using a version of 3D printing. Building on this example, we expect that additive manufacturing methods, supported by many of the same conceptual toolbox components that enabled continuous manufacturing, will emerge, supplemented by a collection of applications that are best described as "precision manufacturing" or "personalized manufacturing." In all of these applications, the scientist/engineer combines relevant knowledge of material properties, sensing, modeling, and process control to design manufacturing methods that can create the desired product at minimum cost and with maximum quality.

This has been the dream of many generations of pharmaceutical scientists. It is now "assigned homework" for our students and younger colleagues to bring it to full reality.

Fernando Muzzio.
January 2021.

References

[1] Oka S, Sahay A, Meng W, Muzzio FJ. Diminished segregation in continuous powder mixing. Powder Technol 2017;309:79–88.

[2] Ervasti T, Niinikoski H, Mäki-Lohiluoma E, Leppinen H, Ketolainen J, Korhonen O, Lakio S. The comparison of two challenging low dose APIs in continuous direct compression process. Pharmaceutics 2020;12(3):279.

[3] Van Snick B, Holman J, Vanhoorne V, Kumar A, De Beer T, Remon JP, Vervaet C. Development of a continuous direct compression platform for low-dose drug products. Int J Pharm 2017;529:329–46.

Index

Note: 'Page numbers followed by "*f*" indicate figures and "*t*" indicate tables.'